Marine Biology

Seaweed

Ecology, Nutrient Composition and Medicinal Uses

MARINE BIOLOGY

Additional books in this series can be found on Nova's website under the Series tab.

Additional E-books in this series can be found on Nova's website under the E-books tab.

EARTH SCIENCES IN THE 21ST CENTURY

Additional books in this series can be found on Nova's website under the Series tab.

Additional E-books in this series can be found on Nova's website under the E-books tab.

MARINE BIOLOGY

SEAWEED

ECOLOGY, NUTRIENT COMPOSITION AND MEDICINAL USES

VITOR H. POMIN
EDITOR

Nova Science Publishers, Inc.
New York

Copyright © 2012 by Nova Science Publishers, Inc.

All rights reserved. No part of this book may be reproduced, stored in a retrieval system or transmitted in any form or by any means: electronic, electrostatic, magnetic, tape, mechanical photocopying, recording or otherwise without the written permission of the Publisher.

For permission to use material from this book please contact us:
Telephone 631-231-7269; Fax 631-231-8175
Web Site: http://www.novapublishers.com

NOTICE TO THE READER

The Publisher has taken reasonable care in the preparation of this book, but makes no expressed or implied warranty of any kind and assumes no responsibility for any errors or omissions. No liability is assumed for incidental or consequential damages in connection with or arising out of information contained in this book. The Publisher shall not be liable for any special, consequential, or exemplary damages resulting, in whole or in part, from the readers' use of, or reliance upon, this material. Any parts of this book based on government reports are so indicated and copyright is claimed for those parts to the extent applicable to compilations of such works.

Independent verification should be sought for any data, advice or recommendations contained in this book. In addition, no responsibility is assumed by the publisher for any injury and/or damage to persons or property arising from any methods, products, instructions, ideas or otherwise contained in this publication.

This publication is designed to provide accurate and authoritative information with regard to the subject matter covered herein. It is sold with the clear understanding that the Publisher is not engaged in rendering legal or any other professional services. If legal or any other expert assistance is required, the services of a competent person should be sought. FROM A DECLARATION OF PARTICIPANTS JOINTLY ADOPTED BY A COMMITTEE OF THE AMERICAN BAR ASSOCIATION AND A COMMITTEE OF PUBLISHERS.

Additional color graphics may be available in the e-book version of this book.

Library of Congress Cataloging-in-Publication Data

Seaweed : ecology, nutrient composition, and medicinal uses / editor, Vitor H. Pomin.
   p. cm.
 Includes index.
 ISBN 978-1-61470-878-0 (hardcover)
 1. Marine algae--Therapeutic use. 2. Marine algae as food. I. Pomin, Vitor H.
 RS165.A45S43 2011
 615.3'29--dc23
                                        2011028019

Published by Nova Science Publishers, Inc. † New York

Contents

Preface		vii
Chapter 1	Alanine is a Possible Compatible Solute Involved in Cold Acclimation in the Marine Red Alga Porphyra yezoensis *Koji Mikami, Megumu Takahashi, Ryo Hirata, Takehiko Yokoyama, Misako Taniguchi, Naotsune Saga and Tomoko Mori*	1
Chapter 2	A Review of the Nutrient Composition of Selected Edible Seaweeds *Leonel Pereira*	15
Chapter 3	Nutritional Quality and Biological Properties of Brown and Red Seaweeds *P. Rupérez, E. Gómez-Ordóñez and A. Jiménez-Escrig*	51
Chapter 4	Edible Seaweeds: A Functional Food with Organ Protective and other Therapeutic Applications *Suhaila Mohamed, Patricia Matanjun, Siti Nadia Hashim, Hafeedza Abdul Rahman and Noordin Mohamed Mustapha*	67
Chapter 5	Seaweeds, Food, and Industrial Products and Nutrition *Maha Ahmed Mohamed Abdallah*	99
Chapter 6	The Potential Health Benefits of Seaweed and Seaweed Extracts *I. A. Brownlee, A. C. Fairclough, A. C. Hall and J. R. Paxman*	119
Chapter 7	Antioxidative Properties of Seaweed Components *R. Sowmya, N. M. Sachindra, M. Hosokawa and K. Miyashita*	137

Chapter 8	Halogenated Compounds from Seaweed, a Biological Overview *Clara Grosso, Juliana Vinholes, Patrícia Valentão and Paula B. Andrade*	**163**
Chapter 9	Sargassum Wightii – A Nature's Gift from the Ocean *Anthony Josephine and Sekar Ashok Kumar*	**185**
Chapter 10	Comparative Medicinal Properties of Seaweed Sulfated Polysaccharides *Wladimir Ronald Lobo Farias*	**209**
Chapter 11	Structural and Biological Insights into Antitumor Seaweed Sulfated Polysaccharides *Hugo Alexandre Oliveira Rocha, Leandro Silva Costa and Edda Lisboa Leite*	**221**
Index		**233**

PREFACE

Plant extracts have been utilized for treatment in medical conditions since ancient times. The decision to use these natural agents was initially supported only by empirical knowledge and/or common-sense. More recently, a body of scientific literature has been generated which confirms a biological basis for these types of medical effects. This book provides a summary of trends in biochemical/pharmacological research concerning seaweeds; an in-depth discussion of nutritional values and pharmaceutical effects of seaweed components and ecological factors responsible for altering algal systems and the possible anabolisms of these bioactive samples are described. The broad-range medicinal use of some seaweed components is reported as well

Chapter 1 - Acclimation to lowered temperature needs newly de novo synthesis of compatible solutes including free amino acids to protect the functions of membranes and cytoplasmic proteins. Since cold stress is one of the harmful abiotic stresses affecting the growth and development of seaweeds, it has been proposed that these organisms also employ compatible solutes for establishing cold tolerance; however, little is known about the compatible solute in red macroalgae. Thus, the authors focused on free amino acids to identify compatible solutes in the intertidal red macroalga *Porphyra yezoensis* whose life cycle consists of haploid leafy gametophyte (thallus) and diploid filamentous sporophyte (conchocelis). The results indicated that the content of free alanine is increased and decreased under cold and heat conditions, especially in the conchocelis, suggesting its function as a compatible solute in diploid generation. This profile of alanine content led us to speculate the inducible expression of a gene encoding alanine aminotransferase (AlaAT) by low temperature. The results of expression analysis of the *P. yezoensis AlaAT* (*PyAlaAT*) gene indicated the tight kinetic correlation between enhancement of the gene expression and accumulation of alanine under cold stress conditions. Taken together, alanine possibly functions as a compatible solute involved in the cold stress response in conchocelis of *P. yezoensis*.

Chapter 2 - Currently, society lives under a misleading apprehension of there being food abundance. Many people of the west are surrounded by fast food rich in calories and unsaturated fats, high powered advertising and over-consumption. The mass market has actually become accustomed to the expression of "junk food" to designate such offerings, but yet this highly processed "food" is consumed in large amounts. The consequences of consumption of these offerings for the mass (western) the lack of essential nutrients, obesity and diseases related to excessive intake of sugars (diabetes) and fat (arteriosclerosis), among others. It is worrying that the fast food trends of the west are being adopted seemingly without concern in developing countries as they become more prosperous, hence rates of associated disease are increasing.

What roles have the seaweeds in this picture?

Represent exactly the opposite: a natural food that gives us a highly nutritious but low in calories. Algae are therefore the best way to address the nutritional deficiencies of the current food, due to its wide range of constituents: minerals (iron and calcium), protein (with all essential amino acids), vitamins and fiber.

Contrary to what happens in East Asia, the West is more involved with use of seaweed as a source in thickeners and gelling properties of hydrocolloids extracted from seaweeds: carrageenan, agar and alginate (E407, E406 and E400, respectively), which are widely used in food industry, especially in desserts, ice cream, the fresh vegetable gelatin. Perhaps in most cases, the consuming public are blissfully unaware they are consuming seaweed derived products.

However attitudes are quite different in Asian cultures where seaweeds are highly valued and regarded for their appearance, texture, flavour and in a number of cases, beneficial health properties.

Some seaweeds can be rich in polysaccharides which, in the absence of appropriate enzymes, due to their long chain molecules , they are not broken down, nor absorbed by the digestive system and behave as soluble fiber, with no calories, having a positive impact on the regulation of intestinal transit.

From the composition of seaweed highlight: Presence of minerals with values about ten times higher than found in traditional vegetables, such as iron in *Himanthalia elongata* (Sea spaghetti) in comparison with that of *Lens esculenta* (lentils) or in the case of calcium present in *Undaria pinnatifida* (Wakame) and *Chondrus crispus* (Irish Moss), in comparison with milk; presence of proteins containing all essential amino acids, constituting a type of protein of high biological value, comparable in quality to the egg; presence of vitamins in significant quantities, in particular the presence of B_{12} (*Porphyra* spp.), absent in higher plants; *Palmaria palmata* and *Himanthalia elongata* are rich in potassium and, together with the algae of the genus *Porphyra* and *Laminaria*, have a ratio of sodium/potassium ratio considered optimal for human health.

This review aims to describe some of the key nutritional characteristics of the main algae used as human food and their potential in the nutraceuticals industry.

Chapter 3 - Brown and red seaweeds are often regarded as under-exploited marine bio-resources. Specifically, research on edible marine macroalgae is on the increase because they are most interesting as a source of macronutrients and associated bioactive compounds with high potentially economical impact in food and pharmaceutical industry, and public health. The authors' group has worked on the nutritional evaluation, physicochemical and biological properties of edible Spanish seaweeds. Thus, in brown: *Bifurcaria bifurcata*, *Fucus vesiculosus*, *Himanthalia elongata* (Sea spaghetti), *Laminaria digitata* (Kombu), *Saccharina latissima* (Sugar Kombu), *Undaria pinnatifida* (Wakame), and red seaweeds: *Chondrus crispus* (Irish moss), *Gigartina pistillata*, *Mastocarpus stellatus* and *Porphyra tenera* (Nori), total dietary fiber content ranges from 29-50% of which approximately 20-75% is soluble. For brown seaweeds, soluble fiber consists of uronic acids from alginates and neutral sugars from sulfated fucoidan and laminarin. For red seaweeds, main neutral sugars correspond to sulfated galactans such as carrageenan or agar. Insoluble fibers (7.4–40%) are essentially made of cellulose, with an important contribution of Klason lignin, up to 31% in *Fucus*. In Nori, insoluble fiber consists of a mannan and xylan. Protein content is generally higher in red (15–30%), than in brown seaweeds (7–26%), although protein digestibility is apparently low. Ash content is high (21–40%) and sulfate, related to the presence of sulfated polysaccharides, represents 7.4–57% of ash. Except for the brown seaweeds *Fucus* (2.5%) and *Bifurcaria* (5.6%), oil content is usually lower than 1%. Relevant biological properties of seaweeds (such as anticoagulant or antioxidant capacity) seem to be associated to sulfate content in sulfated polysaccharides and to a lesser degree to minor components, such as extractable polyphenols (0.4%). In brown seaweeds, the correlation among reduction power, radical-scavenging activity and total phenolic content would suggest the involvement of phenolic compounds in the antioxidant mechanisms, whereas in the case of red seaweeds, the role of sulfate-containing polysaccharides is presumably evidenced in the reduction power. It should be stressed that processing and storage conditions are essential for the optimal preservation of bioactive compounds (such as phenolic compounds and pigments) and their antioxidant activity. Regarding the main physicochemical properties of dietary fiber in seaweeds, related to the hydrophilic nature of sulfated polysaccharides, oil retention is low, while swelling, water retention, and cation exchange capacity are higher in brown than in red algae. The authors' results indicate that vibrational FTIR-ATR spectroscopy is a useful tool for a rapid and preliminary identification of the main natural phycocolloids (namely alginate, agar and carrageenan) in edible brown and red seaweeds. Moreover, other storage and structural polysaccharides in seaweeds such as laminarin, fucoidan and cellulose, as well as protein or sulfate, can be identified from specific absorption bands in their infrared spectra. Accordingly, alginate is the main polysaccharide in brown seaweeds, whereas all red seaweeds studied are carrageenan producers, except for Nori. In summary, edible seaweeds can be considered as an excellent source of dietary fiber, protein and minerals for human consumption. Moreover, algal sulfated polysaccharides potentially could afford natural antioxidants for the food, pharmaceutical and cosmetics industry.

Chapter 4 - The in vitro and in vivo antioxidant properties, total phenolic and chemical composition of various seaweeds is compiled. Seaweeds medicinal uses include for cardiovascular disease prevention, cholesterol-lowering, anti-diabetes, anti-coagulative, anti-inflammatory, immunomodulating and anti-cancer effects. The nutrients composition, vitamin C, tocopherol, dietary fibers, minerals, fatty acid and amino acid profiles of some tropical seaweeds is presented. Effects of tropical seaweeds in preventing cardiovascular diseases and cancer in animals via assessing the plasma and organs biomarkers will be given as example. Such biomarkers include activities of antioxidant enzymes such as superoxide dismutase (SOD), glutathione peroxidase (GSH-Px) and catalase (CAT); alanine aminotransferase (ALT), aspartate aminotransferase (AST), gama glutamyltransferase (GGT), creatinine kinase (CK), CK-MB isoenzyme, urea, creatinine and uric acid. Positive changes caused by dietary seaweeds on somatic index and histological changes in the liver, heart, kidney, brain, spleen and eye of the experimental animals are shown. The comparative in vivo cardiovascular protective effects of red and green tropical seaweeds in mammals fed on a rich lipogenic or sometimes called Western diet (24% fat and 1% cholesterol) are elaborated as a case. The potential anti-infective, antiviral and tissue healing properties of seaweeds are also incorporated.

Chapter 5 - Algae, as processed and unprocessed food, have a commercial value of several billion dollars annually. Approximately 500 species are used as food or food products for humans and about 160 species are valuable commercially.

Commercially available varieties of marine macroalgae are commonly referred to as "seaweeds". Seaweed is suitable for human and animal feed, as well as for fertilizer, fungicides, herbicides, and Phycocolloids (as Chlorophyta) are commonly used as food due to high contents of vitamins and minerals, Phaeophyta are typical suppliers of alginic acid. Rhodophyta are responsible to produce agar and carrageenan. However, there is a worldwide interest for macroalgal components, since they are used in medicine and in pharmacology for their antimicrobial, antiviral, antitumor, anticoagulant and fibrinolytic properties. Seaweeds contain high amounts of carbohydrates, protein and minerals. Because of their low fat contents and their proteins and carbohydrates, which cannot be entirely digested by human intestinal enzymes, they contribute few calories to the diet. The protein content of seaweed varieties varies greatly and demonstrates a dependence on such factors as season and environmental growth conditions. Thus edible marine seaweeds may be an important source of minerals, since some of these trace elements are lacking or very minor in land vegetables. About 221 seaweeds are utilized commercially worldwide of which 65% are used as human food. Most recently seaweeds have been utilized in Japan as raw materials in the manufacture of many seaweed food products, such as jam, cheese, wine, tea, soup and noodles. Currently, human consumption of green algae (5%), brown algae (66%) and red algae (33%) is high in Asia. However demand for seaweed as food has now also extended to other parts of the world.

Seaweeds have a wide range in mineral content, not found in edible land plants, is related to factors such as seaweed phylum, geographical origin and seasonal environmental and physiological variations. The mineral fraction of some seaweed even accounts for up to 40% of dry matter.

Algal products are used in the preparation or manufacture of many nonfood products. The agars, carrageenans, and alginates, collectively termed hycocolloids, are a major source of industrially important algal products. Agar and agaroses are used in medical and biological sciences for culture media and for gel electrophoresis. Agars are also used in many other products, including ion-exchange and affinity chromatography, pharmaceutical products, and fruit fly foods. Carrageenans are used as binders and thickeners in a wide variety of pastes, lotions, and water-based paints. Alginates are used to bind textile printing dyes, to stabilize paper products during production, to coat the surfaces of welding rods, to serve as binders and thickeners in numerous pharmaceutical products, and to act as binders in animal feed products

Chapter 6 - Edible seaweeds have historically been consumed by coastal populations across the globe. Today, seaweed is still part of the habitual diet in many Asian countries. Seaweed consumption also appears to be growing in popularity in Western cultures, due both to the influx of Asian cuisine as well as notional health benefits associated with consumption. Isolates of seaweeds (particularly viscous polysaccharides) are used in an increasing number of food applications in order to improve product acceptability and extend shelf-life.

Epidemiological evidence suggests regular seaweed consumption may protect against a range of diseases of modernity. The addition of seaweed and seaweed isolates to foods has already shown potential to enhance satiety and reduce the postprandial absorption rates of glucose and lipids in acute human feeding studies, highlighting their potential use in the development of anti-obesity foods. As seaweeds and seaweed isolates have the potential to both benefit health and improve food acceptability, seaweeds and seaweed isolates offer exciting potential as ingredients in the development of new food products.

This review will outline the evidence from human and experimental studies that suggests consumption of seaweeds and seaweed isolates may impact on health (both positively and negatively). Finally, this review will highlight current gaps in knowledge in this area and what future strategies should be adopted for maximising seaweed's potential food uses.

Chapter 7 - Seaweeds constitute one of the major components of diet in several Asian countries, particularly Japan, China and Korea. The consumption of seaweed as a part of diet has been shown to be one of the reasons for low incidence of breast and prostrate cancer in Japan and China compared to North America and Europe. Seaweeds are rich source of variety of nutrients and bioactive components. *In vivo* studies have demonstrated the anti-cancerous, anti-obesity, anti-inflammatory and anti-proliferative effects of seaweeds and their components. Reactive oxygen species (ROS) that include free radicals such as superoxide, hydroxyl and peroxyl are responsible for manifestation of oxidative stress related diseases like cancer and arteriosclerosis. ROS are the important initiators of lipid oxidation in biological membranes, which lead to many diseases. Thus, it is believed that the antioxidative components in seaweeds are responsible for their effects on disease protection. As the ROS are implicated in several diseases, antioxidants play an important role in preventing the interaction of reactive oxygen species with biological system. In addition to being implicated in diseases and aging, ROS also play an important role in chemical deterioration of food. The lipid oxidation initiated by ROS can lead to unacceptability of food products in addition to formation of harmful lipid oxidation products. Studies using model systems also reported that free radicals induce protein oxidation and prevention of protein oxidation by antioxidants has protective effects on lipid fractions. Antioxidants or ingredients having antioxidative properties are used extensively for improvement of food stability. With the focus being shifting towards finding alternatives for synthetic food ingredients, natural substances having antioxidative properties need to be further explored. Studies have been carried out on the antioxidative potential of different seaweeds mainly from the waters of China, Korea and Japan. Few studies have been reported on the antioxidative potential of extracts from Indian seaweeds. The presence of antioxidative substances in seaweeds is suggested to be an endogenous defence mechanism as a protection against oxidative stress due to extreme environmental conditions. The antioxidative components in seaweeds include chlorophyll and carotenoid pigments, vitamins like α-tocopherol and phenolic substances. The antioxidant activity of extracts from red algae, 'dulse' (*Palmaria palmate*) is associated with aqueous/alcohol soluble compounds characterized by phenolic functional groups with reducing activity. Antioxidant activity of red seaweed extracts correlates with their polyphenol content. Fucoxanthin, isolated from the brown seaweed Wakame exhibits various radical scavenging activities. The free radical scavenging activity of polysaccharide extracts from seaweeds, particularly the sulphated polysaccharides laminarin and beta-glucans, fucoidan, from seaweeds has been demonstrated. This chapter highlights the information available on the antioxidative potential of polyphenols and polysaccharides from different types of seaweeds and their potential health benefits.

Chapter 8 - Seaweeds are a renewable marine resource recognized as a rich provider of valuable compounds. Since they live in a competitive environment, marine algae developed different strategies to survive, including the biosynthesis of a variety of compounds from different metabolic pathways. Indeed, more than 15000 metabolites were already determined, belonging to several groups of primary and secondary metabolites. For this fact the interest of pharmacologists, physiologists and chemists in this group of living organisms has risen. The research on natural products of marine origin has shown that seaweeds are of unlimited potential for textile, fuel, plastics, paint, varnish, cosmetics, pharmaceutical and food industries. The last three take advantage from the several biological properties attributed to seaweeds.

The present chapter will focus on seaweeds' halogenated compounds that are naturally produced by Phaeophyta, Chlorophyta and Rhodophyta. Halogenation can occur in several classes of metabolites, like indoles, terpenes, acetogenins, phenols, fatty acids and volatile hydrocarbons. The incorporation of a halogen in such compounds can induce/increase their biological properties, like antimicrobial and antitumor. Consequently, the consumption of seaweeds and their based products constitute a benefit to human health and therefore they are seen as a potential medicinal food of the 21st century.

Chapter 9 - Seaweeds or marine algae have long been made up a key part of the Asian diet and are also consumed in other parts of the world. The relative longevity and health of Okinawan Japanese population have been attributed to the consumption of marine algae in their diet. The antique tradition and daily routine of consuming seaweeds in Asian countries has made possible a huge number of epidemiological researches to screen the health benefits coupled to seaweed consumption and for centuries, brown seaweeds has been hailed as a natural answer to a lengthier and a healthier life. Brown algae belong to a very large group called the heterokonts, most of which are colored flagellates. A notable example is Sargassum, which creates unique habitats in the Sargasso Sea (hence the name Sargassum). *Sargassum wightii* is one such species with diverse biological and pharmacological properties. Recently, sulphated polysaccharides from marine brown algae are receiving continuous attention, and especially as an antioxidant, sulphated polysaccharides have piqued the interest of many scientists and researchers as one of the ocean's greatest treasures. In line with this scenario, this review highlights the habitat, economical value and medicinal importance of *Sargassum wightii*, with special emphasis on the significance of sulphated polysaccharides from *Sargassum wightii*.

Chapter 10 - In recent years, much attention has been focused on sulfated polysaccharides (SP) isolated from natural sources. They are found in various marine organisms and their biological activities are of great interest in the medical sciences and animal health. Anticoagulant and antithrombotic properties have been the most widely exploited SP biological's activities but these molecules also show other activities such as antitumor and antioxidant properties, inhibition of the human complement system, and immunomodulatory and antinociceptive actions. These biological activities appear to be dependent on the sulfation content and/or position of the sulfate groups. A repeating structure (4-α-D-Galp-1→3-β-D-Galp-1→) with a variable sulfation pattern was found for a sulfated galactan purified from the red algae *Botryocladia occidentalis*. The presence of two sulfate esters to a single alpha-galactose residue enhances the anticoagulant action. The antithrombotic activity of these sulfated galactans was investigated on an experimental thrombosis model and in contrast with heparin, the sulfated galactans showed a dual dose-response curve preventing thrombosis at low doses but losing the effect at higher doses. The SP isolated from the seaweed *Champia feldmannii* promotes an antitumor activity with high inhibition rates of sarcoma 180 tumor development. It was demonstrated that this polymer acts as an immunomodulatory agent, raising the production of specific antibodies. This SP was not antiinflammatory, but rather induced maximal edematogenic activity, increased vascular-permeability and stimulated neutrophil migration. The polymer was also antinociceptive and extended human plasma coagulation time by 3 times, suggesting that this molecule may be an important immunostimulant. Immunomodulatory agents such as SP from marine algae have also been widely used to minimize stress in cultivated aquatic organisms. Stress is the most powerful immunesuppressor agent on aquaculture, causing the decrease of animal's natural defenses, leaving them weakened and susceptible to contaminations by pathogens. The administration of SP from the seaweeds *B. occidentalis*, *H. pseudofloresii* and *Spatoglossum schroederi* enhances *Litopenaeus vannamei* shrimp survival after stress and SP from *B. occidentalis* improved tilapia, *Oreochromis niloticus*, growth. Fish survival was also increased after the administration of SP extracted from the red marine algae *Gracilaria caudata* to *O. niloticus* post-larvae. This large range of biological activities expressed by SP from seaweed in human or animal biological systems turns these molecules important tools to promote human and animal health.

Chapter 11 - Sulfated polysaccharides comprise a complex group of macromolecules. These anionic polymers are widespread in nature, occurring in a great variety of organisms such as mammals and invertebrates. Seaweeds are the most important source of non-animal sulfated polysaccharides. Furthermore, the structure of algal sulfated polysaccharides varies according to the species of seaweed. Thus, each new sulfated polysaccharide purified from a seaweed is a new compound with unique structures and, consequently, with potential novel biological activities. Sulfated polysaccharides are found in varying amounts in three major divisions of marine algal groups, Rhodophyta, Phaeophyta and Chlorophyta. These compounds found in Rhodophyta are manly galactans consisting entirely of galactose or modified galactose units. The general sulfated polysaccharides of Phaeophyta are called fucans, which comprise families of polydisperse molecules based on sulfated L-fucose. Heterofucans are also called fucoidans. The major polysaccharides in Chlorophyta are polydisperse heteropolysaccharides, although, homopolysaccharides also may be found. The seaweeds are an untapped source of bioactive sulfated polysaccharides and the marine pharmacology research during the last years, with researchers from several international research institutes, contributing to the preclinical pharmacology of several polysaccharides which are part of the preclinical marine pharmaceuticals. Here the authors have reviewed publications regarding the bioactivity of seaweed polysaccharides-rich extracts or as yet structurally uncharacterized sulfated polysaccharides that showed several promising biological activities, with emphasis on antitumor. In addition, the present paper will also review the recent progress in research on structural features and the major antitumor effect of the marine algal biomaterials in a comprehensive manner. Moreover, when possible, the relationship between structure and antitumoral action of algal sulfated polysaccharides will also be reviewed.

Chapter 1

ALANINE IS A POSSIBLE COMPATIBLE SOLUTE INVOLVED IN COLD ACCLIMATION IN THE MARINE RED ALGA PORPHYRA YEZOENSIS

Koji Mikami[1,*], Megumu Takahashi[1], Ryo Hirazi[2], Takehiko Yokoyama[3], Misako Taniguchi[4], Naotsune Saga[1], and Tomoko Mori[4]

[1]Faculty of Fisheries Sciences, Hokkaido University, Hakodate, Japan
[2]Graduate School of Fisheries Sciences, Hokkaido University, Hakodate, Japan
[3]School of Fisheries Sciences, Kitasato University, Sagamihara, Japan
[4]NIBB Core Research Facilities, National Institute for Basic Biology, Okazaki, Japan

ABSTRACT

Acclimation to lowered temperature needs newly de novo synthesis of compatible solutes including free amino acids to protect the functions of membranes and cytoplasmic proteins. Since cold stress is one of the harmful abiotic stresses affecting the growth and development of seaweeds, it has been proposed that these organisms also employ compatible solutes for establishing cold tolerance; however, little is known about the compatible solute in red macroalgae. Thus, we focused on free amino acids to identify compatible solutes in the intertidal red macroalga *Porphyra yezoensis* whose life cycle consists of haploid leafy gametophyte (thallus) and diploid filamentous sporophyte (conchocelis). The results indicated that the content of free alanine is increased and decreased under cold and heat conditions, especially in the conchocelis, suggesting its function as a compatible solute in diploid generation. This profile of alanine content led us to speculate the inducible expression of a gene encoding alanine aminotransferase

---

[*] Email: komikami@fish.hokudai.ac.jp.

(AlaAT) by low temperature. The results of expression analysis of the *P. yezoensis AlaAT* (*PyAlaAT*) gene indicated the tight kinetic correlation between enhancement of the gene expression and accumulation of alanine under cold stress conditions. Taken together, alanine possibly functions as a compatible solute involved in the cold stress response in conchocelis of *P. yezoensis*.

1. INTRODUCTION

Seasonal and daily changes in environmental conditions create abiotic stresses affecting the growth and development of sessile plants. Since low temperature is one of the most harmful stresses, plants generally acclimate to non-freezing low temperatures to achieve freeze tolerance [1-3]. To respond and acclimate to non-freezing low temperatures, metabolic activities are changed by increasing the contents of protective compounds, called compatible solutes, via de novo synthesis. Compatible solutes, which are soluble in water and non-toxic at high concentrations, have an ability to protect membranes and proteins against toxic effects such as membrane disorder and protein denaturation by the production of free radicals and reactive oxygen species (ROS) [2]. It is well known that free amino acids including their derivatives, like proline and glycinebetaine, and carbohydrates including sugars and polyols, such as glycerol, mannitol, trehalose and inositols, act as compatible solutes in both prokaryotic and eukaryotic organisms [2,4,5]. For example, the cytoplasmic concentrations of proline, glycinebetaine and mannitol are increased in rye, barley and brown algae under low but non-freezing temperature [6-8].

The accumulation of compatible solute requires the new expression of genes involved in their biosynthesis. Increases in the proline content in land plants is regulated by the activation of a gene encoding the Δ^1-pyrroline-5-carboxylase (P5C) synthase (P5CS) and inactivation of a gene for P5C reductase [9,10]. Moreover, activation of a gene encoding mannitol-1-phosphate dehydrogenase is responsible for the accumulation of mannitol in the brown alga *Ectocarpus siliculosus* [11]. Thus, analysis of the stress-inducible expression of genes involved in the biosynthesis of compatible solutes is essential for understanding how plants respond and acclimate to abiotic stresses.

The red alga *Porphyra yezoensis*, an important cultivar in Japan for producing nori, is a sessile seaweed living at the intertidal zone and therefore usually exposed to abiotic stresses such as temperature, salinity and drought stresses, suggesting the presence of compatible solutes in intertidal red algae including *P. yezoensis*. In fact, it has been observed that salinity tolerance requires accumulation of the photosynthetic product floridoside in *P. purpurea* [12]; however, the compatible solutes involved in the establishment of cold tolerance remain to be identified. We here summarize recent progress in our research on the free amino acids that probably functions in cold acclimation as compatible solutes in *P. yezoensis*.

2. AMINO ACID COMPOSITIONS IN HAPLOID AND DIPLOID GENERATIONS

Bangiophycean algae including *P. yezoensis* represent a bi-phase life cycle that consists of a haploid leafy gametophyte (thallus) and a diploid filamentous sporophyte (conchocelis), whose appearance depends on the season; the thallus and conchocelis appear in winter and summer, respectively [13-15]. Free amino acid composition has been analyzed in the thallus [16,17] but not yet in the conchocelis. Therefore, we compared free amino acid compositions between two morphologically different generations, both of which were grown at 15°C under a 10 h light/14 h dark illumination cycle. Table 1 showes the free amino acid composition in the thallus and conchocelis of the laboratory culture strain *P. yezoensis* TU-1, indicating the presence of four major amino acids such as glutamic acid, alanine, taurine and aspartic acid in conchocelis as found in the thallus.

Table 1. Comparison of free amino acid composition at 15°C between the thallus and conchocelis in *P. yezoensis*

Amino acid	Thallus (μmol/g)	Conchocelis (μmol/g)
Taurine	12.2	15.7
Alanine	8.8	22.5
Glutamic acid	7.9	17.4
Aspartic acid	3.1	4.0
Proline	0.2	1.6
Phosphoserine	0.2	0.7
Threonine	0.4	1.5
Serine	1.0	1.2
Asparagine	0.9	1.6
Glutamine	0.2	1.5
Glycine	0.6	0.9
Citrulline	1.2	2.0
α-aminobutanoic acid	0.1	0.3
Valine	<0.1	0.2
β-alanine	0.1	0.8
Lysine	<0.1	<0.1
Hydroxylysine	0.6	n.d.
Methionine	n.d.	<0.1
Cystathione	n.d.	<0.1
Isoleucine	n.d.	<0.1
Gamma aminobutyric acid	n.d.	<0.1
Arginine	n.d.	<0.1
Total	37.6	73.8

n.d. Not detected.

[Methods] To determine the free amino acid contents, the thallus and conchocelis dried by evaporation were dissolved in 0.02N HCl to a final concentration of 1 g (fresh weight)/ml and extracted using the vortex mixer. Extracts were filtered through a 0.22 μm Millipore filter by centrifugation at 17,800 x *g* for 10 min. The resultant filtrates were diluted 10 or 100 fold by 0.02N HCl and subjected to analysis by the ninhydrin method with a Hitachi L-8500A amino acid analyzer, in which free amino acids (proline and hydroxyproline) and other free amino acids were detected at 440 nm and 570 nm, respectively.

3. TEMPERATURE-DEPENDENT CHANGES IN CONTENTS OF MAJOR FREE AMINO ACIDS

To investigate the temperature-dependency of changes in the free amino acid composition, *P. yezoensis* strain TU-1, normally cultured at 15°C, was incubated at 5, 10, 15, 20 and 25°C for 7 days. Amino acid analyses indicated that the total amino acid content was increased and decreased under cold and heat stress conditions, respectively, in conchocelis, whereas no change in amino acid content was observed in the thallus (Figure 1A). There is thus a relationship between the increase in free amino acid content and the cold stress response in the conchocelis.

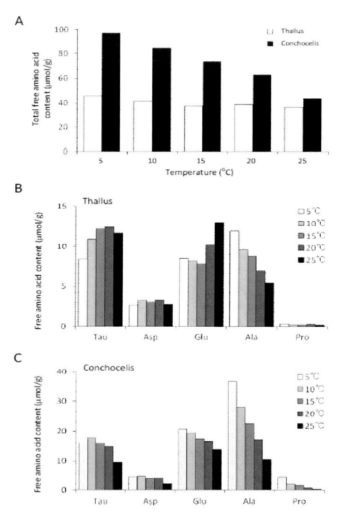

Figure 1. Effects of temperature changes on contents of free amino acids in *P. yezoensis*. The thallus and conchocelis were transferred from 15°C to 5, 10, 15, 20 and 25°C and incubated for 7 days. Amino acid analysis was performed as mentioned in Table 1. (A) Comparison of total amino acid contents under various temperatures between thallus and conchocelis. White and black boxes represent thallus and conchocelis, respectively. (B) Effects of temperature changes in the thallus on the contents of four major amino acids and proline. (C) Effects of temperature changes in the conchocelis on the contents of four major amino acids and proline.

Changes in the contents of the major amino acids were analyzed. In the thallus, alanine was increased and decreased under cold and heat stress conditions, respectively, although glutamic acid and taurine showed a reciprocal pattern (Figure 1B). However, there was not much variation in the contents of these free amino acids and that this lack of variation may be a responsible for keeping the level of total amino acid contents under temperature changes.

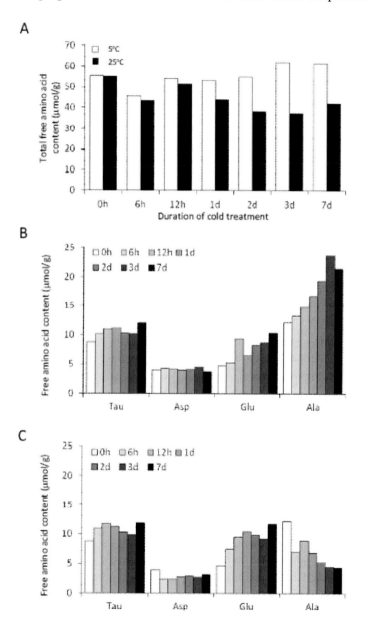

Figure 2. Kinetic analysis of temperature-dependent changes in the contents of four major free amino acids in the thallus. Amino acid analysis was performed as mentioned in Table 1. (A) Effects of incubation at 5°C (white bar) and 25°C (black bar) on accumulation kinetics of total amino acid contents in the thallus. (B) Changes in amino acid contents in the thallus during incubation at 5°C. (C) Changes in amino acid contents in the thallus during incubation at 25°C.

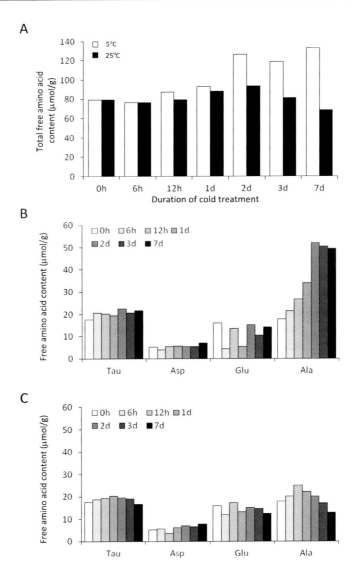

Figure 3. Kinetic analysis of temperature-dependent changes in the contents of four major free amino acids in the conchocelis. Amino acid analysis was performed as mentioned in Table 1. (A) Effects of incubation at 5°C (white bar) and 25°C (black bar) on accumulation kinetics of total amino acid contents in the conchocelis. (B) Changes in amino acid contents in the conchocelis during incubation at 5°C. (C) Changes in amino acid contents in the conchocelis during incubation at 25°C.

In contrast, the accumulation of a large amount of alanine was observed under cold stress conditions in the conchocelis, in which the alanine content increased from 22% to 37% of the total amino acid content by a temperature downshift from 15°C to 5°C (Figure 1C). Since glutamic acid and taurine showed only a small increase under cold stress conditions (Figure 1C), alanine is a possible compatible solute involved in the cold stress response in the conchocelis.

Interestingly, the cold-dependent accumulation of proline was observed in the conchocelis, although no significant variation of this amino acid was detected in the thallus (Figures 1A and 1B). Proline is a major compatible solute in land plants for establishing salinity, drought and cold tolerances [6,8,18]. However, since the content of proline was very low, the significance of the proline accumulation under cold stress in *P. yezoensis* is not clear.

Next, kinetic analysis of the temperature-dependent changes in the amino acid content was performed. In the thallus, the total amino acid content increased and decreased slightly under 5°C and 25°C, respectively (Figure 2A). However, by incubation at 5°C, the accumulation of alanine began within 6 h after cold treatment and the alanine content was increased from 22% to 35% of the total amino acid after 7 days, although only a small accumulation of glutamic acid and taurine was found (Figure 2B). In conchocelis, cold treatment resulted in a remarkable accumulation of alanine and total amino acid contents, an accumulation which was also observed within 6 h after a temperature downshift; however, there were no significant changes in the contents of the other major amino acids (Figures 3A and 3B). Although the ratio of the increase in alanine content in the conchocelis was similar to that in the thallus, the absolute amount of alanine was 2 fold higher in the conchocelis than in the thallus (compare Figures 2B and 3B). Taken together, alanine accumulation is responsible for the increase in the total amino acid content in the conhocelis under low temperature conditions.

The above results show there is a close reciprocal relationship between the contents of free alanine and changes in the temperature in the conchocelis of *P. yezoensis*, suggesting the ability of alanine to function as a compatible solute under cold stress conditions in diploid generation. This proposal is supported by a decrease in the alanine content by a temperature upshift (Figure 3C). At present, the functional significance of alanine in the thallus is unknown.

4. ABSENCE OF D-AMINO ACIDS UNDER COLD STRESS CONDITIONS

Several species of marine algae and invertebrates are known to contain free D-amino acids [19-23]. Among them, D-alanine plays an important role as the compatible solute in marine invertebrate which acclimate them to hyperosmotic and anoxic stresses [21,23]. In fact, D-alanine has been identified in brown algae such as *Sargassum fusiforme* (formerly *Hizikia fusiformis*), *Heterochordaria abietina* and *Sargassum nigrifolium*, red algae like *Rhodomela larix* and *Gloiopeltis furcata*, and green algae such as *Monostroma nitidum* and *Ulva pertusa* [19,20]. Although the functions of these D-amino acids in seaweeds are largely unknown, it is possible that D-amino acids act as compatible solutes in these macroalgae as is in marine vertebrates.

To test this possibility, the presence of D-amino acids was examined in *P. yezoensis* under cold and heat stress conditions. The results indicated the complete absence of any D-amino acids in *P. yezoensis* (data not shown), indicating that D-alanine is not employed as a compatible solute and that the de novo synthesis of L-alanine is responsible for the accumulation of alanine under cold stress in *P. yezoensis*. Thus, it is possible that the mechanisms for establishing cold tolerance through the de novo synthesis of amino acids as compatible solutes differ by species in seaweeds.

5. Genes Involved in Alanine Biosynthesis

Alanine biosynthesis is mainly regulated by alanine aminotransferase (AlaAT) that catalyzes the reversible reaction of the interconversion of pyruvate and glutamic acid to alanine and 2-oxoglutarate.

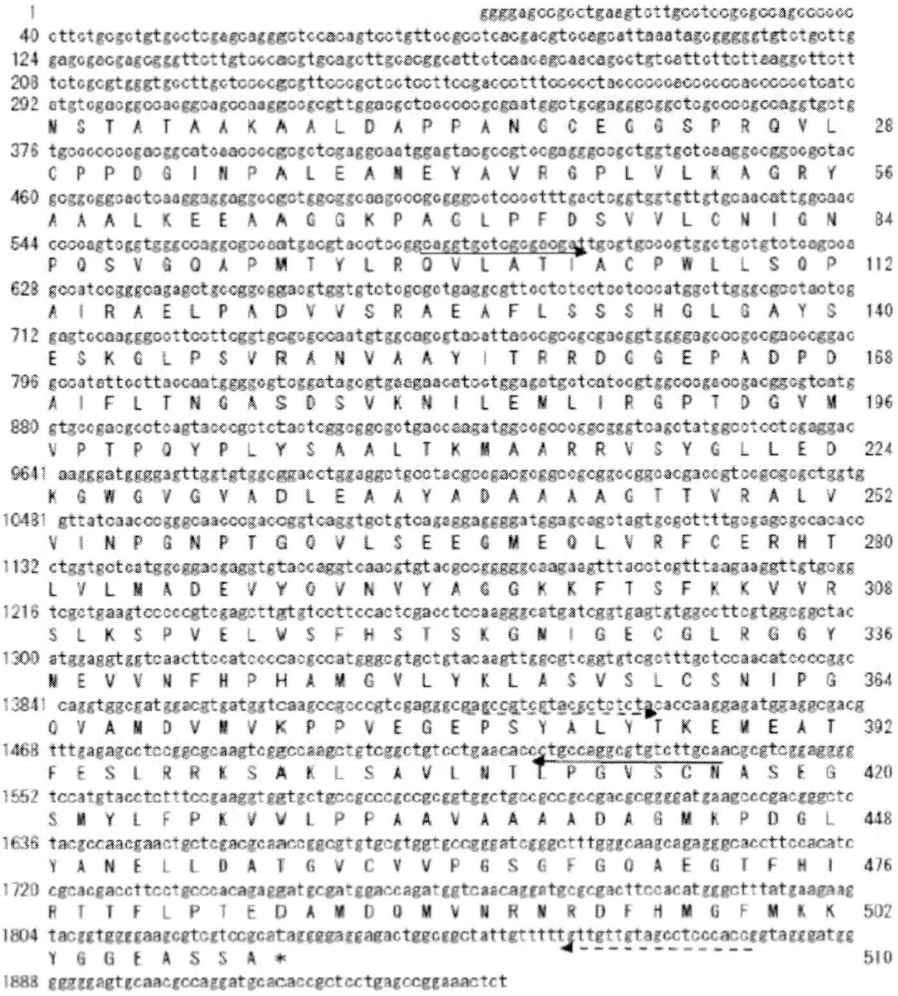

Figure 4. Nucleotide and deduced amino acid sequences of the *PyAlaAT* gene. From the EST database (http://est.kazusa.or.jp/en/plant/porphyra/EST/index.html), the N-terminal part was found as a contig of AU187060 and AV188833, whereas the other contig of AU191646 and AV429515 corresponded to the C-terminal part. Since these contigs were not overlapped, a missing part of the *PyAlaAT* cDNA was obtained by RT-PCR with primers 5'-CAGGTGCTCGCGACGATTG-3' and 5'-TTGCAAGACACGCCTGGCAG-3', for which total RNA was extracted using an RNeasy plant mini kit (Qiagen) from the thallus and further purified with a TURBO DNA-free kit (Ambion, TX, USA). The conditions of RT-PCR were as follows: 94°C for 1 min, followed by 30 cycles of 94°C for 30 sec, 60°C for 30 sec, 72°C for 2 min, and at 72°C for 7 min with LA Taq with GC buffer (TaKaRa Bio. Inc., Japan). Numbers on the left and right refer to nucleotide and amino acid positions relative to the first residues. Positions corresponding to the primers using amplification of the missing region and the expression analysis shown in Figure 6 are indicated by solid and dotted arrows, respectively.

In land plants, the close relationship between increase in AlaAT activity based on the *AlaAT* gene expression and alanine accumulation has been established under hypoxic conditions [24-27]. However, it is unknown if activation of the *AlaAT* gene is responsible for the accumulation of alanine under cold stress in *P. yezoensis*. To address this question, the full-length *P. yezoensis AlaAT* (*PyAlaAT*) cDNA was isolated (Figure 4). The deduced amino acid sequence of PyAlaAT shows high homology to those of AlaATs from bacteria, fungi, vertebrates and land plants (data not shown), suggesting that the *PyAlaAT* gene encodes a functional enzyme.

Alanine-glyoxylate transaminase (AGT) produces pyruvate and glycine from alanine and glyoxylate, which is considered to be a detoxification reaction by negative regulation of alanine production [28]. Thus, a full-length cDNA encoding *P. yezoensis* AGT (*PyAGT1*) was isolated (Figure 5). The amino acid sequences of PyAGT1 demonstrated the complete conservation of four residues identified in the aminotransferase domain as AGT1 from *Arabidopsis thaliana* [29], indicating that PyAGT1 is functional enzyme with aminotrasferase activity.

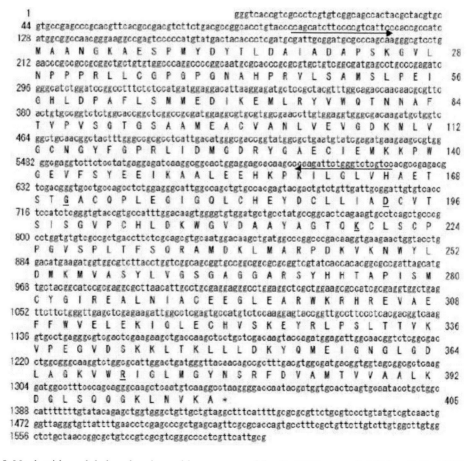

Figure 5. Nucleotide and deduced amino acid sequences of the *PyAGT1* gene. A full-length *PyAGT1* cDNA was obtained from unpublished full-length cDNA information of *P. yezoensis*. Four amino acid residues conserved in an aminotransferase domain are indicated as underlined bold characters. Numbers on the left and right refer to the nucleotide and amino acid positions relative to the first residues. Positions corresponding to primes for expression analysis are indicated by arrows.

6. COLD-INDUCIBLE EXPRESSION OF THE *PYALAAT* GENE IN *P. YEZOENSIS*

To examine the possibility that the cold stress-dependent accumulation of alanine requires the expression of the *AlaAT* gene in *P. yezoensis*, the expression kinetics of the *PyAlaAT* and *PyAGT1* genes were investigated by RT-PCR after temperature downshift of the culture conditions from 15°C to 5°C. In the thallus, expression of the *PyAlaAT* gene was gradually but weakly increased by temperature downshift, although no significant induction of the *PyAGT1* gene was observed (Figure 6). On the other hand, strong induction of the *PyAlaAT* gene expression by cold treatment was observed in the conchocelis (Figure 6), which is consistent with the fact that alanine is the most accumulating free amino acid under low temperature conditions in the conchocelis (Figure 3B). The *PyAGT1* gene maintained a constitutive level of expression, as in the thallus (Figure 6). Importantly, alanine accumulation by temperature downshift occurred early in both thallus and conchocelis (Figures 2 and 3), there is therefore good correlation between enhancement of the *PyAlaAT* gene expression and alanine accumulation. Together, these results show that the cold-inducible expression of the *PyAlaAT* gene is responsible for accumulation of alanine by low temperature in the conchocelis of *P. yezoensis*; however, the possibility of functional involvement of alanine as the compatible solute in thallus cannot be ruled out.

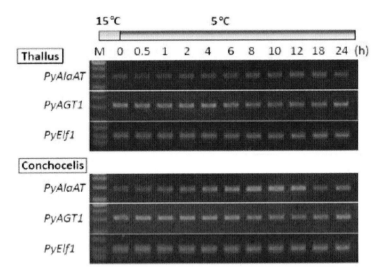

Figure 6. Cold-inducible expression of the *PyAlaAT* gene. Both the thallus and conchocelis were treated with 5°C for 0.5, 1, 2, 4, 6, 8, 10,12, 18 and 24 h, and then total RNA was extracted using an RNeasy plant mini kit (Qiagen) from each sample and further purified with a TURBO DNA-free kit (Ambion, TX, USA). RT-PCR with LA Taq with GC buffer (TaKaRa Bio Inc., Japan) was performed as follows: 94°C for 1 min followed by 28 cycles of 94°C for 30 sec, 60°C for 30 sec and 72°C for 30 sec, and at 72°C for 7 min with two gene-specific primer sets, 5'-AGCCGTCGTACGCTCTCTACAC-3' and 5'-CTACCGGTGGGAGGCTACAA-3', and 5'-AGCATCTTCCCCGTCATTCC-3' and 5'-TGGACGAGACCCAGAATCTTG-3' to detect expression of the *PyAlaAT* and *PyAGT1* genes, respectively. To make a reference, RT-PCR was performed for the *PyElf1* gene [33] with primers 5'-CACCATGGGCAAGCGCAAA-3' and 5'-CTCCTCCCCCCCATTCACCT-3', under the same conditions as those for the *PyAlaAT* and *PyAGT1* genes. Amplified fragments were 455, 524 and 259 bp for the *PyAlaAT*, *PyAGT1* and *PyElf1* genes, respectively.

7. DIFFERENCES IN THE STRATEGIES FOR USING FREE AMINO ACIDS BETWEEN SEA AND LAND PLANTS

It is clear that *P. yezoensis* accumulates free alanine under cold stress conditions (Figures 1-3). Non-freezing low temperature mainly causes a decrease in membrane fluidity and formation of radical producing oxidative stress [2]. Although the abilities of proline for membrane protection and detoxification of ROS have been demonstrated [30,31], it is not well known that alanine can act as a protector of membrane disorder and a radical and ROS scavenger. Thus, the functional significance of alanine under cold stress conditions in *P. yezoensis* is unknown at present. However, land plants use alanine under hypoxic stress conditions, in which close correlation between the accumulation of alanine and activation of the *AlaAT* gene are observed [24-27]. Since hypoxic stress by flooding depletes oxygen and nitrogen through inhibition of their entry from the environment into plant cells, alanine seems to play a role in enhancing the recycling efficiency of oxygen and nitrogen under hypoxic conditions. The above findings suggest that strategies to use alanine as a compatible solute are different between sea and land plants.

Employment of compatible solutes in the cold stress response is also different in sea and land plants. In land plants, proline functions under salinity, hyperosmotic, drought and cold stresses [3,4,9,10,18] and hypoxic stress requires alanine [24-27], indicating that the signal transduction pathways for abiotic stresses can selectively target the *P5CS* or *AlaAT* gene to archive a correct response to each stress, in which the *AlaAT* gene is not targeted by cold stress signaling. However, *P. yezoensis* synthesizes alanine under cold stress via the cold-enhanced expression of the *PyAlaAT* gene (Figure 6). Thus, since glutamic acid is a precursor of both alanine and proline, the difference in the fate of glutamic acid seems to be responsible for the differential use of alanine and proline between sea and land plants under low temperature. To confirm this possibility, it would be interesting to test whether artificial accumulation of proline in *P. yezoensis* and alanine in land plants establishes cold tolerance.

CONCLUSION

A cold-dependent increase and heat-dependent decrease in alanine contents are remarkable in the conchocelis (Figures 1 and 3) strongly suggesting that alanine is a possible compatible solute for cold acclimation in the diploid generation of *P. yezoensis*. However, since a similar but weak pattern in the changes in the contents of free amino acids was observed in the thallus (Figures 1 and 2), it is possible that alanine also functions in haploid generation as the compatible solute. These proposals are further supported by the fact that both accumulation of alanine and enhancement of the *PyAlaAT* gene expression begin within 6 h after a temperature downshift, especially in the conchocellis and slightly in the thallus (Figure 6). This work is the first to identify a candidate of the compatible solute involved in cold acclimation in red macroalgae.

The functional significance of alanine as the compatible solute is still largely unknown in *P. yezoensis*. Elucidation of the mechanisms regulating the cold-enhanced expression of the *PyAlaAT* gene is therefore important for understanding the metabolic alteration required for establishment of cold-tolerance via de novo synthesis of the compatible solute in *P. yezoensis*.

Thus, it is necessary to examine what happens by genetic inactivation of the *PyAlaAT* gene under cold stress conditions, which provides direct evidence of the importance of alanine as the compatible solute. In addition, it is unknown whether *P. yezoensis* properly use different free amino acids as compatible solutes for other stresses. To resolve this question, it is also necessary to examine whether alanine is accumulated via the activation of the *PyAlaAT* gene under salinity, drought and other stress conditions and whether other free amino acids and/or carbohydrates act as compatible solutes in *P. yezoensis*. Since a gene manipulation technique has not yet been established in any seaweeds including *P. yezoensis* [32], establishment of a stable transformation system is an important challenge for advancing the study of the molecular mechanisms of environmental stress responses through the identification of compatible solutes in *P. yezoensis*.

ACNOWLEDGEMENTS

This work was supported by the Joint Studies Program on Model Organisms and Technical Advancement (2009-2010) of the National Institute for Basic Biology (no. 9-201) to K.M. and in part by a grant from the Regional Innovation Cluster Program (Global Type) from the Ministry of Education, Culture, Sports, Science, and Technology, Japan to N. S. and the Hokusui Foundation to K.M.

REFERENCES

[1] Thomashow MF. Plant cold acclimation: freezing tolerance genes and regulatory mechanisms. *Annual Review of Plant Physiology and Plant Molecular Biology*, 1999, 50, 571-599.

[2] Beck EH, Fettig S, Katja C, Hartig K, Bhattarai T. Specific unspecific responses of plants to cold and drought stress. *Journal of Bioscience*, 2007, 32, 501-510.

[3] Janská A, Maršík P, Zelenková S, Ovesná J. Cold stress and acclimation – what is important for metabolic adjustment? *Plant Biology*, 2010, 12, 395-405.

[4] Rathinasabapathi B. Metabolic engineering for stress tolerance: installing, osmoprotectant synthesis pathways. *Annals of Botany*, 2000, 86, 709-716.

[5] Empadinhas N, de Costa MS. Osmoprotection mechanisms in prokaryotes: distribution of compatible solutes. *Internal Microbiology*, 2008, 11, 151-161.

[6] Reed RH, Davison IR, Chudek JA, Foster R. The osmotic role of mannitol in the Phaeophyta: an appraisal. *Phycologia*, 1985, 24, 35-47.

[7] Koster KL, Lynch DV. Solute accumulation and compartmentation during the cold acclimation of Puma rye. *Plant Physiology*, 1992, 98, 108-113.

[8] Nomura M, Morimoto Y, Yasuda S, Takebe T, Kishitani S. The accumulation of glycincebetaine during cold acclimation in early and late cultivars of barley. *Euphytica*, 1995, 83, 247-250.

[9] Y Yoshiba Y, Kiyosue T, Nakashima K, Yamaguchi-Shinozaki K, Shinozaki K. Regulation of levels of proline as an osmolyte in plants under water stress. *Plant Cell Physiology*, 1997, 38, 1095-1102.

[10] Verbruggen N, Hermans C. Proline accumulation in plants: a review. *Amino acids,* 2008, 35, 753-759.

[11] Rousvoal S, Groisiller A, Dittami SM, Michel G, Boyen C, Tonon T. Mannitol-1-phosphate dehydrogenase activity in Ectocarpus silicuosus, a key role for mannitol synthesis in brown algae. *Planta*, 2011, 233, 261-273.

[12] Reed RH, Collins JC, Russel G. The effects of salinity upon galactosyl-glycerol content and concentration of the marine red alga *Porphyra purpurea* (Roth) C.Ag. *Journal of Experimental Botany*, 1980, 31, 1539-1554.

[13] Blouin NA, Brodie JA, Grossman AC, Xu P, Brawley SH. *Porphyra*: a marine crop shaped by stress. *Trends in Plant Science,* 2011, 16, 29-37.

[14] Mikami K. Migrating plant cell: F-actin asymmetry directed by phosphoinositide signaling. In: Lansing S and Rousseau T (eds). *Cytoskeleton: Cell Movement, Cytokinesis and Organelles Organization.* New York, NY: Nova Science Publishers, 2010. p. 205-218.

[15] Mikami K, Li L, Takahashi M. Monospore-based asexual life cycle in *Porphyra yezoensis*. In: Mikami K (ed). *Porphyra yezoensis: Frontiers in Physiological and Molecular Biological Research.* New York, NY: Nova Science Publishers, 2011. in press.

[16] Niwa K, Furuita H, Aruga Y. Free amino acid contents of the gametophytic blades from the green mutant conchocelis and the heterozygous conchocelis in *Porphyra yezoensis* Ueda (Bangioles, Rhodophyta). *Journal of Applied Phycology*, 2003, 15, 407-413.

[17] Niwa K, Furuita H, Yamamoto T. Changes of growth characteristics and free amino acid content of cultured *Porphyra yezoensis* Ueda (Bangioles Rhodophyta) blades with the progression of the number of harvests in a nori farm. *Journal of Applied Phycology*, 2008, 20, 687-693.

[18] Yuanyuan M, Yaki Z, Jiang L, Hongbo S. Roles of plant soluble sugars and their responses to plant cold stress. *African Journal of Biotechnology*, 2009, 8, 2004-2010.

[19] Nagahisa E, Kan-no N, Sato M, Sato Y. Occurrence of free D-aspartic acid in marine macroalgae. *Biochemistry International*, 1992, 28, 11-19.

[20] Nagahisa E, Kan-no N, Sato M, Sato Y. Occurrence of free D-alanine in marine macroalgae. *Bioscience, Biotechnology, and Biochemistry*, 1995, 59, 2176-2177.

[21] Fujimori T, Abe H. Physiological roles of free D- and L-alanine in the crayfish Procambarus clarkia with special reference to osmotic and anoxic stress responses. *Comparative Biochemistry and Physiology*, 2002, 131A, 893-900.

[22] Yokoyama T, Kan-no N, Ogata T, Kotaki Y, Sato M, Nagahisa E. Presence of free D-amino acids in microalgae. *Bioscience, Biotechnology, and Biochemistry*, 2003, 67, 388-392.

[23] Abe H, Yoshikawa N, Sarower MG, Okuda S. Physiological function and metabolism of free D-alanine in aquatic animals. *Biologica. Pharmaceutical and Bulletin*, 2005, 28, 1571-1577.

[24] de Sousa CAF, Sodek L. Alanine metabolism and alanine aminotransferase activity in soybean (Glycine max) during hypoxia of the root system and subsequent return to normoxia. *Environmental and Experimental Botany*, 2003, 50, 1-8.

[25] Ricoult C, Echeverria LO, Cliquet J-B, Limami AM. Characterization of alanine aminotransferase (AlaAT) multigene family and hypoxic response in young seedlings

of the model legume *Medicago truncatula*. *Journal of Experimental Botany*, 2006, 57, 3079-3089.

[26] Miyashita Y, Dolferus R, Ismond KP, Good AG. Alanine aminotransferase catalyses the breakdown of alanine after hypoxia in *Arabidopsis thaliana*. *Plant Journal*, 2007, 49, 1108-1121.

[27] Rocha M, Sodek L, Licausi F, Hameed MW, Dornelas MC, van Dongen JT. Analysis of alanine aminotransferase in various organs of soybean (Glycine max) and in dependence of different nitrogen fertilizers during hypoxic stress. *Amino Acid*, 2010, 39, 1043-1053.

[28] Danpure CJ, Lumb MJ, Birdsey GM, Zhang X. Alanine: glyoxylate aminotransferase peroxisome-to-mitochondrion mistargeting in human hereditary kidney stone disease. *Biochimica et Biophysica Acta,* 2003, 1647, 70-75.

[29] Liepman AH, Olsen LJ. Peroxisomal alanine: glyoxylate aminotransferase (AGT1) is a photorespiratory enzyme with multiple substrates in *Arabidopsis thaliana*. *Plant Journal,* 2001, 25, 487-498.

[30] Kaul S, Sharma SS, Mehta IK. Free radical scavenging potential of L-proline: evidence from in vitro assays. *Amino Acids*, 2008, 34, 315-320.

[31] Banu MNA, Hoque MA, Watanabe-Sugimoto M, Islam MM, Uraji M, Matsuoka K, Nakamura Y, Murata Y. Proline and glycinebetaine ameliorated NaCl stress via scavenging of hydrogen peroxyide and methylglyoxal but not superoxide or nitric oxide in Tobaaco cultured cells. *Bioscience, Biotechnology, and Biochemistry*, 2010, 74, 2043-2049.

[32] Mikami K, Uji T. Transient gene expression systems in *Porphyra yezoensis*: Establishment, application and limitation. In: Mikami K (ed). *Porphyra yezoensis: Frontiers in Physiological and Molecular Biological Research*. New York, NY: Nova Science Publishers, 2011. in press.

[33] Uji T, Takahashi M, Saga N, Mikami K. Visualization of nuclear localization of transcription factors with cyan and green fluorescent proteins in the red alga *Porphyra yezoensis*. *Marine Biotechnology*, 2010, 12, 150-159.

In: Seaweed
Editor: Vitor H. Pomin

ISBN 978-1-61470-878-0
© 2012 Nova Science Publishers, Inc.

Chapter 2

A REVIEW OF THE NUTRIENT COMPOSITION OF SELECTED EDIBLE SEAWEEDS

Leonel Pereira[*]

Institute of Marine Research, Department of Life Sciences,
Faculty of Sciences and Technology, University of Coimbra,
Coimbra, Portugal

ABSTRACT

Currently, our society lives under a misleading apprehension of there being food abundance......etc, etc...... Many people of the west are surrounded by fast food rich in calories and unsaturated fats, high powered advertising and over-consumption. The mass market has actually become accustomed to the expression of "junk food" to designate such offerings, but yet this highly processed "food" is consumed in large amounts. The consequences of consumption of these offerings for the mass (western) the lack of essential nutrients, obesity and diseases related to excessive intake of sugars (diabetes) and fat (arteriosclerosis), among others. It is worrying that the fast food trends of the west are being adopted seemingly without concern in developing countries as they become more prosperous, hence rates of associated disease are increasing.

What roles have the seaweeds in this picture?

Represent exactly the opposite: a natural food that gives us a highly nutritious but low in calories. Algae are therefore the best way to address the nutritional deficiencies of the current food, due to its wide range of constituents: minerals (iron and calcium), protein (with all essential amino acids), vitamins and fiber [1,2].

Contrary to what happens in East Asia, the West is more involved with use of seaweed as a source in thickeners and gelling properties of hydrocolloids extracted from seaweeds: carrageenan, agar and alginate (E407, E406 and E400, respectively), which are widely used in food industry, especially in desserts, ice cream, the fresh vegetable gelatin. Perhaps in most cases, the consuming public are blissfully unaware they are consuming seaweed derived products.

However attitudes are quite different in Asian cultures where seaweeds are highly valued and regarded for their appearance, texture, flavour and in a number of cases, beneficial health properties.

---

[*] E-mail: leonel@bot.uc.pt

Some seaweeds can be rich in polysaccharides which, in the absence of appropriate enzymes, due to their long chain molecules , they are not broken down, nor absorbed by the digestive system and behave as soluble fiber, with no calories, having a positive impact on the regulation of intestinal transit.

From the composition of seaweed highlight: Presence of minerals with values about ten times higher than found in traditional vegetables, such as iron in *Himanthalia elongata* (Sea spaghetti) in comparison with that of *Lens esculenta* (lentils) or in the case of calcium present in *Undaria pinnatifida* (Wakame) and *Chondrus crispus* (Irish Moss), in comparison with milk; presence of proteins containing all essential amino acids, constituting a type of protein of high biological value, comparable in quality to the egg; presence of vitamins in significant quantities, in particular the presence of B_{12} (*Porphyra* spp.), absent in higher plants; *Palmaria palmata* and *Himanthalia elongata* are rich in potassium and, together with the algae of the genus *Porphyra* and *Laminaria*, have a ratio of sodium/potassium ratio considered optimal for human health.

This review aims to describe some of the key nutritional characteristics of the main algae used as human food and their potential in the nutraceuticals industry.

1. INTRODUCTION

Seaweeds are used in many maritime countries as a source of food, for industrial applications and as a fertilizer. The major utilization of these plants as food is in Asia, particularly Japan, Korea and China, where seaweed cultivation has become a major industry. In most western countries, food and animal consumption is restricted and there has not been any major pressure to develop seaweed cultivation techniques. Industrial utilization is at present largely confined to extraction for phycocolloids and, to a much lesser extent, certain fine biochemical. Fermentation and pyrolysis are not been carried out on an industrial scale at present but are possible options for the 21st century.

The present uses of seaweeds are as human foods, cosmetics, fertilizers, and for the extraction of industrial gums and chemicals. They have the potential to be used as a source of long- and short-chain chemicals with medicinal and industrial uses [3].

Worldwide only about 221 species of algae: 125 Rhodophyta (Red algae), 64 Phaeophyceae (Brown algae) and 32 Chlorophyta (Green algae) are used. Of these, about 145 species are used (66%) directly in food: 79 Rhodophyta, 38 Phaeophyceae and 28 Chlorophyta. In phycocolloid industry, 101 species are used: 41 alginophytes (algae that produce alginic acid), 33 agarophytes (algae producing agar) and 27 carrageenophytes (algae producing carrageenan). Other activities will use: 24 species in traditional medicine, 25 species in agriculture, animal feed and fertilizers and about 12 species are cultivated in "marine agronomy" [4,5].

The species *Alaria esculenta* (Linnaeus) Greville, *Codium fragile* (Suhr) Hariot, *Caulerpa lentillifera* J.Agardh, *Caulerpa racemosa* (Forsskål) J.Agardh, *Dilsea carnosa* (Schmidel) Kuntze, *Eisenia bicyclis* (Kjellman) Setchell, *Fucus vesiculosus* Linnaeus, *Fucus spiralis* Linnaeus, *Gelidium* spp., *Gracilaria changii* (B.M.Xia and I.A.Abbott) I.A.Abbott, J.Zhang and B.M.Xia, *Gracilaria chilensis* C.J.Bird, McLachlan and E.C.Oliveira, *Laminaria digitata* (Hudson) J.V.Lamouroux, *Laminaria ochroleuca* Bachelot de la Pylaie, *Porphyra leucosticta* Thuret, *Porphyra tenera* Kjellman, *Porphyra umbilicalis* Kützing, *Porphyra yezoensis* Ueda, *Saccharina japonica* (Areschoug) C.E.Lane, C.Mayes, Druehl and G.W.Saunders, *Saccharina latissima* (Linnaeus) C.E.Lane, C.Mayes, Druehl and G.W.Saunders, *Sargassum fusiformes* (Harvey) Setchell, *Ulva compressa* Linnaeus, *Ulva lactuca* Linnaeus, *Ulva pertusa* Kjellman, *Ulva rigida* C.Agardh and *Ulva rotundata* Bliding are analyzed in this chapter.

2. WORLD PRODUCTION OF SEAWEED

The world seaweed production reached in 2000 around 10 millions tons including wild and maricultured. The top 12 main producing countries are: China, France, UK, Japan, Chile, Philippines, Korea, Indonesia, Norway, USA, Canada and Ireland. The wild seaweed harvesting did not change much the last 12 years but aquaculture (including integrated mariculture) is increasing incessantly [3,6].

3. USE OF VARIOUS SEAWEEDS AS HUMAN FOOD

Seaweed as a staple item of diet has been used in Japan, Korea and China since prehistoric times. In 600 BC, *Sze Teu* wrote in China, "Some algae are a delicacy fit for the most honored guests, even for the King himself." Some 21 species are used in everyday cookery in Japan, six of them since the 8th century. Seaweed (Kaiso) accounted for more than 10% of the Japanese diet until relatively recently, and seaweed consumption reached an average of 3.5 kg per household in 1973, a 20% increase in 10 years [1,7]. Although there is little tradition of using seaweed in Western cuisine, there is now renewed interest in Western countries in the use of seaweed as sea vegetables [2,8,9].

In recent years, there has been a growing interest in so-called functional food groups, amongst which seaweeds would seem to be able to play an important role since they can provide physiological benefits, additional to nutritional as, for instance, anti-hypertensive, anti-oxidant or anti-inflammatory [10,11]. A functional food can be defined as a food that produces a beneficial effect in one or more physiological functions, increases the welfare and or decreases the risk of suffering from the onset or development of a particular disease. The functionalities are far more preventative than curative. Furthermore, new types of products, derived from food, often referred to as nutraceuticals have recently been developed and marketed extensively. These products are usually employed as food supplements, rather than whole foods and are marketed as tablets and pills and can provide important health benefits. Frequently, functional foods are obtained from traditional foods enriched with an ingredient which is able to provide or promote a beneficial action for human health. These are the so-called functional ingredients. According Madhusudan et al. [11], many biologically active compounds are present in seaweed, which can be used as therapeutic agents (see Table 4) in dietary supplements.

4. EXAMPLES OF SEAWEEDS USED AS HUMAN FOOD

4.1. Chlorophyta (Green Algae)

Dichotomous sponge tang or shui-sung (*Codium fragile*, Bryopsidophyceae) – The marine green alga *Codium fragile* is a invasive species (in particular the subspecies *tomentosoides*), widely distributed in temperate areas throughout the world and is eaten in Korea, China and Japan [1, 12, 13].This alga is an additive of Kinchi, a traditional fermented vegetable [14]. The nutrient composition and vitamin content of this species [15,16] are shown in Table 1 and 3, respectively.

Table 1. Nutrient composition of selected edible seaweed (% dry weight)

Species	Protein	Ash	Dietary fiber	Carbohydrate	Lipid	Reference
Chlorophyta (Green seaweed)						
Caulerpa lentillifera	10 - 13	24 - 37	33	38 - 59	0.86 - 1.11	[18,19,20]
C. racemosa	17.8 - 18.4	7 -19	64.9	33 - 41	9.8	[95,96,97,98]
Codium fragile	8 - 11	21 - 39	5.1	39 - 67	0-5 - 1.5	[15,16]
Ulva compressa	21 - 32	17 - 19	29 – 45	48.2	0.3 - 4.2	[21,23,28,29]
U. lactuca	10 - 25	12.9	29 – 55	36 - 43	0.6 - 1.6	[21,22,24,26,56,99]
U. pertusa	20 - 26	-	-	47.0	-	[22,27]
U. rigida	18 - 19	28.6	38 – 41	43 - 56	0.9 - 2.0	[31,56,97,100]
U. reticulata	17 - 20	-	65.7	50 - 58	1.7 - 2.3	[100,101]
Phaeophyceae (Brown seaweed)						
Alaria esculenta	9 - 20	-	42.86	46 - 51	1 - 2	[26,12]
Eisenia bicyclis	7.5	9.72	10 – 75	60.6	0.1	[21,35,103]
Fucus spiralis	10.77	-	63.88	-	-	[29]
F. vesiculosus	3 - 14	14 - 30	45-59	46.8	1.9	[8,22,37,102,104,105]
Himanthalia elongata	5 - 15	27 - 36	33 – 37	44 - 61	0.5 - 1.1	[8,21, 23,62,106,107]
Laminaria digitata	8 - 15	38	36 – 37	48	1.0	[21,22,23,26,37]
L. ochroleuca	7.49	29.47	-	-	0.92	[21]
Saccharina japonica	7 - 8	27 - 33	10 – 41	51.9	1.0 – 1.9	[21,34,35,103]
S. latissima	6 - 26	34.78	30	52 - 61	0.5 - 1.1	[8,26,107]
Sargassum fusiforme	11.6	19.77	17 – 69	30.6	1.4	[21,34,35,103]
Undaria pinnatifida	12 - 23	26 - 40	16 – 51	45 - 51	1,05 - 4.5	[8,23,34,35,37,39,62, 103,108]
Rhodophyta (Red seaweed)						
Chondrus crispus	11 - 21	21	10 – 34	55 - 68	1.0 - 3.0	[8,21,26,37,39,51]
Gracilaria changii	6.9	22.7	24.7		3.3	[21]
G. chilensis	13.7	18.9	-	66.1	1.3	[15]
Palmaria palmata	8 - 35	12 - 37	29 – 46	46 - 56	0.7 - 3	[8,21,22,26,29,51]
Porphyra tenera	28 - 47	8 - 21	12 – 35	44.3	0.7 - 1.3	[21,22,23,35,37,103]
P. umbilicalis	29 - 39	12	29 – 35	43	0.3	[8,62]
P.yezoensis	31 - 44	7.8	30 – 59	44.4	2.1	[21,34,51,63]

Sea grapes or Green caviar (*Caulerpa* spp., Bryopsidophyceae) – There are many species of the genus *Caulerpa*, but *Caulerpa lentillifera* and *C. racemosa* are the two most popular edible ones. Both have a grape-like appearance and due to their grass-green in color, soft and succulent texture, are usually consumed in the form of fresh vegetable or salad. They are commonly found on sandy or muddy sea bottoms in shallow protected, sub-tropical areas.

Table 2. Mineral composition of some edible seaweeds (mg.100 g^{-1} DW)

Species	Na	K	P	Ca	Mg	Fe	Zn	Mn	Cu	I	Reference
Chlorophyta (Green seaweed)											
Caulerpa lentillifera	8917	700 - 1142	1030	780 - 1874	630 - 1650	9.3 - 21.4	2.6 - 3.5	7.9	0.11 - 2.2	-	[18,19,21]
C. racemosa	2574	318	29.71	1852	384 - 1610	30 - 81	1 - 7	4.91	0.6 - 0.8	-	[97,100]
Ulva lactuca	-	-	140	840	-	66	-	-	-	-	[25]
U. rigida	1595	1561	210	524	2094	283	0.6	1.6	0.5	-	[31]
Phaeophyceae (Brown seaweed)											
Fucus vesiculosus	2450 - 5469	2500 - 4322	315	725 - 938	670 - 994	4 - 11	3.71	5.50	<0.5	14.5	[8,37]
Himanthalia elongata	4100	8250	240	720	435	59	-	-	-	14.7	[8]
Laminaria digitata	3818	11,579	-	1005	659	3.29	1.77	<0.5	<0.5	-	[37]
Saccharina japonica	2532 - 3260	4350 - 5951	150 - 300	225 - 910	550 - 757	1.19 - 43	0.89 - 1.63	0.13 - 0.65	0.25 - 0.4	130 - 690	[44,109,110]
S. latissima	2620	4330	165	810	715	-	-	-	-	15.9	[8]
Sargassum fusiforme	-	-	-	1860	687	88,6	1.35	-	-	43.6	[21,42]
Undaria pinnatifida	1600 - 7000	5500 - 6810	235 - 450	680 - 1380	405 - 680	1.54 - 30	0.94	0.33 2	0.185	22 - 30	[8,21,44]
Rhodophyta (Red seaweed)											
Chondrus crispus	1200 -4270	1350 - 3184	135	420 - 1120	600 -732	4 - 17	7.14	1.32	<0.5	24.5	[8,37]
Gracilaria spp.	5465	3417	-	402	565	3.65	4.35	-	-	-	[111]
Palmaria palmata	1600 - 2500	7000 - 9000	235	560 - 1200	170 - 610	50	2.86	1.14	0.376	10 - 100	[8,21]
Porphyra tenera	3627	3500	-	390	565	10 - 11	2 - 3	3	<0.63	1.7	[21,37]
P. umbilicalis	940	2030	235	330	370	23	-	-	-	17.3	[8]
P. yezoensis	570	2400	-	440	650	13	10	2	1.47	-	[63]

The pond cultivation of *C. lentillifera* has been very successful on Mactan Island, Cebu, in the central Philippines, with markets in Cebu and Manila and some exports to Japan [1,17,18]. Compared to those reported in other seaweeds, the protein content of *C. lentillifera* (12.49%) was comparable to the red algae *Palmaria* sp. (13.87%), and was notably higher than some other brown algae tested, e.g. *Himanthalia elongata* (7.49%) and *Laminaria ochroleuca* (7.49%) (see Table 1) [19,20]. Apart from iodine, *C. lentillifera* is also rich in phosphorus, calcium, cooper and magnesium (Table 2) [19,21]. This species is also rich in vitamin E with moderate amount of vitamin B1, vitamin B2 and niacin (Table 3) [18].

Sea lettuce or Ao-Nori (*Ulva* spp., Ulvophyceae) – The sea lettuces comprise the genus *Ulva*, a group of edible green algae that are widely distributed along the coasts of the world's oceans. The type species within the genus *Ulva* is *Ulva lactuca* (see http://macoi.ci.uc.pt/imagem.php?id=247andtp=7), "lactuca" meaning lettuce. Sea lettuce as a food for humans is eaten raw in salads and cooked in soups. It is high in protein (level between 10 and 25% of dry mass; see Table 1) [22-24], soluble dietary fibers, and a variety of vitamins and minerals, especially iron (Table 2 and 3) [25,26].

The species *Ulva pertusa* (see http://www.algaebase.org/_mediafiles/algaebase/ 5B7BE95A076ca2C19Dsxv2CAFF8E/k857fWdJXeJD.jpg), which is frequently consumed under the name of "ao-nori" by the Japanese people, has a high protein level between 20 and 26% (dry product) (see Table 1) [22,27,159]. According Pengzhan et al. [159] the sulfated polysaccharide (ulvan) extracted from this species has antilipidemic effects.

The species *Ulva compressa* (formerly *Enteremorpha compressa*) (see http://macoi.ci.uc.pt/imagem.php?id=940andtp=7) is used dried in cooking, particularly with eggs [3]. Is used to as an ingredient in the preparation of a high fibre snack, namely Pakoda, a common Indian product made from chickpea flour [28] and their crude protein levels ranging from 21 and 32% (see Table 1) [21,28,29].

The level of aspartic and glutamic acids can represent up to 26 and 32% of the total amino acids of the edible species *Ulva rigida* (see http://macoi.ci.uc.pt/ imagem.php?id=1508andtp=7) and *Ulva rotundata* (see http://www. algaebase.org/_mediafiles/algaebase/3EE735B10772e14708IjI34FDB98/dgduKPpgE9Zn.jpg), respectively [30,31].

4.2. Phaeophyceae (Brown Algae)

Arame (*Eisenia bicyclis*, Lessoniaceae) – Is a brown alga or kelp (see http://www.algaebase.org/_mediafiles/algaebase/5B7BE95A076ca284FAXLt2BF7E95/x7Zsj f19i3Bh.jpg) that is also known as "sea oak" because of the shape of its leaves. It grows wild attached to stable on rock at a depth of a few meters on many coasts of the Pacific Ocean. This alga is one of the most nutritious of all plants [32]. It is a species of kelp best known for its use in many Japanese dishes. Arame is high in calcium, iodine, iron, magnesium, and vitamin A as well as being a good dietary source for many other minerals. It also is harvested for alginate. It contains the storage polysaccharide laminarin and the tripeptide eisenin, a peptide with immunological activity [33-35], and phlorotannins with antioxidant activity [134] (see Table 4).

Table 3. Vitamin content of some edible seaweeds (mg/100 g edible portion)

Species	A	B₁ (Thiamin)	B₂ (Riboflavin)	B₃ (Niacin)	B₅ (Pantothenic Acid)	B₆ (Pyridoxine)	B₈ (Biotin)	B₆ (Cobalamin)	C (Ascorbic Acid)	E	Folic acid	Reference
Chlorophyta (Green seaweed)												
Caulerpa lentillifera	-	0.05	0.02	1.09	-	-	-	-	1.00	2.22	-	[18]
Codium fragile	0.527	0.223	0.559	-	-	-	-	-	<0.223	-	-	[56]
Ulva lactuca	0.017	<0.024	0.533	98*	-	-	-	6*	<0.242	-	-	[26,56]
Ulva pertusa	-	-	-	-	-	-	-	-	30 – 241**	-	-	[112]
Ulva rigida	9581	0.47	0.199	<0.5	1.70	<0.1	0.012	6	9.42	19.70	0.108	[31]
Phaeophyceae (Brown seaweed)												
Alaria esculenta	-	-	0.3 - 1*	5*	-	0.1*	-	-	100 - 500*	-	-	[26]
Fucus vesiculosus	0.307	0.02	0.035	-	-	-	-	-	14.124	-	-	[8,56]
Himanthalia elongata	0.079	0.020	0.020	-	-	-	-	-	28.56	-	0.176 - 0.258	[8,47,56]
Laminaria digitata	-	1.250	0.138	61.2	-	6.41	6.41	0.0005	35.5	3.43	-	[113]
Laminaria ochroleuca	0.041	0.058	0.212	-	-	-	-	-	0.353	-	0.479	[47,56]
Saccharina japonica	0.481	0.2	0.85	1.58	-	0.09	-	-	-	-	-	[44]
Saccharina latissima	0.04	0.05	0.21					0.0003	0.35	1.6		[8]
Undaria pinnatifida	0.04 - 0.22	0.17 - 0.30	0.23 - 1.4	2.56	-	0.18	-	0.0036	5.29	1.4 - 2.5	0.479	[44,47,56]
Rhodophyta (Red seaweed)												
Chondrus crispus	-	-	-	-	-	-	-	0.6 -4*	10 - 13*	-	-	[26,37]
Gracilaria spp.	-	-	-	-	-	-	-	-	16 - 149**	-	-	[112]
Gracilaria changii	-	-	-	-	-	-	-	-	28.5	-	-	[21]
Palmaria palmata	1.59	0.073 - 1.56	0.51 - 1.91	1.89	-	8.99	-	0.009	6.34 - 34.5	2.2 - 13.9	0.267	[8,26,47]
Porphyra umbilicalis	3.65	0.144	0.36	-	-	-	-	0.029	4.214	-	0.363	[47,56]
Porphyra yezoensis	16000***	0.129	0.382	11.0	-	-	-	0.052	-	-	-	[63,114]

*expressed as *ppm*; **expressed as *mg%*; ***expressed as *I.U.*

Fucus (Fucus vesiculosus and F. spiralis, Fucaceae) – Members of this genus (see http://macoi.ci.uc.pt/imagem.php?id=242andtp=7 for F. vesiculosus photo and http://macoi.ci.uc.pt/imagem.php?id=2493andtp=7 for F. spiralis photo) are not commonly used as food, but their extracts are reported to be useful as anti-inflammatory and anti-cellulite and weight loss treatments. Fucus species has are reported to contain (see Table 1, 2 and 3): polysaccharides mucilage with algin, fucoidan and laminarin; polyphenols, trace elements and minerals (iodine in the form of salts and attached to proteins and lipids), potassium, bromine, chlorine, magnesium, calcium, iron and silicon, mannitol, vitamins and pro-vitamins A and D, ascorbic acid and lipids (glycosylglycerides) [36-39].

Hiziki or Hijiki (*Sargassum fusiforme*, Sargassaceae) – The species *Sargassum fusiforme* (formerly *Hizikia fusiformis*) (see http://www.algaebase.org/_mediafiles/algaebase/5B7BE95A076ca2541CiyH2B27FDF/mmQhPonex6Cw.jpg) is a common, edible alga which is widely consumed and used as a medicinal herb in China, Japan, Korea and Southeast Asia [22,40]. It is collected from the wild in Japan and cultivated in the Republic of Korea. The alga naturally grows at the bottom of the eulittoral and top of the sublittoral zones, and is found on the southern shore of Hokkaido, all around Honshu, on the Korean peninsula and most coasts of the China Sea. About 90 percent of the Republic of Korea production is processed and exported to Japan [17].

Hiziki contains potential and intensively investigated bioactive compounds especially fucoxanthin pigments and phlorotannins, a polyphenolic secondary metabolite (see Table 4) [34,41]. The protein, fat, carbohydrate and vitamin contents (see Table 1 and 3) are similar to those found in Kombu (formerly *Laminaria japonica*), although most of the vitamins are destroyed in the processing of the raw seaweed. The iron, copper and manganese contents (Table 2) are relatively high, certainly higher than in Kombu [21, 42]. Like most brown seaweeds, its fat content is low (1.5%) but 20-25% of the fatty acid is eicosapentaenoic acid (EPA) [17,35].

According to the Canadian Food Inspection Agency (CFIA) reports, this seaweed contains inorganic arsenic that can exceed the tolerable daily intake levels considered safe for safe human consumption. Even though, inorganic arsenic has been linked with gastrointestinal effects, anemia and liver damage, no evidence of such health complications reported to date due to direct consumption of Hiziki [41].

Kombu or Haidai (*Laminari*a spp. and *Saccharina* spp., Laminariaceae) – *Saccharina japonica* (formerly *Laminaria japonica*) (see http://www.algaebase.org/_mediafiles/algaebase/3EE735B10772e11C1FMvk34536ED/YSqPfyD87qHw.jpg) is perhaps the best known species of kelp. It has broad, shiny leaves and flourishes in cool waters off the coasts of Japan and Korea. It has been cultivated in Japan for about 300 years and elsewhere on a large scale for about forty years. A rich stock (Dashi) can be prepared from kelp because of its concentration of the flavor-enhancer glutamic acid. It is considered that the best varieties of Kombu grow in the cool coastal waters of the northern-most Japanese island of Hokkaido [32]. Haidai is the Chinese name for *Saccharina japonica*, seaweed that was introduced to China accidentally from Japan in the late 1920s. Previously, China had imported all of its requirements from Japan and the Republic of Korea. This alga is now cultivated on a large scale in China. *Saccharina japonica* grows naturally in the Republic of Korea and is also cultivated, but on a much smaller scale; the demand is lower because Koreans prefer Wakame (*Undaria pinnatifida*) [17].

The species *Saccharina latissima* (formerly *Laminaria saccharina*), despite being a deep seaweed (see http://macoi.ci.uc.pt/imagem.php?id=1506andtp=7), prefers areas with calm waters, being present in the North Atlantic from Norway to northern Portugal. Commercially this seaweed is called "Royal Kombu" and its composition is very similar to that of *Laminaria ochroleuca* (see http://macoi.ci.uc.pt/imagem.php?id=1638andtp=7), known commercially as "Atlantic Kombu" and of *L. digitata* (see http://www.algaebase.org/_mediafiles/algaebase/3EE735B10772e033A6jpH30F0391/2rgkFQ1L8AyP.jpg), known commercially as "Kombu Breton". The Atlantic Kombu is a rather tougher than the Kombu from Japan and is distributed in Iberian Peninsula from Santander, in Cantabria (Spain), to Cape Mondego in Portugal [8,38,43].

Kombu stands out for its high mineral content (particularly magnesium, calcium and iodine). Calcium and magnesium regulate together many functions, including the nervous system and muscles. The various species of the genera *Laminaria* and *Saccharina* have been used as a source of iodine in the industry, mineral with a role in thyroid function, as noted above (see Table 2 and 4). The alginic acid present in these algae has shown preventive effects against contamination by heavy metals and radioactive substances, especially Strontium 90. Among the properties of these seaweeds, we highlight the following: anti-rheumatic, anti-inflammatory, regulators of body weight and blood pressure (due to the presence of laminarin and laminin). These Laminariaceae also prevent atherosclerosis and other vascular problems due to its bloodstream fluidifying effects [8,34,39].

Sea spaghetti or Haricot vert de mer (*Himanthalia elongata*, Himanthaliaceae) - Is long, dark (see http://macoi.ci.uc.pt/imagem.php?id=276andtp=7), and rich in trace elements and vitamins. It is successfully cultivated in Brittany, France, and increasingly exported fresh for the Japanese restaurant trade. The long strands must first have its furry layer removed by hand under cold running water before it is prepared for eating [32].

Little known in Asian countries, it is increasingly valued in Europe, both in restaurants and in specialty bakeries. For several years they have manufactured specialty pies, pizzas, pastas, pates, breads, and snacks, since its taste is reminiscent of some cephalopods (squid and cuttlefish) [8].

This species is characterized in particular by its high iron content (59 mg per 100 g of algae) and the simultaneous presence of vitamin C, which facilitates the absorption of this trace element (see Table 2 and 3) [47]. Sea spaghetti is rich in phosphorus, a mineral known to enhance brain function, helping to preserve memory, concentration and mental agility [8,38].

Wakame or Quandai-cai (*Undaria pinnatifida*, Alariaceae) – Is a invasive brown seaweed (see http://macoi.ci.uc.pt/imagem.php?id=1146andtp=7) originating from the Pacific, which lives in deep waters (up to 25 m) and can reach 1.5 m in length and is one of the most important species of commercial seaweed, next to nori, on the Japanese menu and is eaten both dried and fresh [8, 38].

The nutritional value is high, as the leaves consist of 13% protein, as well as containing substantial amounts of calcium (see Table 1 and 2) [21,44]. Traditionally Wakame is harvested from wild populations by boats by means of long hooks and then sold fresh or sun dried. Since this seaweed is salted for transport, certain cleansing must take place before eating. Wakame must be thoroughly rinsed under running water, then placed in boiling water for thirty seconds, then rinsed in ice water. The leaves are then spread out and the hard midrib is removed [32]. Wakame has relatively high total dietary fibre content; it is higher than Nori or Kombu (see Table 1). Consumption of dietary fibre has a positive influence on several aspects related to health such as reducing the risk of suffering from colon cancer, constipation, hypercholesterolemia, obesity and diabetes. Besides, many constituents of dietary fibre show antioxidant activity as well as immunological activity [45]. In this sense, *U. pinnatifida* (Wakame) showed some positive effect on cardiovascular diseases (hypertension and hypercholesterolemia) [46]; this alga contains basically dietetic fibre, being its principal component alginate. This alginic acid has demonstrated to reduce hypertension in hypertensive rates [46].

Like other brown seaweeds, the fat content is quite low (see Table 1). Air-dried Wakame has a similar vitamin content to the wet seaweed and is relatively rich in the vitamin B group, especially niacin (see Table 3) [44,47]; however, processed Wakame products lose most of their vitamins. Wakame contains appreciable amounts of essential trace elements (see Table 2) such as manganese, copper, cobalt, iron, nickel and zinc, similar to Kombu and Hiziki [8,7,44].

Wakame is one of the most popular edible seaweed in Japan and has been found to contain 5–10% fucoxanthin [48] apart from containing polar lipids such as glycolipids. Health benefits of fucoxanthin are anticancer effect — it is evaluated that neoxanthin and fucoxanthin were reported to cause a remarkable reduction in growth of prostate cancer cells, and also demonstrated anti-obesity activity and anti-inflammatory activity [49]. Fucoxanthin (see Table 4) is other major biofunctional pigment of brown seaweeds and the content in various edible seaweeds including *U. pinnatifida* has been reviewed by Hosakawa et al. [50].

Winged kelp, Edible kelp or Atlantic wakame (*Alaria esculenta*, Alariaceae) – This is a large brown kelp (see http://macoi.ci.uc.pt/imagem.php?id=2115andtp=7) which grows in the upper limit of the sublittoral zone. It has a wide distribution in cold waters and does not survive above 16°C. It is found in areas such as Ireland, Scotland (United Kingdom), Iceland, Brittany (France), Norway, Nova Scotia (Canada), Sakhalin (Russia) and northern Hokkaido (Japan). The seaweed is eaten in Ireland, Scotland (United Kingdom) and Iceland either fresh or cooked, and it is said to have the best protein among the kelps and is also rich in trace metals and vitamins (see Table 1 and 3), especially niacin and contains up to 42% alginic acid [17,26,51,52]. The species is used for a cultivar of purposes from value-added sea-vegetables to fodder and body care products. Recently, it has become of economic interest as a foodstuff in aquaculture for herbivorous mollusks, urchins, shrimp and fish [53].

4.3. Rhodophyta (Red Algae)

Kanten (Japan) or Agar-Agar, Dai choy goh (China), Gulaman (Philippines) (Agarophytes, Florideophyceae) – *Ahnfeltia*, *Gelidiella*, *Gelidium*, *Gracilaria* and *Pterocladiella* are the major sources of raw materials used for the commercial extraction. Agar-Agar is the Malay name for a gum discovered in Japan that had been extracted from a red seaweed of the genus *Eucheuma* (see Phycocolloid "Agar").

With common names such as Kanten (see http://www.mitoku.com/products/seavegetables/img/kanten001_s.jpg) in Japan , but can also be referred to by many names including 'Grass jelly', 'Seaweed jelly', and 'Vegetable gelatin' (true gelatin is an animal by-product and as such can be unacceptable on account of dietary or religious preferences).

Table 4. Summary of nutraceutical value of some seaweed compounds

Category	Compounds	Seaweed source	Potential health benefit	Reference
Lipids and fatty acids	Omega 3 and omega 6 acids	*Porphyra* spp. Brown algae	Prevention of cardio-vascular diseases, osteoarthritis and diabetes	[8,23,39]
Carotenoids	β-carotene, lutein	*Chondrus crispus* *Porphyra yezoensis* Red algae	Antimutagenic; protective against breast cancer	[94,115,116,117,118,119]
	β-carotene, lycopene	*Porphyra* spp. Red algae	Recent studies have shown the correlation between a diet rich in carotenoids and a diminishing risk of cardio-vascular disease, cancers	[39,120,121,161]
	Lutein, zeaxanthin	Red algae Brown algae	Diminishing risk of ophthalmological diseases	
	Fucoxanthin	*Undaria pinnatifida* Brown algae	Antiangiogenic; protective effects against retinol deficiency; anticancer effect; anti-obesity and anti-inflammatory activity	[39,120,121,122] [1,48,123,124,125,126]
Minerals	Iodine	*Fucus vesiculosus* *Laminaria* spp. *Undaria pinnatifida*	The brown seaweeds have traditionally been used for treating thyroid goiter.	[8,39,127]
	Calcium	*Undaria pinnatifida* *Laminaria* spp. *Saccharina* spp.	Seaweed consumption may thus be useful in the case of expectant mothers, adolescents and elderly that all exposed to a risk of calcium deficiency.	[8,23]
Phycobilin pigments	Phycoerythrin, Phycocyanin	Red algae	Antioxidant properties, which could be beneficial in the prevention or treatment of neuro-degenerative diseases caused by oxidative stress (Alzheimer's and Parkinson's) as well as in the cases of gastric ulcers and cancers Amelioration of diabetic complications	[39,128,129,130]
	Phycoerythrin	Red algae		[131]

Table 4. (Continued).

Category	Compounds	Seaweed source	Potential health benefit	Reference
Polyphenols	Flavonoids	*Palmaria palmata*	At high experimental concentrations that would not exist *in vivo*, the antioxidant abilities of flavonoids *in vitro* are stronger than those of vitamin C and E	[132,133]
	Phlorotannins	Brown algae	Antioxidant activity of polyphenols extracted from brown and red seaweeds has already been demonstrated by *in vitro* assays; anti-inflammatory effect

Algicidal and bactericidal effect | [134,135,136,137,138]

[138,139,140] |
| Polysaccharides and dietary fibers | Agars, carrageenans, ulvans and fucoidans | Red, green and brown algae | These polysaccharides are not digested by humans and therefore can be regarded as dietary fibers.

Antihyperlipidemic effects | [141,142,143]

[27,159] |
	Ulvan	*Ulva pertusa*	Antitumor and anti-viral	[1,138,144,145,146,147,148]
	Carrageenan, fucoidan	Red algae (carrageenophytes), *Undaria pinnatifida*, brown algae	Anti-viral, anti-HSV and anti-HIV	[60,71,149]
		Red algae (carrageenophytes)	Anticoagulant and antithrombotic activity	[60,138,150]
	Carrageenan (lambda, iota and nu variants)	Brown algae	Antitumor and immunomodulatory activity	[146,150,151]
	Fucoidan		Antiviral and anti-HIV	[60,94,150,152,153,154,155]
	Fucoidan	*Fucus vesiculosus Saccharina japonica*	Hypolipidemic effect	[156,157]

Category	Compounds	Seaweed source	Potential health benefit	Reference
Proteins and amino acids	Proteins	*Palmaria palmata* *Porphyra tenera*	Higher protein contents are recorded in green and red seaweeds (on average 10-30 % of the dry weight). In some red seaweed, such as *Palmaria palmata* (dulse) and *Porphyra tenera* (nori), proteins can represent up to 35 and 47% of the dry matter, respectively.	[23]
	Proteins, amino acids	*Undaria pinnatifida*	*Undaria pinnatifida* (wakame) has a high balance between the essential amino acids, which gives a high biological value to their proteins. Proteins, in addition, with a high bioavailability (85-90%)	[8,39]
Vitamins	Vitamin B_{12}	*Porphyra* spp.	Is particularly recommended in the treatment of the effects of ageing, of CFS and anemia.	[114]
	Vitamin C	*Himanthalia elongata Palmaria palmata*	Strengthens the immune defense system, activates the intestinal absorption of iron, controls the formation of conjunctive tissue and the protidic matrix of bony tissue, and also acts in trapping free radicals and regenerates Vitamin E.	[8,23]
	Vitamin E	*Fucus* spp.	Due to its antioxidant activity, vitamin E inhibits the oxidation of the low-density lipoproteins. It also plays an important part in the arachidonic acid chain by inhibiting the formation of prostaglandins and thromboxan.	[23]

Abbreviations: CFS - Chronic fatigue syndrome; HIV - Human immunodeficiency virus; HSV - Herpes simplex virus.

Agar-agar is a powerful gel-forming of all gums because of the unusual length of its carbohydrate molecules. It is also unique in its ability to withstand near boiling-point temperatures, making it ideal for use in jellied confections in tropical countries since the ingredients can be treated at high temperatures and then cooled [32,54,55].

The agarophytes, from which this gum is extracted, are gathered and left on the beach to dry and bleach before being sold to a factory where it is cleaned, washed, and boiled to extract the gum. Traditionally the, water soluble extract is it is frozen and thawed. More recently precipitation methods have been developed, which alternative process relies on synaeresis [17]. As the water runs out of it, so do any of the impurities, leaving the purified gum to be dried. This method of purifying (freezing and thawing) is said to have been discovered accidentally by a Japanese innkeeper during a frosty winter of 1658. Since then, the product has gained in popularity in Japanese cuisine, not only for making jellies, but also as a general thickener for soups and sauces [32,38,54].

A popular Japanese sweet dish is mitsumame; this consists of cubes of agar gel containing fruit and added colors. It can be canned and sterilized without the cubes melting. Agar is also used in gelled meat and fish products, and is preferred to gelatin because of its higher melting temperature and gel strength. In combination with other gums, agar has been used to stabilize sherbets and ices. It improves the texture of dairy products such as cream cheese and yoghurt. Agar has been used to clarify wines, especially plum wine, which can prove difficult by traditional methods. Unlike starch, agar is not readily digested and so adds little calorific value to food. It is used in vegetarian foods such as meat substitutes. There is an increased recent interest in agar as used in dedicated Kanten restaurants catering for modern weight conscious Japanese consumers [17].

Dulse or Dilisk (*Palmaria palmata*, Florideophyceae) – Is a relatively common Atlantic seaweed (see http://macoi.ci.uc.pt/imagem.php?id=854andtp=7). It is comparatively small (up to 50 cm long), and can occupy a wide range of habitats from the intertidal, with brief exposure to relatively deep niches, in cold and turbulent waters. The name "dulse" comes from the Irish vocabulary (dils = edible seaweed) and has little to do with the Latin dulce meaning tasting good or sugary or sweet. In fact eating dried dulse may be described as being an acquired taste which can be quite strong and distinctive [38]. Dulse was prized by the Celts and the Vikings and has been harvested on beaches at low tide, air-dried, and boiled in soups from Ireland to Iceland well into the 20th century. The people of Scotland, Ireland, and Iceland have been using Dulse for centuries, and collect it off their coasts. Many consider it to be the most delectable of all seaweeds [32,56]. Today, this species is successfully cultivated along the coast of Brittany in France, Ireland and northern Spanish coast [57,58,59].

Dilsea carnosa is another type of edible seaweed (http://macoi.ci.uc.pt /imagem .php?id=370andtp=7), unrelated to the regular Dulse, but identical in taste, appearance, and nutritional value. Dried Dulse is a popular food in Canada, where much of the world's current supply is harvested in New Brunswick and Nova Scotia. From there, it is exported to Scotland, Ireland and the US. Dulse is extremely rich in iodine, phosphorus, calcium, and contains more potassium than any other food. In Canada, Dulse is available in many major coastal food outlets and supermarkets and can be served in a variety of ways: as a side dish, in soups and salads, as a sandwich ingredient or in powdered form to be used as a spice or condiment flavoring [17,32].

About 30% of the dry weight of dulse comprises minerals (e.g. iron, iodine and potassium) and proteins of high nutritive value (18%). *Palmaria palmata* also has relatively high amounts of vitamin C, which facilitates the absorption of iron (see Table 2 and 3). This seaweed is ideal as a restorative in states of anemia and asthenia (weakness). Strengthens vision (vitamin A) and is recommended for treatment of gastric and intestinal problems and for regeneration of the mucous membranes (respiratory, gastric, and vaginal). Like other few red algae [60], *Palmaria palmata* has anthelmintic effect and acts as an antiseptic and parasites control, cleaning up the gut [8,56].

Irish moss or Carrageen moss (*Chondrus crispus*, Florideophyceae) – This species (see http://macoi.ci.uc.pt/imagem.php?id=305andtp=7 for hand harvest photo, http://macoi.ci.uc.pt/imagem.php?id=1522andtp=7 for Irish moss pudding photo and http://macoi.ci.uc.pt/imagem.php?id=1020andtp=7 for habit photo) is found along the coasts of the North Atlantic in both Europe and North America [61]. It can either be reddish-purple or green in color. Ireland is a major source of the world's supply and where this vegetable is steamed and eaten with potatoes or cabbage. Its most common use outside of Ireland is in the making of rennet-free gelatin (carrageen). This is preferred by full vegetarians and on certain religious grounds since true gelatin is a product of animal processing. One example of its traditional use is in the production of blancmange (literally white jelly), a traditional vanilla-flavoured pudding. In eastern Canada, a company is cultivating a strain of *Chondrus crispus* in on land tanks and marketing it as Hana Tsunomata, for seaweed salad (see www.acadianseaplants.com), a yellow variant that resembles traditional Japanese seaweed that is in limited supply from natural resources [17,38,56].

Mastocarpus stellatus is frequently collected with *C. crispus* and sold as a *mixture* under the name Carrageen or Irish moss [17].

Carrageenan is extensively used in the manufacture of various soft cheeses, ice cream, aspics and jellies (see "Phycocolloids" and Table 6).

Nori or Purple laver (a large number of species including *Porphyra yezoensis*, *P. tenera*, *P. umbilicalis* and *Porphyra* spp., Bangiophyceae) – The original and traditional Nori is produced from *Porphyra yezoensis* (see http://upload.wikimedia.org/wikipedia/commons/5/5f/Porphyra_yezoensis.jpg) and *P. tenera* cultivated in Japan. The word "nori" originally means "all seaweed"; however the modern application of the word is taken to include the purplish-black seaweed sheets often seen wrapped around rice in sushi cuisine. Nori sheets come largely from cultivation in Japan, the Republic of Korea and China. In Japan's list of products from marine culture, Nori has the highest production volume, followed by oysters, yellowtails and Wakame, the last being another seaweed used as food. In traditional way, to obtain Nori, freshly harvested fronds of *Porphyra* are chopped, pressed between bamboo mats, and dried either in drying rooms or in the sun. Good quality Nori is mild-tasting and black in color, but having a purple sheen. It should be packed airtight since it is very hygroscopic However, today the production of Nori is more mechanical [17,38].

There is an Atlantic Nori (see http://macoi.ci.uc.pt/imagem.php?id=569andtp=7), which is produced from wild algae of the genus *Porphyra* (e.g. *P. umbilicalis*, *P. leucosticta* and others), which is traditionally consumed in Celtic countries and in the Azores archipelago. In Wales and Ireland it is still used in preparing the dish called "laverbread" [8,38].

Many species of the genus *Porphyra* are rich in amino acids. Nori is exceptionally rich in provitamin A (see Table 3), surpassing the vegetables and also seafood and fish. Nori has a low percentage of fats and these are of great nutritional value because more than 60% of them are polyunsaturated fatty acids omega 3 and 6. This dried seaweed contains large amounts of protein, ash, vitamins and carbohydrate (see Table 1) [21,62]. The levels of taurine (> 1.2%) are notable as this compound aids enterohepatic circulation of bile acid, thus preventing gallstone through controlling blood-cholesterol levels. Relatively high levels ofeicosapentanoic acid, choline, inositol and other B-group vitamins are regarded as beneficial to health. The occurrence of porphyosins and betaines that prevent respectively, gastric ulcers and lower blood-cholesterol levels are particular interest (see Table 4) [8,38,63].

Ogo, Ogonori or Sea moss (*Gracilaria* spp., Florideophyceae) – Fresh *Gracilaria* species have been collected and sold as a salad vegetable in Hawaii (United States of America) for several decades. The mixture of ethnic groups in Hawaii (Hawaiians, Filipinos, Koreans, Japanese and Chinese) creates an unusual demand and supply has, at times, been limited by the availability of stocks natural sources. The alga is being successfully cultivated in Hawaii using an aerated tank system, producing up to 6 tones fresh weight per week. In Indonesia, Malaysia, the Philippines and Vietnam, species of *Gracilaria* are collected by coastal people for food [64]. In southern Thailand, an education program was undertaken to show people how it could be used to make jellies by boiling and making use of the extracted agar (See Phycocolloid "Agar" and Table 6). In the West Indies, *Gracilaria* is sold in markets as "Sea moss" and in some locations is marketed also as 'Irish moss"; it is reputed to have aphrodisiac properties and is also used as a base for a non-alcoholic drink. It has been successfully cultivated for this purpose in St Lucia and adjacent islands. *Gracilaria changii* (see http://www.naturia.per.sg/cjsurvey/vegetative/text/gracilaria%20changii.htm) is consumed in certain coastal areas especially along the east coast of Peninsula Malaysia and in East Malaysia, where it is occasionally eaten as a salad dish [17,65].

The red alga *Gracilaria chilensis* (see http://www.algaebase.org/_mediafiles/algaebase/3EE735B1076ca33F0Bquh2E9B16C/f8jBtU6jij4V.jpg), belonging to the Gracilariaceae, is known as "Pelillo" in Chile, on account of its appearance [15]. It has a long and filamentous thallus. It is a reddish brown alga, with variable branching reaching 2 m. It grows in bunches or isolated, in habitats with solid substrates [66]. This alga is almost entirely used in the domestic and foreign industry for the development of agar, and is one of the most exported (126,000 tones/year) [15].

5. PHYCOCOLLOIDS

What Are Phycocolloids?

Colloids are compounds that form colloidal solutions, an intermediate state between a solution and a suspension, and are used as thickeners, gelling agents, and stabilizers for suspensions and emulsions (see Table 6). Hydrocolloids are carbohydrates that when dissolved in water form viscous solutions. The phycocolloids are hydrocolloids extracted from algae and represent a growing industry, with more than 1 million tons of seaweeds extracted annually for hydrocolloid production [67-69].

Many seaweeds produce hydrocolloids, associated with the cell wall and intercellular spaces. Members of the red algae (Rhodophyta) produce galactans (e.g. carrageenans and agars) and the brown algae (Heterokontophyta, Phaeophyceae) produce uronates (alginates) [68,70-72].

The different phycocolloids used in food industry as natural additives are (European codes of phycocolloids):

- Alginic acid – E400
- Sodium alginate – E401
- Potassium alginate – E402

- Ammonium alginate – E403
- Calcium alginate – E404
- Propylene glycol alginate – E405
- Agar – E406
- Carrageenan – E407
- Semi-refined carrageenan or processed eucheuma seaweed – E407A

Agar (Agarophytes, Rhodophyta) – Most Agar is extracted from species of *Gelidium* and *Gracilaria*. Closely related to *Gelidium* are species of *Pterocladiella* (see http://macoi.ci.uc.pt/imagem.php?id=571andtp=7), and small quantities of these are collected, mainly in the Azores (Portugal) and New Zealand. *Gelidiella acerosa* is the main source of agar in India. *Ahnfeltia* species have been used in both Russia and Japan, one source being the island of Sakhalin (Russia) [17, 38]. *Gelidium* spp. and *Gracilaria* spp. are collected in Morocco and Tunisia and Chile for Agar production [15,73-76].

Agar is a phycocolloid the name of which comes from Malaysia and means "red alga" in general and has traditionally been applied to what we now know taxonomically as – *Eucheuma* (see "Agar-Agar"). Ironically we now know this to be the commercial source of iota carrageenan. Agar is composed of two polysaccharides: namely agarose and agaropectin. The first is responsible for gelling, while the latter has thickening properties [77].

Agar is a relatively mature industry in terms of manufacturing methods and applications. Today most processors are using press/syneresis technology; although some still favor freeze/thaw technology or a mixture of these technologies. While the basic processes may not have changed, improvements in presses and freezing equipment must be noted. High-pressure membrane presses have greatly improved dewatering of agar and thereby reducing energy requirements for final drying before powder milling. Average prices of this phycocolloid were US$ 18 kg^{-1} and global sales in 2009 were US$ 173 million [70].

The origin of agar as a food ingredient is in Asia where it has been consumed for several centuries. Its extraordinary qualities as a thickening, stabilizing and gelling agent make it an essential ingredient for preparing processed food products. Furthermore, its satiating and gut regulating characteristics make it an ideal fiber ingredient in the preparation of low calorie food products. The principal applications of agar food grade are (see Table 6): fruit jellies, milk products, fruit pastilles, caramels, chewing gum, canned meat, soups, confectionery and baked goods, icing, frozen and salted fish [77].

About 80 percent of the agar produced globally is for food applications (see Table 5 and 6), the remaining 10 percent is used for bacteriological plates and other biotechnology uses (in particular agarose electrophoresis). Agar has been classified as GRAS (Generally Recognized as Safe) by the United States of America Food and Drug Administration, which has set maximum usage levels depending on particular applications. In the baked goods industry, the ability of agar gels to withstand high temperatures allows for its use as a stabilizer and thickener in pie fillings, icings and meringues. Cakes, buns, etc., are often pre-packed in various kinds of modern wrapping materials and often stick to them, especially in hot weather; by reducing the quantity of water and adding some agar, a more stable, smoother, non-stick icing may be obtained [17,68]. Some agars, especially those extracted from *Gracilaria chilensis*, can be used in confectionery with very high sugar content, such as fruit candies. These agars are said to be "sugar reactive" because the sugar (sucrose) increases

Table 5. Agar grades depending on their final use (Adapted from Armisen [78])

	Agar type	Source
Natural agar	Strip	Only *Gelidium* by old traditional methods
	Square	
Industrial agar	Food grade	*Gelidium, Gracilaria, Pterocladiella, Gelidiella, Ahnfeltia*
	Pharmacological grade	Only *Gelidium*
	Clonic plants production grade	*Gelidium, Pterocladiella*
	Bacteriological agar	Only *Gelidium, Pterocladiella*
	Purified agar	*Gelidium*

the strength of the gel. Since agar is tasteless, it does not interfere with the flavors of foodstuffs; this is in contrast to some of its competitive gums which require the addition of calcium or potassium salts to form gels. In Asian countries, it is a popular component of jellies; this has its origin in the early practice of boiling seaweed, straining it and adding flavors to the liquid before it cooled and formed a jelly [17].

The remaining 20 percent is accounted for biotechnological applications [77]. A list of different uses and the corresponding type of algae required can be found in Table 5 [78]. Agar is fundamental in biotechnology studies, and is used in the preparation of inert, solidified culture media for bacteria, microalgae, fungi, tissue culture. It is also used to obtain monoclonal antibodies, interferons, steroids and alkaloids. The biotechnological applications of agar are increasing – it essential for the separation of macromolecules by electrophoresis, chromatography and DNA sequencing [38,69].

Alginate (Alginophytes, Phaeophyceae) – "Alginate" is the term usually used for the salts of alginic acid, but it can also refer to all the derivatives of alginic acid and alginic acid itself; in some publications the term "algin" is used instead of alginate. Alginate is a linear copolymer of β-D-mannuronic acid (M) and α-L-guluronic acid (G) (1→4)-linked residues, arranged either in heteropolymeric (MG) and/or homopolymeric (M or G) blocks [54,79,80].

Alginic acid is present in the cell walls of brown seaweeds, and it is partly responsible for the flexibility of the seaweed. Consequently, brown seaweeds that grow in more turbulent conditions usually have higher alginate content than those in calmer waters. While any brown seaweed could be used as a source of alginate, the actual chemical structure of the alginate varies from one genus to another, and similar variability is found in the properties of the alginate that is extracted from the seaweed. Since the main applications of alginate are in thickening aqueous solutions and forming gels, its quality is judged on how well it performs in these uses [17].

Twenty-five to 30 years ago almost all extraction of alginates took place in Europe, USA, and Japan. The major change in the alginates industry over the last decade has been the emergence of producers in China in the 1980s. Initially, production was limited to low cost, low quality alginate for the internal, industrial markets produced from the locally cultivated *Saccharina japonica*. By the 1990s, Chinese producers were competing in western industrial markets to sell alginates, primarily based on low cost. Average prices of alginates were 12 US$ kg^{-1} and global sales in 2009 were 318 million US$ [70].

Table 6. Applications of seaweed phycocolloids as food additives (Adapted from van de Velde and de Ruiter [86], Dhargalkar and Pereira [158] and Pereira [38])

Use	Phycocolloid	Function
Baked food	Agar, Kappa, Iota, Lambda	Improving quality and controlling moisture
Beer and wine	Alginate, Kappa	Promotes flocculation and sedimentation of suspended solids
Canned and processed meat	Alginate, Kappa	Hold the liquid inside the meat and texturing
Cheese	Kappa	Texturing
Chocolate milk	Kappa, lambda	Keep the cocoa in suspension
Cold preparation puddings	Kappa, Iota, Lambda	Thicken and gelling
Condensed milk	Iota, lambda	Emulsify
Dairy Creams	Kappa, iota	Stabilize the emulsion
Fillings for pies and cakes	Kappa	Give body and texture
Frozen fish	Alginate	Adhesion and moisture retention
Gelled water-based desserts	Kappa + Iota Kappa + Iota + CF	Gelling
Gums and sweets	Agar, Iota	Gelling, texturing
Hot preparation flans	Kappa, Kappa + Iota	Gelling and improve the mouth-feel
Jelly tarts	Kappa	Gelling
Juices	Agar, Kappa, Lambda	Viscosity, emulsifier
Low calorie gelatins	Kappa + Iota	Gelling
Milk ice-cream	Kappa + GG, CF, X	Stabilize the emulsion and prevent ice crystals formation
Milkshakes	Lambda	Stabilize the emulsion
Salad dressings	Iota	Stabilize the suspension
Sauces and condiments	Agar, Kappa	Thicken
Soymilk	Kappa + iota	Stabilize the emulsion and improve the mouth-feel

Non-seaweed colloids: CF - Carob flour; GG - Guar gum; X – Xanthan.

A high quality alginate forms strong gels and gives thick, aqueous solutions. A good raw material for alginate extraction should also give a high yield of alginate. Brown seaweeds that fulfill the above criteria are species of *Ascophyllum, Durvillaea, Ecklonia, Fucus, Laminaria, Lessonia, Macrocystis* and *Sargassum*. However, *Sargassum*, is only used when nothing else is available: its alginate is usually borderline quality and the yield usually low [38, 81].

The goal of the extraction process is to obtain dry, powdered, sodium alginate. The calcium and magnesium salts do not dissolve in water; the sodium salt does. The rationale behind the extraction of alginate from the seaweed is to convert all the alginate salts to the sodium salt, dissolve this in water, and remove the seaweed residue by filtration [17].

Water-in-oil emulsions such as mayonnaise and salad dressings are less likely to separate into their original oil and water phases if thickened with alginate. Sodium alginate is not useful when the emulsion is acidic, because insoluble alginic acid forms; for these applications propylene glycol alginate (PGA) is used since this is stable in mild acid conditions. Alginate improves the texture, body and sheen of yoghurt, but PGA is also used in the stabilization of milk proteins under acidic conditions, as found in some yoghurts. Some fruit drinks have fruit pulp added and it is preferable to keep this in suspension; addition of sodium alginate, or PGA in acidic conditions, can prevent sedimentation of the pulp and to create foams. In chocolate milk, the cocoa can be kept in suspension by an alginate/phosphate mixture, although in this application it faces strong competition from carrageenan (see Table 6). Small amounts of alginate can thicken and stabilize whipped cream [82,83].

Carrageenan (Carrageenophytes, Rhodophyta) – Carrageenans represent one of the major texturising ingredients used by the food industry; they are natural ingredients, which have been used for decades in food applications and are generally regarded as safe (GRAS). The phycocolloid "carrageen*in*", as it was first called, was discovered by the British pharmacist, Stanford in 1862 who extracted it from Irish moss (*Chondrus crispus*).The name was later changed to "carrageen*an*" so as to comply with the '-an' suffix for the names of polysaccharides. The modern carrageenan industry dates from the 1940s, receiving its impetus from the dairy applications (see the carrageenan applications in Table 6) where carrageenan was found to be the ideal stabilizer for the suspension of cocoa in milk chocolate [68].

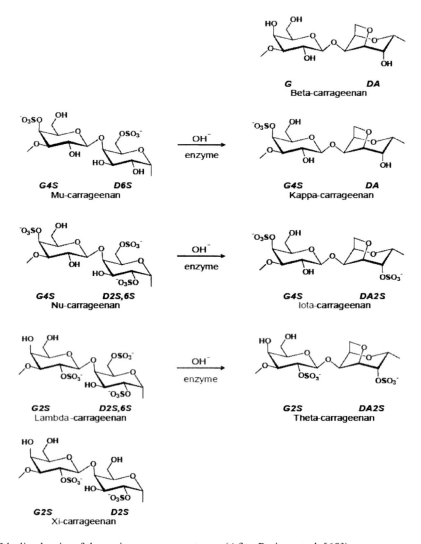

Figure 1. Idealized units of the main carrageenan types (After Periera et al. [68]).

The commercial carrageenans are normally divided into three main types: kappa-, iota- and lambda-carrageenan. The idealized disaccharide repeating units of these carrageenans are given in Figure 1. Generally, seaweeds do not produce these idealized and pure carrageenans, but more likely a range of hybrid structures and or precursors (see Table 7). Several other carrageenan repeating units exist: e.g. xi, theta, beta, mu and nu (Figure 1). The precursors (mu and nu), when exposed to alkali conditions, are modified into kappa and iota, respectively, through formation of the 3,6-anhydrogalactose bridge [68,72,84,85]. This is a feature used extensively in extraction and industrial modification.

Carrageenans are the third most important hydrocolloid in the food industry, after gelatin (animal origin) and starch (plant origin) [86]. The most commonly used, commercial carrageenans are extracted from *Kappaphycus alvarezii* and *Eucheuma denticulatum* [17].

Primarily, wild-harvested genera such as *Chondrus*, *Furcellaria*, *Gigartina*, *Chondracanthus*, *Sarcothalia*, *Mazzaella*, *Iridaea*, *Mastocarpus*, and *Tichocarpus* are also mainly cultivated as carrageenan raw materials and producing countries include Argentina, Canada, Chile, Denmark, France, Japan, Mexico, Morocco, Portugal, North Korea, South Korea, Spain, Russia, and the USA [4,70].

The original source of carrageenans was from the red seaweed *Chondrus crispus*, which continues to be used, but in limited quantities. *Betaphycus gelatinum* is used for the extraction of beta (β) carrageenan. Some South American red algae used previously only in minor quantities have, more recently, received attention from carrageenan producers, as they seek to increase diversification of raw materials in order to provide for the extraction of new carrageenan types with different physical functionalities and therefore increased product development, which in turn stimulates demand [17]. *Gigartina skottsbergii*, *Sarcothalia crispata*, and *Mazzaella laminaroides* are currently the most valuable species and all are harvested from natural populations in Chile and Peru. We can not let to mention the recent earthquake in Chile (February 27th, 2010), which caused the elevation of intertidal areas and the consequent large losses of harvestable biomass. Small quantities of *Gigartina canaliculata* are harvested in Mexico and *Hypnea musciformis* has been used in Brazil [87]. The use of high value carrageenophytes as a dissolved organic nutrient sink to boost economic viability of integrated multitrophic aquaculture (IMTA) operations has been considered [88,160].

Large carrageenan processors have fuelled the development of *Kappaphycus alvarezii* (which goes by the name "cottonii" to the trade) and *Eucheuma denticulatum* (commonly referred to as "spinosum" in the trade) farming in several countries including the Philippines, Indonesia, Malaysia, Tanzania, Kiribati, Fiji, Kenya, and Madagascar [17]. Indonesia has recently overtaken the Phils as the world's largest producer of dried carrageenophyte biomass.

Shortages of carrageenan-producing seaweeds suddenly appeared in mid-2007, resulting in doubling of the price of carrageenan; some of this price increase was due to increased fuel costs and a weak US dollar (most seaweed polysaccharides are traded in US dollars). The reasons for shortages of the raw materials for processing are less certain: perhaps it is a combination of environmental factors, sudden increases in demand, particularly from China, and some market manipulation by farmers and traders. Most hydrocolloids are experiencing severe price movements. Average prices of carrageenans were 10.5 US$ kg−1 and the global sales in 2009 were 527 million US$ [4,70].

Table 7. Industrial carrageenophytes: composition as determined by FTIR-ATR and FT-Raman (After Pereira et al. [4])

Family	Species	Lifecycle phase	Harvest Season	Origin	Yield [1]	Carrageenan Alkali-extracted	Iota/kappa ratio	Native [2]
Gigartinaceae	Chondracanthus chamissoi	NF	Summer	Chile (W)	13.5	kappa/iota	0.77	kappa/iota (mu/nu)
	C. chamissoi	T	Late Spring	Chile (W)	24.6	xi/theta	-	xi/theta
	C. chamissoi	FG	Summer	Chile (W)	14.2	kappa/iota	0.79	kappa/iota (mu/nu)
	Chondrus crispus	G + T	Late Spring	Canada (W)	33.8	kappa/iota lambda	-	kappa/iota (mu/nu) lambda/alpha
	Sarcothalia crispata	NF	Late Winter	Chile (W)	14.6	kappa/iota	0.81	kappa/iota (mu/nu)
	S. crispata	NF	Spring	Chile (W)	16.7	kappa/iota	0.79	kappa/iota (mu/nu)
	S. crispata	FG	Late Winter	Chile (W)	5.4	kappa/iota	0.81	kappa/iota (mu/nu)
Petrocelidaceae	Mastocarpus papillatus	G	Winter	Chile (W)	5.4	kappa/iota	0.83	kappa/iota (mu/nu)
Solieriaceae	Betaphycus gelatinum	-	June - October	Philippines (W)	71.0	kappa/beta	1.004 (4)	kappa/beta (mu/gamma)
	Eucheuma denticulatum	-	October - February	Philippines (F)	39.7	iota	0.92	iota (nu)
	E. denticulatum	-	Late Spring	Madagascar (F)	35.3	iota	0.93	iota (nu)
	E. denticulatum	-	Spring	Tanzania (F)	31.5	iota/kappa	0.88	iota/kappa (nu)
	Eucheuma isiforme	-	Late Summer	Colombia (F)	20.4	kappa/iota	0.71	kappa/iota (mu)
	Kappaphycus alvarezii	-	October – February	Indonesia (F)	20.0	kappa/iota	0.64	kappa/iota (mu)
	K. alvarezii (Ph)	-	October - February	Philippines (F)	30.4	kappa/iota	0.72	kappa/iota (mu)
	K. alvarezii	-	June-October	Philippines (F)	68.0	kappa/iota	0.70	kappa/iota (mu)
	K. alvarezii (Tz)	-	Winter	Tanzania (F)	18.7	kappa/iota	0.69	kappa/iota (mu)
	K. alvarezii (M1)	-	2 weeks [3]	Mexico (C)	58.1	kappa/iota	0.80	kappa/iota (mu/nu)
	K. alvarezii	-	4	Mexico	60.2	kappa/iota	0.75	kappa/iota

Family	Species	Lifecycle phase	Harvest Season	Origin	Carrageenan Yield (1)	Alkali-extracted	Iota/kappa ratio	Native (2)
	(M2)		weeks (3)	(C)				(mu)
	K. alvarezii (M3)	-	6 weeks (3)	Mexico (C)	62.4	kappa/iota	0.76	kappa/iota (mu)
	K. alvarezii (M4)	-	8 weeks (3)	Mexico (C)	48.0	kappa/iota	0.80	kappa/iota (mu/nu)
	K. alvarezii (P1)	-	2 weeks (3)	Panama (C)	-	-	0.59	kappa/iota (mu)
	K. alvarezii (P2)	-	3 weeks (3)	Panama (C)	-	-	0.66	kappa/iota (mu)
	K. alvarezii (P3)	-	4 weeks (3)	Panama (C)	-	-	0.65	kappa/iota (mu)
	K. alvarezii (P4)	-	5 weeks (3)	Panama (C)	-	-	0.70	kappa/iota (mu)
	K. alvarezii (P5)	-	6 weeks (3)	Panama (C)	-	-	0.67	kappa/iota (mu)
	K. alvarezii (P6)	-	7 weeks (3)	Panama (C)	-	-	0.71	kappa/iota (mu)
	K. alvarezii (P7)	-	8 weeks (3)	Panama (C)	-	-	0.60	kappa/iota (mu)
	Kappaphycus striatum	-	Late Spring	Madagascar (F)	75.6	kappa/iota	0.66	kappa/iota (mu)

Ph – the Philippines; Tz – Tanzania; M – Mexico; P – Panama; C – Experimental Cultivation; F – Farmed; W – Wild; T – Tetrasporophyte; FG – Female Gametophyte; G – Gametophyte; NF – Non-fructified thalli; 1 – Yield expressed as percentage of dry weight; 2 – Composition determined by FTIR-ATR and FT-Raman analysis of native carrageenan or ground seaweed samples; the carrageenans are identified according to the Greek lettering system; the letters between parentheses ( ) correspond to the biological precursors of the carrageenans, present in native samples (or ground seaweed); 3 – Carrageenophytes subjected to increasing duration of culture; 4 – The ratio between 845 and 890 cm^{-1} absorption bands in FTIR spectra was calculated and used as a parameter to determine the degree of the kappa/beta hybridization.

However, the monocultures of some carrageenophytes (namely *Kappaphycus alvarezii*) have several problems due to environmental change and also diseases. The problems with ice-ice and epiphytes have resulted in large scale crop losses [89-91].

CONCLUSION

In addition to their ecological importance, seaweeds exhibit original and interesting nutritional properties. From a nutritional standpoint, the main properties of seaweeds are their high mineral (iodine, calcium) and soluble dietary fibre contents, the occurrence of vitamin B$_{12}$ and specific components such as fucoxanthin, fucosterol, phlorotannin. If more research is needed to evaluate the nutritional value of other marine algae (e.g. *Grateloupia* spp., *Bonnemaisonia* spp., *Delesseria* spp., etc.) seaweeds can be regarded as an under-exploited source of health benefit molecules for food processing and nutraceuticals industry.

The potential for commercialization of seaweed based, antioxidant compounds as food supplements or nutraceuticals ensures continued dedicated efforts to eventually develop functional, condition-specific, antioxidant products. Seaweeds are indeed suitable natural agents for producing and delivering these products based on the multi-functional aspects of secondary seaweed metabolites and the presence of a wide variety of associated non-toxic antioxidants [60, 92]. Such relatively non-toxic associations can enhance the synergistic effects of multiple antioxidants and provide buffering capacity if necessary for those compounds which may have been intentionally increased. Algae are efficient harvesters and proficient managers of electromagnetic energy and as highly nutritional foodstuffs, can be regularly consumed without fear of metabolic toxicities. As part of a balanced diet, seaweeds can provide fibre, protein, minerals, vitamins and low fat carbohydrate content [93]. The versatility of algae as food allows consumption in fresh, dried, pickled or cooked forms and as a component in a wide assortment of other products. Cornish and Garbary [94], in the review "Antioxidants from macroalgae: potential applications in human health and nutrition", advocates the regular consumption of a variety of marine algae, primarily for their anticipated *in vivo* antioxidant capacities and associated synergistic effects.

REFERENCES

[1] Nisizawa K. Seaweeds Kaiso – Bountiful harvest from the seas. In: Critchley A, Ohno M, Largo D, editors. World Seaweed Resources - An authoritative reference system: ETI Information Services Ltd.; 2006. *Hybrid Windows and Mac DVD-ROM*; ISBN: 90-75000-80-4.

[2] Hotchkiss S, Trius A. Seaweed: the most nutritious form of vegetation on the planet? *Food Ingredients – Health and Nutrition*, 2007, January/February, 22-33.

[3] Guiry M. *Seaweed site*. 2011. URL: http://www.seaweed.ie/.

[4] Pereira L, Critchley AT, Amado AM, Ribeiro-Claro PJA. A comparative analysis of phycocolloids produced by underutilized versus industrially utilized carrageenophytes (Gigartinales, Rhodophyta). *Journal of Applied Phycology*, 2009, 21, 599-605.

[5] Zemke-White WL, Ohno M. World seaweed utilisation: An end-of-century summary. *Journal of Applied Phycology*, 1999, 11, 369-76.

[6] Alga-Net. *Seaweeds, alga-net seaweed resources and artwork*. 2010. URL: http://www.alga-net.com/.

[7] Indergaard M. The aquatic resource. I. *The wild marine plants: a global bioresource*. In: Cote W, editor. Biomass utilization. New York: Plenum Publishing Corporation; 1983. p. 137-68.

[8] Saá CF. *Atlantic Sea Vegetables - Nutrition and Health: Properties, Recipes, Description*. Redondela - Pontevedra: Algamar; 2002.

[9] Pereira L. Seaweed: an unsuspected gastronomic treasury. *Chaîne de Rôtisseurs Magazine*, 2010, 2, 50.

[10] Goldberg I. Introduction. In: Goldberg I, editor. *Functional Food: Designer Foods, Pharmafoods, Nutraceuticals*. London: Chapman and Hall; 1994. p. 3-16.

[11] Madhusudan C, Manoj S, Rhaul K, Rishi C. Seaweeds: A Diet with nutritional, medicinal and industrial value. *Research Journal of Medicinal Plant*, 2011, 5, 153-7.

[12] Abbott IA. *Food and food products from seaweeds*. In: Lembi CA, Waaland JR, editors. Algae and Human Affairs. New York: Cambridge University Press; 1988. p. 135-147.

[13] Thomsen MS, McGlathery KJ. Stress tolerance of the invasive macroalgae Codium fragile and Gracilaria vermiculophylla in a soft-bottom turbid lagoon. *Biological Invasions*, 2007, 9, 499-513.

[14] Hwang EK, Baek JM, Park CS. Cultivation of the green alga, Codium fragile (Suringer) Hariot, by artificial seed production in Korea. *Journal of Applied Phycology*, 2007, 20, 19-25.

[15] Ortiz J, Uquiche E, Robert P, Romero N, Quitral V, Llantén C. Functional and nutritional value of the Chilean seaweeds Codium fragile, Gracilaria chilensis and Macrocystis pyrifera. *European Journal of Lipid Science and Technology*, 2009, 111, 320-327.

[16] Guerra-Rivas G, Gómez-Gutiérrez CM, Alarcón-Arteaga G, Soria-Mercado IE, Ayala-Sánchez NE. Screening for anticoagulant activity in marine algae from the Northwest Mexican Pacific coast. *Journal of Applied Phycology*, 2010, DOI 10.1007/s10811-010-9618-3.

[17] McHugh DJ. A guide to the seaweed industry. *FAO, Fisheries Technical Paper*, 2003, 441, 73-90.

[18] Pattama RA, Chirapart A. Nutritional evaluation of tropical green seaweeds Caulerpa lentillifera and Ulva reticulata. *Kasetsart Journal: Natural Science*, 2006, 40, 75-83.

[19] Matanjun P, Mohamed S, Mustapha NM, Muhammad K. Nutrient content of tropical edible seaweeds, Eucheuma cottonii, Caulerpa lentillifera and Sargassum polycystum. *Journal of Applied Phycology*, 2009, 21, 75-80.

[20] Saito H, Xue CH, Yamashiro R, Moromizato S, Itabashi Y. High polyunsaturated fatty acid levels in two subtropical macroalgae, Cladosiphon okamuranus and Caulerpa lentillifera. *Journal of Phycology*, 2010, 46, 665-673.

[21] Yuan YV. Marine algal constituents. In: Barrow C, Shahidi F, editors. *Marine Nutraceuticals and Functional Foods*. New York: CRC Press, Taylor and Francis Group; 2008. p. 259-296.

[22] Fleurence J, Le Coeur C, Mabeau S, Maurice, M, Land-rein, A. Comparison of different extractive procedures from the edible seaweeds Ulva rigida and Ulva rotundata. *Journal of Applied Phycology*, 1995, 7, 577-582.

[23] Burtin P. Nutritional value of seaweeds. *Electronic Journal of Environmental, Agricultural and Food Chemistry*, 2003, 2, 498-503.

[24] Kumar IJN, Kumar RN, Manmeet K, Bora A, Sudeshnachakraborty. Variation of biochemical composition of eighteen marine macroalgae collected from Okha coast, Gulf of Kutch, India. *Electronic Journal of Environmental Agricultural and Food Chemistry*, 2010, 9, 404-410.

[25] Castro-Gonzalez MI, Romo FPG, Perez-Estrella S, Carrillo-Dominguez S. Chemical composition of the green alga Ulva lactuca. *Ciencias Marinas*, 1996, 22, 205-213.

[26] Morrissey J, Kraan S, Guiry MD. *A guide to commercially important seaweeds on the Irish coast.* Dun Laoghaire: Bord Iascaigh Mhara; 2001.

[27] Fujiwara-Arasaki T, Mino N, Kuroda M. The protein value in human nutrition of edible marine algae in Japan. *Hydrobiologia*, 1984, 116/117, 513-516.

[28] Mamatha BS, Namitha KK, Senthil A, Smitha J, Ravishankar GA. Studies on use of Enteromorpha in snack food. *Food Chemistry*, 2007, 101, 1707-1713.

[29] Patarra RF, Paiva L, Neto AI, Lima E, Baptista J. Nutritional value of selected macroalgae. *Journal of Applied Phycology*, 2010, DOI 10.1007/s10811-010-9556-0.

[30] Fleurence J. Seaweed proteins: biochemical, nutritional aspects and potential uses. *Trends in Food Science and Technology*, 1999, 10, 25-28.

[31] Taboada C, Millán R, Míguez I. Composition, nutritional aspects and effect on serum parameters of marine algae Ulva rigida. *Journal of The Science of Food and Agriculture,* 2010, 90, 445-449.

[32] Duff D, Duff P, Duff S, Duff MA, Duff C. *Sea Vegetables. Innvista*, 2010, 1-4. PDF URL: http://www.innvista.com/health.

[33] Waley SG. Naturally occurring peptides. *Advances in Protein Chemistry*, 1966, 21, 1-112.

[34] Dawczynski C, Schubert R, Jahreis G. Amino acids, fatty acids, and dietary fibre in edible seaweed products. *Food Chemistry*, 2007, 103, 891-899.

[35] Mišurcová L, Kráčmar S, Klejdus B, Vacek J. Nitrogen content, dietary fiber, and digestibility in algal food products. *Czech Journal of Food Sciences*, 2010, 28, 27-35.

[36] Guiry MD, Blunden G. *Seaweed Resources in Europe: Uses and Potential.* New York: John Wiley and Sons; 1991.

[37] Rupérez P. Mineral content of edible marine seaweeds. *Food Chemistry*, 2002, 79, 23-26.

[38] Pereira L. As algas marinhas e respectivas utilidades. Monografias.com, 2008. PDF URL: http://br.monografias.com/trabalhos913/algas-marinhas-utilidades/algas-marinhas-utilidades.pdf.

[39] Holdt SL, Kraan S. Bioactive compounds in seaweed: functional food applications and legislation. *Journal of Applied Phycology*, 2011, DOI: 10.1007/s10811-010-9632-5.

[40] Mabeau S, Fleurence J. Seaweed in food products: biochemical and nutritional aspects. *Trends in Food Science and Technology*, 1993, 4, 103-107.

[41] Siriwardhana N. Hizikia: A popular edible brown algae with great health benefits. Herbication, 2009, 1-3, PDF URL: http://www.herbication.com/Herebilife/ Siriwardhana%20Hiziki%20Article/HFP-01%20Siriwardhana%20et%20al.pdf.

[42] Sugawa-Katayama Y, Katayama M. Release of minerals from dried Hijiki, Sargassum fusiforme (Harvey) Setchell, during water-soaking. *Trace Nutrients Research*, 2009, 24, 106-109.

[43] Pérez-Ruzafa I, Izquierdo JL, Araújo R, Pereira L, Bárbara I. Distribution map of marine algae from the Iberian Peninsula and the Balearic Islands. XVII. Laminaria

rodriguezii Bornet and additions to the distribution maps of L. hyperborea (Gunner.) Foslie, L. ochroleuca Bach. Pyl. and L. saccharina (L.) Lamour. (Laminariales, Fucophyceae). *Botanica Complutensis*, 2003, 27, 155-164.

[44] Kolb N, Vallorani L, Milanovi N, Stocchi V. Evaluation of marine algae Wakame (Undaria pinnatifida) and Kombu (Laminaria digitata japonica) as food supplements. *Food Technology and Biotechnology*, 2004, 42, 57-61

[45] Suzuki N, Fujimura A, Nagai T, Mizumoto I, Itami T, Hatate H. Antioxidative activity of animal and vegetable dietary fibers. *Biofactors*, 2004, 21, 329-333.

[46] Ikeda K, Kitamura A, Machida H, Watanabe M, Negishi H and Hiraoka J. Effect of Undaria pinnatifida (Wakame) on the development of cerebrovascular diseases in stroke-prone spontaneously hypertensive rats. *Clinical and Experimental Pharmacology and Physiology*, 2003, 30, 44-48.

[47] Quirós ARB, Ron C, López-Hernández J, Lage-Yusty MA. Determination of folates in seaweeds by high-performance liquid chromatography. *Journal of Chromatography* A, 2004, 1032, 135-139.

[48] Kadam SU, Prabhasankar P. Marine foods as functional ingredients in bakery and pasta products. *Food Research International*, 2010, 43, 1975-1980.

[49] Miyashita K, Hosokawa M. Beneficial health effects of seaweed carotenoid, fucoxanthin. In: Barrow C, Shahidi F, editors. *Marine Nutraceuticals and Functional Foods*. Boca Raton, USA: CRC Press; 2008. p. 297-320.

[50] Hosokawa M, Wanezaki S, Miyauchi K, Kunihara H, Kohno H, Kawabata J, Odashima S, Takahashi K. Apoptosisinducing effect of fucoxanthin on human leukemia cell line HIL-60. *Food Science* and *Technology*, 1999, 5, 243-246.

[51] Indergaard M, Minsaas J. Animal and human nutrition. In: Guiry MD, Blunden G, editors. *Seaweed Resources in Europe*. Chichester: Wiley; 1991. p. 21-64.

[52] Lewallen E, Lewallen J. *Sea Vegetable Gourmet Cookbook and Wildcrafter's Guide*. CA, USA: Mendocino Sea Vegetable Company; 1996.

[53] Kraan S, Tramullas AV, Guiry MD. The edible brown seaweed Alaria esculenta (Phaeophyceae, Laminariales): hybridization, growth and genetic comparisons of six Irish populations. *Journal of Applied Phycology*, 2000, 12, 577-583.

[54] Pereira L, Sousa A, Coelho H, Amado AM, Ribeiro-Claro PJA. Use of FTIR, FT-Raman and 13C-NMR spectroscopy for identification of some seaweed phycocolloids. *Biomolecular Engineering*, 2003, 20, 223-228.

[55] Pereira L. Identification of Phycocolloids by Vibrational Spectroscopy. In: Critchley AT, Ohno M, Largo DB, editors. World Seaweed Resources - An Authoritative Reference System V1.0, ETI Information Services Ltd.; 2006. *Hybrid Windows and Mac DVD-ROM*; ISBN: 90-75000-80-4.

[56] García I, Castroviejo R, Neira C. Las algas en Galicia: *Alimentación y Otros Usos. La Coruña: Consellería de Pesca,* Marisqueo e Acuicultura - Xunta de Galícia; 1993.

[57] Le Gall L, Pien S, Rusig AM. Cultivation of Palmaria palmata (Palmariales, Rhodophyta) from isolated spores in semi-controlled conditions. *Aquaculture*, 2004, 229, 181-91.

[58] Pang S, Lüning K. Tank cultivation of the red alga Palmaria palmata: effects of intermittent light on growth rate, yield and growth kinetics. *Journal of Applied Phycology*, 2004, 16, 93-99.

[59] Martínez B, Viejo RM, Rico JM, Rødde RH, Faes VA, Oliveros J, Álvarez D. Open sea cultivation of Palmaria palmata (Rhodophyta) on the northern Spanish coast. *Aquaculture,* 2006, 254, 376-87.

[60] Smit AJ. Medicinal and pharmaceutical uses of seaweed natural products: A review. *Journal of Applied Phycology*, 2004, 16, 245-262.

[61] Dixon PS, Irvin LM. *Seaweeds of the British Isles: Volume I - Rhodophyta, Part* 1 - Introduction, Nemaliales, Gigartinales. London: The Natural History Museum; 1995.

[62] López-López I, Bastida S, Ruiz-Capillas C, Bravo L, Larrea MT, Sánchez-Muniz F, Cofrades S, Jiménez-Colmenero F. Composition and antioxidant capacity of low-salt meat emulsion model systems containing edible seaweeds. *Meat Science*, 2009, 83, 492-498.

[63] Noda H. Health benefits and nutritional properties of nori. *Journal of Applied Phycology*, 1993, 5, 255-258.

[64] Barsanti L, Gualtieri P. Algae and Men, Chapter 7. In: *Algae - Anatomy, Biochemistry, and Biotechnology.* LLC: CRC Press, Taylor and Francis Group; 2006. p. 251-291.

[65] Norziah MH, Ching YC. Nutritional composition of edible seaweed Gracilaria changii. *Food Chemistry*, 2000, 68, 69-76.

[66] Bird C, McLachlan J, Oliveira E. Gracilaria chilensis sp. nov. (Rhodophyta, Gigartinales), from Pacific South America. *Canadian Journal of Botany*, 1987, 64, 2928-2934.

[67] Ioannou E, Roussis V. Natural products from seaweeds. In: Osbourn AE, Lanzotti V, editors. *Plant-Derived Natural Products*. LLC: Springer Science + Business Media; 2009. p. 51-81.

[68] Pereira L, Amado AM, Critchley AT, van de Velde F, Ribeiro-Claro PJA. Identification of selected seaweed polysaccharides (phycocolloids) by vibrational spectroscopy (FTIR-ATR and FT-Raman). *Food Hydrocolloids*, 2009, 23, 1903-1909.

[69] Pereira L. Algae: *Uses in Agriculture, Gastronomy and Food Industry*. VN: CM Viana do Castelo; 2010.

[70] Bixler HJ, Porse H. A decade of change in the seaweed hydrocolloids industry. Journal *of Applied Phycology*, 2010, DOI: 10.1007/s10811-010-9529-3.

[71] Jiao G, Yu G, Zhang J, Ewart HS. *Chemical structures and bioactivities of sulfated polysaccharides from marine algae.* Marine Drugs, 2011, 9, 196-223.

[72] Pereira L, van de Velde F. Portuguese carrageenophytes: Carrageenan composition and geographic distribution of eight species (Gigartinales, Rhodophyta). *Carbohydrate Polymers*, 2011, 84, 614-623.

[73] Mouradi-Givernaud A, Amina Hassani L, Givernaud T, Lemoine Y, Benharbet O. Biology and agar composition of Gelidium sesquipedale harvested along the Atlantic coast of Morocco. *Hydrobiologia*, 1999, 398-399, 391-395.

[74] Givernaud T, Mouradi A. Seaweed resources of Morocco. In: Critchley AT, Ohno M, Largo DB, editors. World Seaweed Resources - An Authoritative Reference System. ETI Information Services Ltd.; 2006. *Hybrid Windows and Mac DVD-ROM*; ISBN: 90-75000-80-4.

[75] Givernaud T, Sqali N, Barbaroux O, Orbi A, Semmaoui Y, Rezzoum NE, Mouradi A, Kaas R. Mapping and biomass estimation for a harvested population of Gelidium sesquipedale (Rhodophyta, Gelidiales) along the Atlantic coast of Morocco. *Phycologia*, 2005, 44, 66-71.

[76] Krisler AV. Seaweed resources of Chile. In: Critchley AT, Ohno M, Largo DB, editors. World Seaweed Resources - An authoritative reference system. ETI Information Services Ltd.; 2006. *Hybrid Windows and Mac DVD-ROM*; ISBN: 90-75000-80-4.

[77] Armisen R, Galatas F. Agar. In: Phillips G, Williams P, editors. *Handbook of Hydrocolloids*. Boca Raton, FL: CRC Press; 2000. p. 21-40.

[78] Armisen R. World-wide use and importance of Gracilaria. *Journal of Applied Phycology*, 1995, 7, 231-243.

[79] Larsen B, Salem DMSA, Sallam MAE, Mishrikey MM, Beltagy AI. Characterization of the alginates from algae harvested at the Egyptian red sea coast. *Carbohydrate Research,* 2003, 338, 2325-2336.

[80] Leal D, Matsuhiro B, Rossi M, Caruso F. FT-IR spectra of alginic acid block fractions in three species of brown seaweeds. *Carbohydrate Research*, 2008, 343, 308-316.

[81] Draget KI, Smidsrød O, Skjåk-Broek S. Alginates from algae. In: Baets SD, Vandamme E, Steinbüchel A, editors. Biopolymers, v6, Polysaccharides II: *Polysaccharides from Eukaryotes*. Weinheim: Wiley; 2004. p. 215-224.

[82] Nussinovitch A. *Hydrocolloid Applications: Gum Technology in the Food and Other Industries.* London: Chapman and Hall; 1997.

[83] Onsoyen E. Alginates. In: Imeson A, editor. *Thickening and Gelling Agents for Food*. London: Blackie Academic and Professional; 1997. p. 22-44.

[84] Myslabodski DE. Red-Algae Galactans: Isolation and Recovery Procedures - Effects on the Structure and Rheology. *Doctoral Dissertation*. Trondheim: Norwegian Institute of Technology; 1990.

[85] Rudolph B. Seaweed product: Red Algae of Economic Significance. In: Martin RE, Carter EP, Davis LM, Flich GJ, editors. *Marine and Freshwater Products Handbook.* Lancaster, USA: Technomic Publishing Company Inc.; 2000. p. 515–529.

[86] van de Velde F, de Ruiter GA. Carrageenan. In: Vandamme EJ, Baets SD, Steinbèuchel A, editors. Biopolymers V6, Polysaccharides II, *Polysaccharides from Eukaryotes*. Weinheim: Wiley; 2002. p. 245-274.

[87] Furtado MR. Alta lucratividade atrai investidores em hidrocolóides. *Química e Derivados*, 1999, 377, 20-29.

[88] Pereira L. Estudos em Macroalgas Carragenófitas (Gigartinales, Rhodophyceae) da Costa Portuguesa - Aspectos Ecológicos, Bioquímicos e Citológicos. *Ph.D. Thesis,* Coimbra: Departamento de Botânica, FCTUC, Universidade de Coimbra; 2004.

[89] Hurtado AQ, Genevieve BL, Critchley AT. Kappaphycus 'cottonii' farming. In: Critchley AT, Ohno M and Largo DB, editors. World Seaweed Resources - An Authoritative Reference System. ETI Information Services Ltd.; 2006. *Hybrid Windows and Mac DVD-ROM*; ISBN: 90-75000-80-4

[90] Vairappan C, Chung C, Hurtado A, Soya F, Lhonneur G, Critchley A. Distribution and symptoms of epiphyte infection in major carrageenophyte-producing farms. *Journal of Applied Phycology*, 2008, 20, 477-483.

[91] Hayashi L, Hurtado AQ, Msuya FE, Bleicher-Lhonneur G, Critchley AT. A review of Kappaphycus farming: prospects and constraints. In: Seckbach J, editor. *Seaweeds and Their Role in Globally Changing Environments* (Cellular Origin, Life in Extreme Habitats and Astrobiology). Netherlands: Springer; 2010. p. 251-283.

[92] Bocanegra A, Bastida S, Benedí J, Ródenas S, Sánchez-Muniz FJ. Characteristics and nutritional and cardiovascular-health properties of seaweeds. *Journal of Medicinal Food*, 2009, 12, 236-258.

[93] Yuan YV, Walsh NA. Antioxidative and antiproliferative activities of extracts from a variety of edible seaweeds. *Food* and *Chemical Toxicology*, 2006, 44, 1144-1150.

[94] Cornish ML, Garbary DJ. Antioxidants from macroalgae: potential applications in human health and nutrition. *Algae*, 2010, 25, 155-171.

[95] El-Sarraf W, El-Shaarawy G. Chemical composition of some marine macroalgae from the Mediterranean Sea of Alexandria, Egypt. *The Bulletin of the High Institute of Public Health,* 1994, 24, 523-534.

[96] Akhtar P, Sultana V. Biochemical studies of some seaweed species from Karachi coast. *Records Zoological Survey of Pakistan*, 2002, 14, 1-4.

[97] Santoso J, Yoshie-Stark Y, Suzuki T. Comparative contents of minerals and dietary fibres in several tropical seaweeds. *Bulletin Teknologi Hasil Perikanan*, 2006, 9, 1-11.

[98] Kumar M, Kumari P, Trivedi N, Shukla MK, Gupta V, Reddy CRK, Jha B. Minerals, PUFAs and antioxidant properties of some tropical seaweeds from Saurashtra coast of India. *Journal of Applied Phycology*, 2010, DOI 10.1007/s10811-010-9578-7.

[99] Manivannan K, Thirumaran G, Devi GK, Hemalatha A, Anantharaman P. Biochemical Composition of Seaweeds from Mandapam Coastal Regions along Southeast Coast of India. *American-Eurasian Journal of Botany*, 2008, 1, 32-37.

[100] Kumar M, Gupta V, Kumari P, Jha B. Assessment of nutrient composition and antioxidant potential of Caulerpaceae seaweeds, *Journal of Food Composition and Analysis,* 2010, doi:10.1016/j.jfca.2010.07.007

[101] Shanmugam A, Palpandi C. Biochemical composition and fatty acid profile of the green alga Ulva reticulata. *Asian Journal of Biochemistry*, 2008, 3, 26-31.

[102] Applegate RD, Gray PB. *Nutritional value of seaweed to ruminants*. Rangifer, 1995, 15, 15-18.

[103] Mitchell K. Keith *Michell's Practically Macrobiotic Cookbook*. Rochester, Vermont: Healing Arts Press; 2000.

[104] Truus K, Vaher M, Taure I. Algal biomass from Fucus vesiculosus (Phaeophyta): investigation of the mineral and alginate components. *Proceedings of the Estonian Academy of Sciences*. Chemistry, 2001, 50, 95-103.

[105] Díaz-Rubio ME, Pérez-Jiménez J, Saura-Calixto F. Dietary fiber and antioxidant capacity in Fucus vesiculosus products. *International Journal of Food Sciences and Nutrition*, 2008, 60, 23-34.

[106] Plaza M, Cifuentes A, Ibáñez E. In the search of new functional food ingredients from algae. *Trends in Food Science and Technology*, 2008, 19, 31-39.

[107] Gómez-Ordóñez E, Jiménez-Escrig A, Rupérez P. Dietary fibre and physicochemical properties of several edible seaweeds from the northwestern Spanish coast. *Food Research International*, 2010, 43, 2289-2294.

[108] Yamada Y, Miyoshi T, Tanada S, Imaki M. Digestibility and energy availability of Wakame (Undaria pinnatifida) seaweed (in Japanese). *Nippon Eiseigaku Zasshi*, 1991, 46, 788-94.

[109] Kaas R, Campello F, Arbault S, Barbaroux O. *La Culture des Algues Marines dans le Monde. Plouzane*, Brest: Institut Français de Recherche pour l'Exploitation de la Mer, IFREMER; 1992.

[110] Funaki M, Nishizawa M, Sawaya T, Inoue S, Yamagishi T. Mineral composition in the holdfast of three brown algae of the genus Laminaria. *Fisheries Science*, 2001, 67, 295-300.

[111] Krishnaiah D, Rosalam S, Prasad DMR, Bono A. Mineral content of some seaweeds from Sabah's South China sea. *Asian Journal of Scientific Research*, 2008, 1, 166-170.

[112] Tsuchiya Y. Physiological studies on the vitamin C content of marine algae. *Tohoku Journal of Agricultural Research*, 1950, 1, 97-102.

[113] MacArtain P, Christopher RG, Brooks M, Campbell R, Rowland IR. Nutritional value of edible seaweeds. *Nutrition Reviews*, 2007, 65, 535-543.

[114] Watanabe F, Takenaka S, Katsura H, Miyamoto E, Abe K, Tamura Y, Nakatsuka T, Nakano Y. Characterization of a vitamin B12 compound in the edible purple laver, Porphyra yezoensis. *Bioscience, Biotechnology, and Biochemistry*, 2000, 64, 2712-2715.

[115] Okaih Y, Higashi-Okaia K, Yanob Y, Otanib S. Identification of antimutagenic substances in an extract of edible red alga, Porphyra tenera (Asadusa-nori). *Cancer Letters*, 1996, 100, 235-240.

[116] Maruyama H, Watanabe K, Yamamoto I. Effect of dietary kelp on lipid peroxidation and glutathione peroxidase activity in livers of rats given breast carcinogen DMBA. *Nutrition* and *Cancer*, 1991, 15, 221-228.

[117] Nam B, Jin H, Kim S, Hong Y. Quantitative viability of seaweed tissues assessed with 2,3,5triphenyltetrazolium chloride. *Journal of Applied Phycology*, 1998, 10, 31-36.

[118] Lohrmann NL, Logan BA, Johnson AS. Seasonal acclimatization of antioxidants and photosynthesis in Chondrus crispus and Mastocarpus stellatus, two co-occurring red algae with differing stress tolerances. *The Biological Bulletin*, 2004, 207, 225-232.

[119] Yang YJ, Nam S-J, Kong G, Kim M K. A case-control study on seaweed consumption and the risk of breast cancer. *British Journal of Nutrition*, 2010, 103, 1345-1353.

[120] Okuzumi J, Takahashi T, Yamane T, Kitao Y, Inagake M, Ohya K, Nishino H, Tanaka Y. Inhibitory effects of fucoxanthin, a natural carotenoid, on N-ethyl-N'-nitro-N nitrosoguanidineinduced mouse duodenal carcinogenesis. *Cancer Letters*, 1993, 68, 159-68.

[121] Yan X, Chuda Y, Suzuki M, Nagata T. Fucoxanthin as the major antioxidant in Hijikia fusiformis, a common edible seaweed. *On the Bioscience, Biotechnology and Biochemistry*, 1999, 63, 605-607.

[122] Hosakawa M, Bhaskar N, Sashima T, Miyashita K. Fucoxanthin as a bioactive and nutritionally beneficial marine carotenoid: A review. *Carotenoid Science*, 2006, 10, 15-28.

[123] Sugawara T, Matsubara K, Akagi R, Mori M, Hirata T. Antiangiogenic activity of brown algae fucoxanthin and its deacetylated product, fucoxanthinol. *Journal of Agricultural and Food Chemistry*, 2006, 54, 9805-9810.

[124] Maeda H, Tsukui T, Sashima T, Hosokawa M, Miyashita K. Seaweed carotenoid, fucoxanthin, as a multi-functional nutrient. *Asia Pacific Journal of Clinical Nutrition*, 2008, 17, 196-199.

[125] Miyashita K, Hosokawa M. Beneficial Health Effects of Seaweed Carotenoid, Fucoxanthin. In: Barrow C, Shahidi F, editors. *Marine Nutraceuticals and Functional Foods.* Boca Raton, FL: CRC Taylor and Francis Press Inc.; 2008. p. 259-296.

[126] Sangeetha RK, Bhaskar N, Baskaran V. Comparative effects of β-carotene and fucoxanthin on retinol deficiency induced oxidative stress in rats. *Molecular and Cellular Biochemistry*, 2009, 331, 59-67.

[127] Suzuki H, Higuchi T, Sawa K, Ohtaki S, Tolli J. Endemic coast goitre in Hokkaido, Japan. *Acta Endocrinologica*, 1965, 50, 161-176.

[128] Gonzalez R, Rodriguez S, Romay C, Ancheta O, Gonzalez A, Armesta J, Remirez D, Merino N. Anti-inflammatory activity of phycocyanin extract in acetic acid-induced colitis in rats. *Pharmacological Research*, 1999, 39, 55-59.

[129] Padula M, Boiteux S. Photodynamic DNA damage induces by phycocyanin and its repair in Saccharomyces cerevisiae. *Brazilian Journal of Medical and Biological Research,* 1999, 32, 1063-1071.

[130] Remirez D, Gonzalez A, Merino N, Gonzalez R, Ancheta O, Romay C, Rodriguez S. Effect of phycocyanin in Zymosan-induced arthritis in mice-phycocyanin as an antiarthritic compound. *Drug Development Research*, 1999, 48, 70-75.

[131] Yabuta Y, Fujimura H, Kwak CS, Enomoto T, Watanabe F. Antioxidant activity of the phycoerythrobilin compound formed from a dried Korean purple laver (Porphyra sp.) during in vitro digestion. *Food Science and Technology Research*, 2010, 16, 347-351.

[132] Bagchi M, Mark M, Casey W, Jaya B, Xumei Y, Sidney S, Debasis. Acute and chronic stress-induced oxidative gastrointestinal injury in rats, and the protective ability of a novel grape seed proanthocyanidin extract. *Nutrition Research*, 1999, 19, 1189-1199.

[133] YuanYV, Bone DE, Carrington MF. Antioxidant activity of dulse (Palmaria palmata) extract evaluated in vitro. *Food Chemistry*, 2005, 91, 485-494.

[134] Nakamura T, Nagayama K, Uchida K, Tanaka R. Antioxidant activity of phlorotannins isolated from the brown alga Eisenia bicyclis. *Fisheries Science*, 1996, 62, 923-926.

[135] Shin HC, Hwang HJ, Kang KJ, Lee BH. An antioxidative and anti-inflammatory agent for potential treatment of osteoarthritis from Ecklonia cava. *Archives of Pharmacal Research,* 2006, 29, 165-171.

[136] Shibata T, Ishimaru K, Kawaguchi S, Yoshikawa H, Hama Y. Antioxidant activities of phlorotannins isolated from Japanese Laminariaceae. *Journal of Applied Phycology*, 2008, 20, 705-711.

[137] Wijesekara I, Yoon NY, Kim S. Phlorotannins from Ecklonia cava (Phaeophyceae): Biological activities and potential health benefits. *BioFactors,* 2010, 36, 408-414.

[138] Ngo DH, Wijesekara I, Vo TS, Ta QV, Kim SK. Marine food-derived functional ingredients as potential antioxidants in the food industry: An overview. *Food Research International,* 2011, doi:10.1016/j.foodres.2010.12.030

[139] Nagayama K, Shibata T, Fujimoto K, Honjo T, Nakamura T. Algicidal effect of phlorotannins from the brown alga Ecklonia kurome on red tide microalgae. *Aquaculture*, 2003, 218, 601-611.

[140] Nagayama K, Iwamura Y, Shibata T, Hirayama I, Nakamura T. Bactericidal activity of phlorotannins from the brown alga Ecklonia kurome. *Journal of Antimicrobial Chemotherapy,* 2002, 50, 889-89.

[141] Lahaye M, Thibault JF. Chemical and physio-chemical properties of fibers from algal extraction by-products. In: Southgate DAT, Waldron K, Johnson IT, Fenwick GR, editors. *Dietary Fibre: Chemical and Biological Aspects*. Cambridge: Royal Society of Chemistry; 1990. p. 68-72.

[142] Lahaye M. Marine algae as sources of fibres: determination of soluble and insoluble dietary fiber contents in some sea vegetables. *Journal of the Science of Food Agriculture*, 1991, 54, 587-594.

[143] Costa LS, Fidelis GP, Cordeiro SL, Oliveira RM, Sabry DA, Camara RBG, et al. Biological activities of sulfated polysaccharides from tropical seaweeds. *Biomedicine and Pharmacotherapy*, 2010, 64, 21-28.

[144] Haijin M, Xiaolu J, Huashi G. A k-carrageenan derived oligosaccharide prepared by enzymatic degradation containing anti-tumor activity. *Journal of Applied Phycology,* 2003, 15, 297-303.

[145] Yuan H, Song J. Preparation, structural characterization and in vitro antitumor activity of kappa-carrageenan oligosaccharide fraction from Kappaphycus striatum. *Journal of Applied Phycology*, 2005, 17, 7-13.

[146] Choosawad D, Leggat U, Dechsukhum C, Phongdara A, Chotigeat W. Anti-tumour activities of fucoidan from the aquatic plant Utricularia aurea lour. *Songklanakarin Journal* of Science and *Technology*, 2005, 27, 799-807.

[147] Hemmingson JA, Falshaw R, Furneaux RH, Thompson K (2006). Structure and antiviral activity of the galactofucan sulfates extracted from Undaria pinnatifida (Phaeophyta). *Journal of Applied Phycology*, 2006, 18, 185-193.

[148] Yuan H, Song J, Li X, Li N, Liu S. Enhanced immunostimulatory and antitumor activity of different derivatives of κ-carrageenan oligosaccharides from Kappaphycus striatum. *Journal of Applied Phycology*, 2010, DOI 10.1007/s10811-010-9536-4

[149] Spieler R. Seaweed compound's anti-HIV efficacy will be tested in southern Africa. *Lancet*, 2002, 359, 16-75.

[150] Li B, Lu F, Wei X, Zhao R. Fucoidan: Structure and Bioactivity. *Molecules*, 2008, 13, 1671-1695.

[151] Kim KJ, Lee OH, Lee HH, Lee BY. A 4-week repeated oral dose toxicity study of fucoidan from the sporophyll of Undaria pinnatifida in sprague-dawley rats. *Toxicology*, 2010, 267, 154-158.

[152] Sugawara I, Itoh W, Kimura S, Mori S, Shimada K. Further characterization of sulfated homopolysaccharides as anti-HIV agents. *Cellular and Molecular Life Sciences*, 1989, 45, 996-998.

[153] Béress A, Wassermann O, Bruhn T, Béress L, Kraiselburd EN, Gonzalez LV, de Motta GE, Chavez PI. A new procedure for the isolation of anti-HIV compounds (polysaccharides and polyphenols) from the marine alga Fucus vesiculosus. *Journal of Natural Products*, 1993, 56, 478-488.

[154] Witvrouw M, De Clercq E. Sulfated polysaccharides extracted from sea algae as potential antiviral drugs. *Geneneral Pharmacology*, 1997, 29, 497-511.

[155] Feldman SC, Reynaldi S, Stortz CA, Cerezo AS, Damont EB. Antiviral properties of fucoidan fractions from Leathesia difformis. *Phytomedicine*, 1999, 6, 335-40.

[156] Vázquez-Freire MJ, Lamela M, Calleja JM. Hypolipidaemic activity of a polysaccharide extract from Fucus vesiculosus L. *Phytotherapy* Research, 1996, 10, 647-650.

[157] Huang L, Wen K, Gao X, Liu Y. Hypolipidemic effect of fucoidan from Laminaria japonica in hyperlipidemic rats. *Pharmaceutical Biology*, 2010, 48, 422-426.

[158] Dhargalkar VK, Pereira N. Seaweed: Promising plant of the millennium. *Science and Culture*, 2005, 71, 60-66.

[159] Pengzhan Y, Quanbin PZ, Ning L, Zuhong X, Yanmei W, Zhi'en L. Polysaccharides from Ulva pertusa (Chlorophyta) and preliminary studies on their antihyperlipidemia activity. *Journal of Applied Phycology*, 2003, 15, 21-27.

[160] Chopin T. Integrated multi-trophic aquaculture. What it is, and why you should care… and don't confuse it with polyculture. *Northern Aquaculture*, 2006, 12, 4.

[161] Bocanegra A, Bastida S, Benedí J, Ródenas, Sánchez-Muniz FJ, Characteristics and Nutritional and Cardiovascular-Health Properties of Seaweeds. *Journal of Medicinal Food*, 2009, 12(2), 236-258.

In: Seaweed
Editor: Vitor H. Pomin

ISBN 978-1-61470-878-0
© 2012 Nova Science Publishers, Inc.

Chapter 3

NUTRITIONAL QUALITY AND BIOLOGICAL PROPERTIES OF BROWN AND RED SEAWEEDS

P. Rupérez[*], E. Gómez-Ordóñez and A. Jiménez-Escrig

Metabolism and Nutrition Department,
Instituto de Ciencia y Tecnología de Alimentos y Nutrición (ICTAN),
Consejo Superior de Investigaciones Científicas (CSIC), José Antonio Novais,
Ciudad Universitaria, Madrid, Spain

ABSTRACT

Brown and red seaweeds are often regarded as under-exploited marine bio-resources. Specifically, research on edible marine macroalgae is on the increase because they are most interesting as a source of macronutrients and associated bioactive compounds with high potentially economical impact in food and pharmaceutical industry, and public health. Our group has worked on the nutritional evaluation, physicochemical and biological properties of edible Spanish seaweeds. Thus, in brown: *Bifurcaria bifurcata*, *Fucus vesiculosus*, *Himanthalia elongata* (Sea spaghetti), *Laminaria digitata* (Kombu), *Saccharina latissima* (Sugar Kombu), *Undaria pinnatifida* (Wakame), and red seaweeds: *Chondrus crispus* (Irish moss), *Gigartina pistillata*, *Mastocarpus stellatus* and *Porphyra tenera* (Nori), total dietary fiber content ranges from 29-50% of which approximately 20-75% is soluble. For brown seaweeds, soluble fiber consists of uronic acids from alginates and neutral sugars from sulfated fucoidan and laminarin. For red seaweeds, main neutral sugars correspond to sulfated galactans such as carrageenan or agar. Insoluble fibers (7.4–40%) are essentially made of cellulose, with an important contribution of Klason lignin, up to 31% in *Fucus*. In Nori, insoluble fiber consists of a mannan and xylan. Protein content is generally higher in red (15–30%), than in brown seaweeds (7–26%), although protein digestibility is apparently low. Ash content is high (21–40%) and sulfate, related to the presence of sulfated polysaccharides, represents 7.4–57% of ash. Except for the brown seaweeds *Fucus* (2.5%) and *Bifurcaria* (5.6%), oil content is usually lower than 1%. Relevant biological properties of seaweeds (such as anticoagulant or antioxidant capacity) seem to be associated to sulfate content in sulfated polysaccharides and to a lesser degree to minor components, such as extractable

---

[*] E-mail: pruperez@ictan.csic.es.

polyphenols (0.4%). In brown seaweeds, the correlation among reduction power, radical-scavenging activity and total phenolic content would suggest the involvement of phenolic compounds in the antioxidant mechanisms, whereas in the case of red seaweeds, the role of sulfate-containing polysaccharides is presumably evidenced in the reduction power. It should be stressed that processing and storage conditions are essential for the optimal preservation of bioactive compounds (such as phenolic compounds and pigments) and their antioxidant activity. Regarding the main physicochemical properties of dietary fiber in seaweeds, related to the hydrophilic nature of sulfated polysaccharides, oil retention is low, while swelling, water retention, and cation exchange capacity are higher in brown than in red algae. Our results indicate that vibrational FTIR-ATR spectroscopy is a useful tool for a rapid and preliminary identification of the main natural phycocolloids (namely alginate, agar and carrageenan) in edible brown and red seaweeds. Moreover, other storage and structural polysaccharides in seaweeds such as laminarin, fucoidan and cellulose, as well as protein or sulfate, can be identified from specific absorption bands in their infrared spectra. Accordingly, alginate is the main polysaccharide in brown seaweeds, whereas all red seaweeds studied are carrageenan producers, except for Nori. In summary, edible seaweeds can be considered as an excellent source of dietary fiber, protein and minerals for human consumption. Moreover, algal sulfated polysaccharides potentially could afford natural antioxidants for the food, pharmaceutical and cosmetics industry.

ABBREVIATIONS

ABTS	2,2′-azinobis(3-ethylbenzothiazolin-6-sulfonate)
AOAC	Association of Official Analytical Chemists
DF	dietary fiber
DPPH	diphenylpicrylhydrazyl
FRAP	ferric reducing/antioxidant power
FTIR-ATR	Fourier transform algorithm with attenuated total reflectanc
IDF	insoluble DF
KL	Klason lignin
NMR	nuclear magnetic resonance
PCL	photoluminiscent assay
PGE	phloroglucinol equivalent
RP	reduction power towards Fe(III)
RSA	radical scavenging activity
SDF	soluble DF

1. NUTRITIONAL EVALUATION OF EDIBLE SEAWEEDS

1.1. Dietary Fiber

Epidemiological studies have shown that consumption of fruits and vegetables is associated with reduced risk of chronic diseases. As a consequence, an increased consumption of products from vegetable origin, which contain high levels of dietary fiber (DF) and phytochemical constituents, has been recommended [1]. Specifically, there is an impressive body of data providing evidence for an important role for dietary fiber (DF) in the prevention and management of chronic diseases, such as diabetes, cardiovascular diseases, and gastrointestinal diseases [2]. However, the per capita intake of DF in Western countries does not cover the recommendations of DF intake. For example adequate DF intake in the U.S./Canada is set at 31 g/day by the Food and Nutrition Board (FNB), whereas the DF daily recommended intake given by the FAO/WHO and the EURODIET project is about 25 g/day [3].

DF components, which are polysaccharides resistant to the human enzymatic digestion, in brown seaweeds are essentially composed of four families of polysaccharides, laminaran, alginate, fucoidan, and cellulose. Laminaran is the storage polysaccharide in brown seaweeds [4]. The major structural component in brown seaweeds is alginate a gelling polyuronide. Its solubility is influenced by factors such as pH, concentration, ions in solution, the presence of divalent ions and ionic force. Alginate is gelling in presence of divalent ions like calcium. The structural polysaccharide fucoidan is soluble in water and in acid solution. Among structural polysaccharides, fucoidans are the most interesting for their potential biological activities whereas alginate is mostly used as a food ingredient [5,6].

Attending the potential health-promoting of seaweeds, our group has focused on the DF component, which is the main nutritional component in seaweeds, and its associated compounds, sulfate and polyphenols. Thus, we have characterized the nutritional components of several edible seaweeds from the Northwestern Atlantic coast of Galicia, Spain, concluding that brown seaweeds showed relatively high DF content (ranged 29.3% to 50.1% dry weight (d.w.)) (Table 1).

Table 1. Soluble, insoluble and total dietary fiber (% dry weight) in edible Spanish seaweeds

Seaweed		SDF	IDF	TDF	Dietary fiber ratio SDF:TDF	TDF:IDF
Brown	*Bifurcaria bifurcata*	14.64±0.68	22.79±0.97	37.42±0.78	0.39	1.64
	Fucus vesiculosus	9.80±0.78	40.29±0.98	50.09±1.77	0.20	1.24
	Himanthalia elongata (Sea spaguetti)	23.63±0.48	13.51±0.45	37.14±0.86	0.64	2.75
	Laminaria digitata (Kombu)	9.15±0.48	26.98±1.97	36.12±2.46	0.25	1.34
	Saccharina latissima (Sugar Kombu)	17.12±0.84	13.11±0.56	30.23±0.85	0.57	2.31
	Undaria pinnatifida (Wakame)	17.31±0.51	16.26±0.79	33.58±1.31	0.51	2.06
Red	*Chondrus crispus* (Irish moss)	22.25±0.99	12.04±2.89	34.29±3.89	0.65	2.85
	Gigartina pistillata	21.90±0.22	7.41±0.12	29.31±0.34	0.75	3.95
	Mastocarpus stellatus	23.10±0.16	8.85±0.67	31.95±0.11	0.72	3.61
	Porphyra tenera (Nori)	14.56±1.33	19.22±2.05	33.78±3.38	0.43	1.76

Data are mean values of triplicate determinations ± standard deviation.
SDF=Soluble dietary fiber; IDF=Insoluble dietary fiber; TDF=Total dietary fiber.
Modified-AOAC method without starch treatment for brown seaweeds [7, 8].

Table 2. Soluble and insoluble dietary fiber fractions (% d.w.) in edible Spanish seaweeds

Seaweed		SDF			IDF		
		NS	UA	Residue	NS	UA	KL
Brown	B. bifurcata	2.35±0.25	7.67±0.69	4.62±0.45	7.04±0.27	4.43±0.87	10.83±0.84
	F. vesiculosus	4.58±0.08	1.31±0.13	3.90±0.57	7.21±0.18	2.00±0.29	31.08±0.52
	H. elongata	5.14±0.75	12.65±0.95	5.84±0.22	3.31±0.64	0.71±0.03	9.54±0.40
	L. digitata	2.74±0.12	2.18±0.10	4.23±0.26	9.34±0.69	4.55±0.44	13.09±0.84
	S. latissima	2.15±0.15	9.64±0.74	4.91±0.32	5.89±0.75	2.14±0.33	3.89±0.12
	U. pinnatifida	3.79±0.14	4.90±0.24	8.62±0.13	4.96±0.29	1.97±0.20	9.32±0.29
Red	C. crispus	21.66±0.96	0.59±0.03	0	5.95±0.93	0.15±0.01	5.93±1.95
	G. pistillata	19.48±0.34	1.27±0.25	0.25±0.12	4.73±0.29	0.21±0.09	3.24±1.13
	M. stellatus	19.00±0.35	2.35±0.28	0.09±0.03	5.18±0.24	0.62±0.24	3.45±0.90
	P. tenera	14.14±1.31	0.42±0.02	0	8.74±1.53	0.09±0.01	10.39±0.52

Data are mean values of triplicate determinations ± standard deviation.
SDF = soluble dietary fiber; IDF = insoluble dietary fiber; NS = neutral sugars; UA = uronic acids; KL = Klason lignin.
AOAC method without starch treatment for brown seaweeds [7, 8].

This DF content is calculated as non-starch polysaccharides plus lignin, according to the AOAC modified-method [7], which includes the structural non-digestible polysaccharides in seaweeds. Among the brown seaweeds tested, *Saccharina latissima* shows lower DF content (30.2% d.w.) in comparison to other seaweeds such as *Himanthalia elongata* (37.1% d.w.), *Bifurcaria bifurcata* (37.4% d.w) or *Fucus vesiculosus* (50.09%), however *Saccharina latissima* shows the lowest Klason lignin value (Table 2), and consequently the highest relative proportion of carbohydrate components in the total DF content (87.1% *versus* 71.1% or 74.4%, for *H. elongata* and *B. bifurcata*, respectively). Regarding red seaweeds DF content is around 30% d.w. in the four seaweeds tested [7,8].

The ratio for soluble to total DF (Table 1) ranged from 0.20 to 0.57 in the case of brown seaweeds, and from 0.43 to 0.75 in the red ones. In red seaweeds, three out of four seaweeds tested showed higher soluble DF than insoluble DF values.

The AOAC modified-method for DF analysis presented some difficulties when applied to seaweeds. Thus, for brown algae after dialysis of the soluble fiber a sulfuric acid-insoluble residue appeared which amounted to 25–50% of the total soluble fiber (Table 2). The uronic acids in soluble DF of brown seaweeds came from alginates and the acid-insoluble residue was probably composed of alginic acid.

**Table 3. Ashes, protein and fat content (% dry weight)
in edible Spanish seaweeds**

Seaweed		Ashes	Sulfate	Protein	Fat
Brown	B. bifurcata	34.31±0.21	13.5[1]	10.92±0.10	5.67±0.32
	F. vesiculosus	30.12±0.48	20.7[2]	6.86±0.01	2.50±0.08
	H. elongata	36.41±0.15	11.6[1]	14.08±0.21	0.94±0.07
	L. digitata	37.60±0.32	7.5[2]	9.99±0.02	0.24±0.04
	S. latissima	34.78±0.08	5.5[1]	25.70±0.11	0.79±0.07
	U. pinnatifida	39.82±0.05	7.4[2]	15.97±0.04	0.52±0.02
Red	C. crispus	21.44±0.14	48.9[2]	20.90±0.14	0.32±0.05
	G. pistillata	34.56±0.47	45.1[1]	15.59±0.28	0.57±0.06
	M. stellatus	24.99±0.12	57.1[1]	21.30±0.18	0.39±0.02
	P. tenera	21.00±0.42	17.6[2]	29.80±0.04	0.46±0.06

Data are mean values of triplicate determinations ± standard deviation.
Sulfate content determined in ashes obtained at 550 °C 16 h and expressed as % ash dry weight: [1]sulfate by ion chromatography; [2]sulfate by AOAC gravimetric method [7, 8].

1.2. Protein, Ashes and Fat

Regarding other nutritional components apart from DF, ashes content is high and ranged from 21.0% d.w. (*Porphyra tenera*) to 39.8% d.w. (*Undaria pinnatifida*) (Table 3). This relatively high ash content is a general feature of seaweeds, and these values are generally much higher than those of terrestrial vegetables other than spinach [9].

Inorganic anion profile of certain of these seaweeds has been determined by ion chromatography and reported recently by our group [10]. Brown seaweeds are characterized by higher chloride content up to 33.7–36.9% ash d.w., while red seaweeds are characterized by higher sulfate content (45–57% ash d.w.). Regarding protein, for brown seaweeds, the contents ranged from 6.9 to 25.7% (Table 3), being much higher for the cultivated seaweed *Saccharina* (25.7%). In red seaweeds the range is comparatively narrower (15.6–29.8%) than in brown seaweeds. Lipid content was very low in all seaweed samples (0.3–0.9%), except for *Bifurcaria bifurcata* and *Fucus vesiculosus* (Table 3).

1.3. Physicochemical or Functional Properties

The physiological effects of DF are related to their physicochemical properties *in vitro*. Marine algae constitute potential sources of DF that differ chemically and physicochemically from those of land plants and thus may have other physiological effects on man [7,8].

The main physicochemical properties of DF in seaweeds are related to the hydrophilic nature of sulfated polysaccharides. In general, swelling and water retention capacities are high in brown seaweeds (Table 4). These properties could be related to high uronic acids content from alginates in these seaweeds. Whereas, oil retention capacity is low in brown and red seaweeds (Table 4) [7,8].

Table 4. Physicochemical properties of edible Spanish seaweeds

Seaweed	SC (mL/g dw)	WRC (g/g dw)	ORC (g/g dw)
B. bifurcata	7.75±0.73	4.89±0.12	1.65±0.11
F. vesiculosus	5.77±0.42	5.48±0.42	0.89±0.01
H. elongata	10.97±0.62	7.26±0.13	1.61±0.07
L. digitata	9.82±0.64	10.33±0.60	1.11±0.05
S. latissima	10.20±0.37	8.93±0.52	1.67±0.11
U. pinnatifida	10.53±0.24	10.96±0.23	0.96±0.03
C. crispus	5.87±0.35	7.29±0.21	0.91±0.06
G. pistillata	11.43±0.63	10.22±0.67	1.32±0.03
M. stellatus	7.20±0.42	5.42±0.06	1.22±0.04
P. tenera	6.08±0.10	5.12±0.15	1.04±0.03

Data are mean values of triplicate determinations ± standard deviation.
SC= Swelling capacity; WRC= Water retention capacity; ORC= Oil retention capacity [7, 8].

2. BIOLOGICAL PROPERTIES OF SEAWEEDS

2.1. Multifunctional Antioxidant Capacity

Epidemiological studies have shown that consumption of fruits and vegetables is associated with reduced risk of chronic diseases. As a consequence, an increased consumption of products from vegetable origin, which contain high levels of DF and phytochemical constituents, has been recommended by the National Health Councils [1]. A possible mechanism for the protective effects of fruits and vegetables with regard to chronic diseases is that their bioactive compounds reduce oxidative stress [11]. The characteristic of oxidative stress is an increased production of reactive oxygen species in amounts that exceed cellular antioxidant defenses. Our organism counteracts the reactive oxygen species helped by the antioxidants from the diet in addition to the natural antioxidant defense system [12]. Increased reactive oxygen species may be detrimental and lead to cell death or to acceleration in ageing and age-related diseases. Recently, emerging findings suggest a variety of potential mechanisms of bioactive compounds in cytoprotection against oxidative stress. These potential mechanisms against oxidative stress include interaction of bioactive compounds with cell signaling and influence of these compounds on gene expression, with the consequent modulation of specific enzymatic activities. This modulation drives the intracellular response against oxidative stress [13].

Seaweeds have to survive in a highly competitive environment and, therefore, they need to develop defense strategies that result in a tremendous diversity of antioxidant compounds from different metabolic pathways -carotenoids, phenols, minerals, sulphur compounds, vitamins, etc [14]. In addition, seaweeds are usually located in the intertidal zones and consequently must be able, through protective antioxidant defense systems, to cope with a constantly changing environment, fluctuations in light and oxygen exposition [15]. In this context, seaweeds have been considered over the past few decades as promising organisms for providing both, novel biologically active substances and essential compounds for human nutrition [14,16,17].

Thus, regarding the health impact of consuming vegetables and fruits, the use of seaweeds as a source of natural antioxidants may be revitalized in the growing public consciousness in Western countries [16]. In this context, the aim of the present review is to describe our group's research dealing with the *in vitro* antioxidant activity of edible seaweeds collected from the Atlantic coast of Spain. This specific research may contribute to rise in value the edible seaweeds as a potential source of health-promoting compounds.

On the frame of different Spanish National research projects, we have evaluated the *in vitro* antioxidant activities of edible seaweeds from the coast of Galicia in the Northwest of Spain. Firstly, we selected three of the seaweeds more commonly used as a food, including brown algae such as *Laminaria* spp. (Kombu) and *Undaria* spp. (Wakame), and species of the red algae *Porphyra* (Nori) [18]. Their Japanese names, used worldwide, are shown within parentheses [19,20]. In addition other edible seaweeds, such as the brown seaweed *Fucus vesiculosus* and the red seaweed *Chondrus chrispus*, which are representative of algal consumption in the Spanish diet, are selected. In a second stage of research other edible seaweeds less commonly used as a food were also subject of study: *Bifurcaria bifurcata*, *Himantalia elongata* and *Saccharina latissima* [formerly *Laminaria saccharina*] as brown seaweeds, and *Chondracanthus acicularis*, *Dumontia contorta*, *Gigartina pistillata*, *Mastocarpus stellatus*, *Nemalion helminthoides* and *Osmundea pinnatifida* as red seaweeds.

The generation of radical oxidative species involves either radical processes or different potential redox systems. Thus, to evaluate this potential antioxidant activity it is necessary to assess several antioxidant assays that include different antioxidant mechanisms [21]. A useful way of viewing the interactions among various antioxidants is to take into account oxidation-reduction potentials, measuring the reduction power (RP) by FRAP assay. This method is based on the standard redox potential of the semi-system Fe(III)/Fe(II) (0.77 V). Another mechanism commonly used is the radical scavenging activity (RSA). These models measure the RP of reductants with an ionization potential above specific value, and the RSA of antioxidants containing a RH group with an adequate enthalpy difference ($\Delta H_1 < 0$) in the scavenging reaction, respectively [22]. In addition, the copper-induced oxidation of lipoprotein model could be used to measure the prevention of lipid peroxidation or metal-catalysed radical reactions [23]. Apart from that, the solubility properties of antioxidant compounds determine their effective antioxidant activity either in aqueous or lipid systems. Therefore, one strategy might be the selective extraction of bioactive compounds using solvents with different polarity.

As mentioned above, the first stage of research in antioxidant activities of seaweeds is focused on the edible seaweeds more commonly consumed in the Spanish diet. With this purpose three brown and two red seaweeds are selected. The order of RSA -measured towards the stable free radical DPPH- in methanol-acetone-water extracts of the selected seaweeds is the following: *Fucus vesiculosus* >> *Laminaria ochroleuca* > *Undaria pinnatifida* > *Porphyra umbilicalis*. In the case of the red seaweed *Chondrus crispus*, no RSA has been detected [18]. These results are in agreement with other published data showing higher antioxidant capacity in brown than in red seaweeds [24]. Otherwise, previous studies in some brown seaweeds from Brittany coasts [25] report the order Fucales as those brown seaweeds with the highest polyphenol content. In this sense, the RSA of the mentioned extracts might be related to the phenolic hydroxyl group, since *Fucus vesiculosus* showed the highest total polyphenol content (4.4 g PGE /100 g dry matter) among seaweed samples tested. Moreover, linear regression analysis of the RSA towards the stable free radical DPPH with the total polyphenol estimation content, as measured by Folin-Cioacalteau method, gives a statistically significant correlation (r = 0.73). This significant correlation between polyphenols and scavenging activity has also been evidenced in terrestrial vegetable foods and natural beverages [26,27].

Regarding RP, the different activities of the seaweed extracts mentioned previously varied markedly. *Fucus vesiculosus* is found to be more bioactive than *Porphyra umbilicalis* and *Laminaria ochroleuca*, while the RP is not detected either in *Undaria pinnatifida* or *Chondrus crispus* [18].

In order to evaluate the prevention of a lipid system from oxidation by the mentioned seaweed extracts, *Fucus vesiculosus* is further selected for testing in an *in vitro* system containing human low-density lipoproteins that are oxidized by copper. As a result it is found that *Fucus* extracts prolong the lag phase in a concentration-dependent manner [18].

In a second stage of research, the *in vitro* antioxidant studies are focused on the screening for the potential antioxidant activity of three brown and six red seaweeds with potential consumption in the Spanish diet. With this purpose RP by the FRAP assay, and RSA by the ABTS and photoluminiscent (PCL) assays, based on the transfer by the tested antioxidant of one hydrogen to the synthetic radical ABTS$^{\bullet+}$ or to the biological superoxide radical O_2^-, respectively, are measured in both organic and water soluble extracts from seaweeds [unpublished data].

Besides polyphenol content, total carbohydrate, sulfate content and sulphation degree of polysaccharides are also determined. As a result, it was emphasized the great RP and RSA of *Bifurcaria bifurcata* -a *Sargassaceae* brown species- in both organic and aqueous extracts. In addition, two *Gigartinaceae* red species, *Gigartina pistillata* and *Mastocarpus stellatus*, evidence a relatively high RP in the aqueous extracts. This work gives a novel approach to the search of nutraceuticals from seaweeds, taking into account new aspects of sulfated polysaccharide content in brown and red seaweeds other than the usual approaches based only on polyphenols and antioxidant properties. As a result, the correlation found between RSA, RP and total polyphenol content strongly suggests the involvement of phenolic compounds in the antioxidant mechanisms for brown seaweeds, whereas, the involvement of sulfate-containing polysaccharides in the RP [unpublished data] is presumably evidenced for red seaweeds (Table 5).

Table 5. Comparison of the antioxidant capacity in selected brown and red seaweeds

Seaweed	Reduction Power		Radical Scavenging Activity	
	Organic Extracts	Aqueous Extracts	Organic Extracts	Aqueous Extracts
Brown	+++	++	+++	++
Red	+	+	N.D.	N.D.

(+++): high; (++): medium; (+): low; N.D. Not detected.

2.2. Anticoagulant Capacity

Anticoagulant activity is among the most widely studied properties of sulfated polysaccharides, and sulfated polysaccharides from seaweeds have been described to possess anticoagulant activity similar to heparin [28]. However, the relationship between structure and biological activities of sulfated polysaccharides is not clearly established [29]. Until now, two types of sulfated polysaccharides are identified with high anticoagulant activity, sulfated galactans, also known as carrageenans, from marine red algae and sulfated fucoidans from marine brown algae [28]. Apart from that, it is known that the presence of sulfate groups in polysaccharides can increase both the specific and non-specific binding of these polysaccharides to a wide range of biologically active proteins involved in the coagulation system [30]. Our group is currently working on the fractionation and isolation of the sulfated polysaccharides from selected brown and red seaweeds [unpublished data], with the purpose to increase the present knowledge on the relationship between structure and antiocoagulant activity of these polysacharides.

3. POLYSACCHARIDE IDENTIFICATION IN SEAWEEDS BY FTIR SPECTROSCOPY

Polysaccharides associated with the cell wall and intercellular spaces of some brown and red seaweed species namely alginate, agar and carrageenan respectively, have been used extensively as gels and thickeners in food and industrial preparations. Currently, there is an increasing demand of these products by the food industry and a quick and reliable non-destructive method to assess the quality of raw algal material is needed.

One of the most useful and rapid techniques for a preliminary identification of polysaccharide structures is infrared (IR) spectroscopy, which is based on the analysis of absorption peaks at certain wave numbers (expressed in cm^{-1}). In the latest years, the combination of Fourier transform algorithm with attenuated total reflectance (ATR) techniques has improved the conventional IR spectroscopy with various and important advantages. Thus, FTIR-ATR (Fourier transformed IR from attenuated total reflectance) spectroscopy is direct and non-destructive, requires only small amounts of dried ground material (just a few milligrams) and is a quick method (a few minutes *vs.* several days), thus avoiding lengthy extractions and further sample preparation as a film or KBr pellet [31,32]. Not only information on polysaccharide composition and structure but protein or sulfate content can be gained from algal infrared spectra. Thus, this technique allows to preliminary identify the main polysaccharides in an unknown seaweed sample [33]. In the structural analysis of carbohydrates five frequency regions [34] can be distinguished in the normal spectra (4000-650 cm^{-1}): (1) region of OH and CH stretching vibrations at 3600-2800 cm^{-1}; (2) region of local symmetry at 1500-1200 cm^{-1}; (3) region of CO stretching vibration at 1200-950 cm^{-1}; (4) fingerprint or anomeric region at 950-700 cm^{-1}; and (5) skeletal region below 700 cm^{-1} (Figures 1 and 2). Therefore, it is in the fingerprint or anomeric region where identification of polysaccharides can be mainly achieved. It is the intention of this section to summarize the results obtained by the application of this technique for the identification of the main polysaccharides in brown and red edible seaweeds.

3.1. Brown Seaweed Polysaccharides

3.1.1. Alginate

The major structural polysaccharide of brown seaweeds (Phaeophyta) is alginate. Alginate is the salt of alginic acid, a linear copolymer of β-D-mannuronic acid (M) and its C5-epimer α-L-guluronic acid (G) (1→ 4)-linked residues, arranged either in heteropolymeric (MG) and/or homopolymeric (M or G) blocks [35,36]. The FTIR spectra obtained directly on dried and ground material from *Saccharina latissima*, *Himanthalia elongata* and *Bifurcaria bifurcata* in the region 4000-650 cm^{-1} are shown in Figure 1.

Sodium alginate shows specific infrared absorption bands: the asymmetric stretching of carboxylate (COO$^-$) vibration at 1600 cm^{-1}, the C-OH deformation vibration with contribution of O-C-O symmetric stretching vibration of carboxylate group at 1406 cm^{-1} [33,36]. In the fingerprint or anomeric region (region 4, Figure 1) three bands appear: the C-O stretching vibration of uronic acid residues at 947.9 cm^{-1}, the C1-H deformation vibration of β-mannuronic acid residues at 878.1 cm^{-1} and finally the band at 817.1 cm^{-1} is characteristic of mannuronic acid residues [33]. Alginates are stable in solution between pH 6 and 9 but they form insoluble precipitates at acid pH; thus soluble sodium alginate in salt form is converted into insoluble alginic acid in free acid form. Accordingly, a specific band appears in the alginic acid form at 1730 cm^{-1} due to the carboxylic acid ester form (C=O). This alginic acid band can be observed in some brown seaweed samples as a small shoulder (Figure 1) showing the presence of alginic acid.

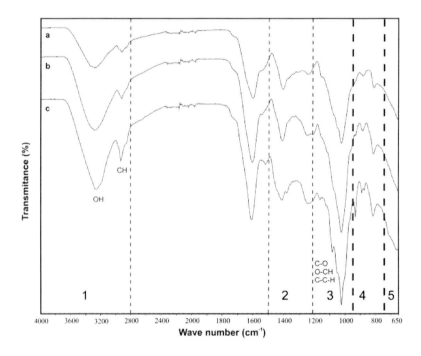

Figure 1. FTIR-ATR spectra of three seaweed samples: (a) *Saccharina latissima*; (b) *Himanthalia elongata* and (c) *Bifurcaria bifurcata*. Also the five frequency (1-5) regions according to Mathlouthi and Koenig [34] are shown.

The content of uronic acids varies with species and tissue types and FTIR spectroscopy can be used for estimation of the composition in hetero- and homopolymeric blocks. Besides, mannuronic to guluronic acid ratio is an index of the nature of gels produced and these different M/G ratios mean that alginates obtained from different seaweeds would have different physicochemical properties. M/G ratio can vary within brown seaweed alginates from 0.5 to 2.5. In general alginic acid with a low M/G ratio (<1) and a large proportion of guluronic acid blocks, form a strong and rigid gel [37] on the contrary alginate with a low number of guluronic acid blocks and a high M/G ratio (>1) produces a soft and elastic gel. A recent research shows how M/G ratio in alginate can be tentatively estimated from specific absorption bands (808/787 cm^{-1} and 1030/1080 cm^{-1}) in infrared spectra [33].

3.1.2. Other Structural and Storage Polysaccharides: Fucoidan, Laminaran and Cellulose

The structural polysaccharide fucoidan, is primarily composed of (1,2)-α-L-fucose-4-sulfate and contains small quantities of D-xylose, D-galactose, D-mannose, and uronic acids [4]. Laminaran is a small β-glucan of 5 kDa, composed of D-glucose with β-(1,3) linkages and β-(1,6) interchain branching [38]. Commercial standards of these polysaccharides can be used for comparison of specific FTIR absorption bands. Table 6 shows the characteristic band assignments found in the spectra of polysaccharide standards from brown seaweeds [5,36,39,40].

Table 6. Identification of main polysaccharide types in brown seaweeds (alginate, fucoidan, laminaran and cellulose) by FTIR spectroscopy

Wave number (cm^{-1})	Assignment [a]	Polysaccharide[b] A	L	F	C
1750-1735	Carboxylic acid (C=O) ester group	-	-	+	-
1650-1640	Amide I bond (N-H_2)	-	-	-	-
1637-1615	Carboxylate anion (COO-)	+	+	+	+
1600-1500	Amide II bond (N-H) (aminosugars and proteins)	-	-	-	-
1425	β-linked glucose	-	-	-	+
1415	Carboxylate group (O-C-O)	+	+	-	-
1260-1225	Sulfate ester (S=O)	+	+	+	-
960-969	(COS); (CH_3) –Fucose, Acetic	-	-	+	-
948	Uronic acid residue (C-O)	+	-	-	-
883	β-mannuronic acid residues (C1-H)	+	-	-	-
890	β-linked polysaccharide	-	+	+	+
845	Sulfate groups at an axial C-4 position	-	-	+	-
820	Sulfate groups at an equatorial position	+	-	+	-
663	β-linked glucose	-	-	-	+

A: alginate; F: fucoidan; L: laminaran; C: cellulose. [a] According to : [5,36,39,40] [b] Standards from Sigma-Aldrich chemicals [38].

Brown seaweeds show an absorption band at 1252 cm^{-1} related to a >S=O stretching vibration of the sulfate group [41]. Also the sulfate band at 845 cm^{-1} is reported to be linked at a C-4 axial position on fucose or galactose [38]. This band appears in *Fucus vesiculosus* [5], and with less intensity in *S. latissima*, *H. elongata* and *B. bifurcata* (Figure 1) indicating the presumable presence of sulfated fucoidans. Also, a band around 963 cm^{-1} is attributed to the presence of fucose (Figure 1). Another two bands attributed to the presence of β-linked glucose and cellulose (at 663 cm^{-1} and 895 cm^{-1}) can be expected in some seaweed.

In addition, protein content can be identified in seaweeds from two specific infrared absorption bands: the N-H$_2$ stretching vibration of the amide I bond at 1650-1640 cm^{-1} (Table 6) and the N-H stretching vibration (amide II bond) at 1550 cm^{-1} (Table 6) [42]. In these seaweeds spectra, the amide I band may appear slightly overlapped with the stronger band of sodium alginate at 1600 cm^{-1} (Figure 1).

To have a more precise characterization of the polysaccharides present in brown algae and their cell-wall fractions it is necessary to perform a sequential extraction of polysaccharides based on their different solubilities [5]. Laminarans are soluble in water, but their solubility depends on branching level; thus, low-branched laminarans are soluble only in warm water (60-80 °C) [5]. Fucans can be extracted with diluted hydrochloric acid; meanwhile alginates can be extracted with alkali [5].

3.2. Red Seaweed Polysaccharides

3.2.1. Carrageenan and Agar

Carrageenan and agar are the principal sulfated polysaccharides produced by red seaweeds (Rhodophyta); the main difference between the highly sulfated carrageenans from the less sulfated agars is the presence of D-galactose and anhydro-D-galactose in carrageenans and of D-galactose, L-galactose or anhydro-L-galactose in agars. Several similar bands can be observed in the FTIR spectra of agar and carrageenan polymers. The characteristic broad band of sulfate esters generally between 1210-1260 cm^{-1} is much stronger in carrageenan than in agar. In the anomeric region, the bands at 930 cm^{-1}, 890 cm^{-1} and bands at 770, 740 and 693 cm^{-1}, assigned to the presence of 3,6-anhydrogalactose residue, β-galactopyranosyl residues and to the skeleton bending of pyranose ring, respectively, are common to agar and carrageenans. Moreover, the second-derivative mode of the FTIR spectra can be applied to distinguish agar-producing from carrageenan-producing red seaweeds [33,43].

The structure of the various types of carrageenans is defined by the number and position of sulfate groups, the presence of 3,6-anhydro-D-galactose and conformation of the pyranosidic ring. There are about fifteen idealized carrageenan structures traditionally identified by Greek letters [44]. Table 7 shows the band assignments of carrageenan types. Commercial carrageenans are normally divided into three main types: kappa (κ), iota (ι)- and lambda (λ)-carrageenan. Their differences on chemical composition and configuration are responsible for their interesting rheological properties as gelling, stabilizing and thickening agents used in the food, pharmaceutical and cosmetics industry. Kappa- and iota-carrageenan contain the 3,6-anhydrogalactose unit and are gel forming, whereas lambda-carrageenan with only galactose residues is a thickener [45,46].

Table 7. Identification of main carrageenan types in red seaweeds by FTIR spectroscopy

Wave number (cm^{-1})	Assignment	κ	ι	λ	μ	ν	θ	ξ
1210-1260	Sulfate ester (S=O)	+	+	+	+	+	+	+
928-933, 1070 (shoulder)	3,6-anhydro-D-galactose	+	-	-	-	-	-	-
840-850	D-galactose-4-sulfate	+	+	-	+	+	-	-
830	D-galactose-2-sulfate	-	-	+	-	-	+	+
820, 825 (shoulder)	D-galactose-2,6-disulfate	-	-	+	-	+	-	-
810-820, 867 (shoulder)	D-galactose-6-sulfate	-	-	-	+	-	-	-
800-805, 905 (shoulder)	3,6-anhydro-D-galactose-2-sulfate	-	+	-	-	-	+	-

(+), Presence of peak; (-), Absence of peak; (κ): Kappa; (ι): Iota; (λ): Lambda; (μ): Mu; (ν): Nu; (θ): Theta; (ξ): Xi.

Generally, red seaweeds do not produce pure carrageenans, but more likely a range of hybrid structures. FTIR spectroscopy is a useful tool in order to identify these hybrid structures of carrageenans in red seaweeds [33]. Moreover, the carrageenan-producing seaweeds seem to present a similar variation related to the season and their reproductive stage. The gametophyte and non-fructified stages produce carrageenan of the kappa family (hybrid kappa/iota/mu/nu-carrageenan), whereas the tetrasporophyte stages produce carrageenans of the lambda family (hybrid xi/theta- or xi/lambda-carrageenan). For example, FTIR spectrum of the red seaweed *M. stellatus* (Figure 2a) shows two specific bands: a strong band at 844.7 and a weak band at 804.3 cm^{-1} that corresponds with a kappa/iota-hybrid carrageenan (Table 7); whereas *G. pistillata*, *C. acicularis* and *D. contorta* (Figure 2b-d) present a broad band at 825-830 cm^{-1} identified with the lambda-family of carrageenans (Table 7).

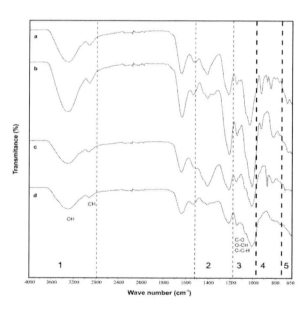

Figure 2. FTIR-ATR spectra of four red seaweed samples: (a) *Mastocarpus stellatus*; (b) *Gigartina pistillata*; (c) *Chondracanthus acicularis* and (d) *Dumontia contorta*. Also the five frequency (1-5) regions according to Mathlouthi and Koenig [34] are shown.

In conclusion, with the FTIR-ATR technique it is possible to identify the principal seaweed polysaccharides and phycocolloids in ground samples as well as in extracted material. Yet further studies are necessary, along with other complementary techniques such as FT-Raman and NMR spectroscopy, in order to characterize completely these phycocolloids in seaweeds.

ACKNOWLEDGMENTS

Financial support given by the Spanish Ministerio de Ciencia e Innovación, through Project AGL2008-00998 ALI is acknowledged.

REFERENCES

[1] P. M. Kris-Etherton, K. D. Hecker, A. Bonanome, S. M. Coval, A. E. Binkoski, K. F. Hilpert, A. E. Griel, T. D. Etherton, *Am. J. Med*. 113, 71S (2002).
[2] J. H. Cummings, J. I. Mann, C. Nishida, H. H. Vorster, *Lancet* 373, 365 (2009).
[3] J. Lunn, J. L. Buttriss, *Nutr. Bull*. 32, 21 (2007).
[4] A. Jiménez-Escrig, F. J. Sánchez-Muniz, *Nutr. Res*. 20, 585 (2000).
[5] P. Rupérez, O. Ahrazem, J. A. Leal, *J. Agric. Food Chem*. 50, 840 (2002).
[6] L. Rioux, S. L. Turgeon, M. Beaulieu, *Carbohydr. Polym*. 69, 530 (2007).
[7] P. Rupérez, F. Saura-Calixto, *Eur. Food Res. Technol*. 212, 349 (2001).
[8] E. Gómez-Ordóñez, A. Jiménez-Escrig, P. Rupérez, *Food Res. Int*. 43, 2289 (2010)
[9] P. Rupérez, *Food Chem*. 79, 23 (2002).
[10] E. Gómez-Ordóñez, E Alonso, P Rupérez, *Talanta* 82, 1313 (2010)
[11] R. L. Prior, *Acta Horticulturae* 841, 75 (2009).
[12] T. Finkel, N. J. Holbrook *Nature* 408, 239 (2000)
[13] R. Masella, R. Di Benedetto, R. Varì, C. Filesi, C. Giovannini C, *J. Nutr. Biochem*. 16, 577 (2005).
[14] K. H. M. Cardozo, T. Guaratini, M. P. Barros, V. R. Falcão, A. P. Tonon, N. P. Lopes, S. Campos, M. A. Torres, A. O. Souza, P. Colepicolo, E. Pinto, *Comp. Biochem. Physiol. C Toxicol Pharmacol* 146, 60 (2007).
[15] C. A. Blanchette, *Ecology* 78, 1563 (1997).
[16] P. MacArtain, C. I. R. Gill, M. Brooks, R. Campbell, I. R. Rowland, *Nutr. Rev*. 65 535, (2007).
[17] A. J. Smit, *J. Appl. Phycol*. 16, 245 (2004).
[18] A. Jiménez-Escrig, I. Jiménez-Jiménez, R. Pulido, F. Saura-Calixto, *J. Sci. Food Agric*. 81, 530 (2001)
[19] *A. Jensen,* Hydrobiologia 260, 15 (1993).
[20] K. Nisizawa, H. Noda, R. Kikuchi, T. Watanabe, *Hydrobiologia* 151, 5 (1987).
[21] C. Sánchez-Moreno, *Food Sci. Technol. Int*. 8, 121 (2002).
[22] P. Trouillas, P. Marsal, A. Svobodová, J. Vostálová, R. Gažák, J. Hrbáč, P. Sedmera, V. Křen, R. Lazzaroni, J. Duroux, D. Walterová, *J. Physical Chem. A* 112, 1054 (2008).
[23] A. Jiménez-Escrig, *Int. J. Food Sci. Nutr*. 58, 629 (2007).

[24] X. Yan, T. Nagata, X. Fan, *Plant Foods Human Nutr.* 52, 253 (1998).
[25] M. Zubia, M. S. Fabre, V. Kerjean, K. L. Lann, V. Stiger-Pouvreau, M. Fauchon, E. Deslandes, *Food Chem.* 116, 693 (2009)
[26] A. Jiménez-Escrig, M. Rincón, R. Pulido R, Saura-Calixto F, *J. Agric. Food Chem.* 49, 5489 (2001).
[27] I. Sánchez-González, A. Jiménez-Escrig, F. Saura-Calixto, *Food Chem.* 90, 133 (2005).
[28] I. Wijesekara I, R. Pangestuti, S. Kim, *Carbohydr. Polym.* 84, 14 (2011).
[29] L. S. Costa, G. P. Fidelis, S. L. Cordeiro, R. M. Oliveira, D. A. Sabry, R. B. G. Câmara, L. T. D. B. Nobre, M. S. S. P. Costa, J. Almeida-Lima, E. H. C. Farias, E. L. Leite, H. A. O. Rocha, *Biomed. Pharmacother.* 64, 21 (2010).
[30] F. R. Melo, M. S. Pereira MS, D. Foguel, P. A. S. Mourão, *J. Biol. Chem.* 279, 20824 (2004).
[31] L. Pereira, J. F. Mesquita, *J. Appl. Phycol.* 16, 369 (2004).
[32] L. Pereira, Identification of phycocolloids by vibrational spectroscopy. In: A. T. Critchley, M. Ohno, and D. B. Largo (Eds.), *World Seaweed Resources - an Authoritative Reference System.* ETI Information Services Ltd. 2006.
[33] E. Gómez-Ordóñez, P. Rupérez, *Food Hydrocoll.* 25, 1514 (2011).
[34] M. Mathlouthi, J. L. Koenig, Advances Carbohydr. *Chem. Biochem.* 44, 7, (1987).
[35] B. Larsen, D. M. S. A: Salem, M. A. E. Sallam, M. M. Mishrikey, A. I. Beltagy, *Carbohydr Res.* 338, 2325 (2003).
[36] D. Leal, B. Matsuhiro, M. Rossi, F. Caruso, *Carbohydr. Res.* 343, 308, (2008).
[37] K. I. Draget, G. Skjåk-Bræk, B. T. Stokke, *Food Hydrocoll.* 20, 170 (2006).
[38] L. Rioux, S. L. Turgeon, M. Beaulieu, *Phytochemistry* 71:1586 (2010)
[39] M. Kačuráková, R. H. Wilson, *Carbohydr. Polym.* 44 291 (2001)
[40] A. Synytsya, W. Kim, S. Kim, R. Pohl, A. Synytsya, F. Kvasnička, J. Čopíková, Y. Park, *Carbohydr. Polym.* 81, 41 (2010)
[41] N. Chattopadhyay, T. Ghosh, S., Sinha, K. Chattopadhyay, P. Karmakar, B. Ray, *Food Chem.* 118, 823 (2010).
[42] A. Femenia, M. García-Conesa, S. Simal, C. Rossello, *Carbohydr. Polym.* 35, 169 (1998).
[43] B. Matsuhiro, *Hydrobiologia* 326-327, 481 (1996).
[44] T. Chopin, B. F. Kerin, R. Mazerolle, *Phycol. Res.* 47, 167 (1999).
[45] E. Tojo, J. Prado, *Carbohydr. Res.* 338, 1309 (2003).
[46] J. Prado-Fernández, J. A. Rodríguez-Vázquez, E. Tojo, J. M. Andrade, Anal. *Chim. Acta* 480, 23 (2003).

In: Seaweed
Editor: Vitor H. Pomin

ISBN 978-1-61470-878-0
© 2012 Nova Science Publishers, Inc.

Chapter 4

EDIBLE SEAWEEDS: A FUNCTIONAL FOOD WITH ORGAN PROTECTIVE AND OTHER THERAPEUTIC APPLICATIONS

Suhaila Mohamed[1],*, Patricia Matanjun[2], Siti Nadia Hashim[1], Hafeedza Abdul Rahman[1], and Noordin Mohamed Mustapha[3]

[1]Institute BioScience, University Putra Malaysia, Serdang, Selangor, Malaysia
[2]School of Food Science and Nutrition, Universiti Malaysia Sabah, Malaysia
[3]Faculty of Veterinary Medicine, Universiti Putra Malaysia, Serdang, Malaysia

ABSTRACT

The in vitro and in vivo antioxidant properties, total phenolic and chemical composition of various seaweeds is compiled. Seaweeds medicinal uses include for cardiovascular disease prevention, cholesterol-lowering, anti-diabetes, anti-coagulative, anti-inflammatory, immunomodulating and anti-cancer effects. The nutrients composition, vitamin C, tocopherol, dietary fibers, minerals, fatty acid and amino acid profiles of some tropical seaweeds is presented. Effects of tropical seaweeds in preventing cardiovascular diseases and cancer in animals via assessing the plasma and organs biomarkers will be given as example. Such biomarkers include activities of antioxidant enzymes such as superoxide dismutase (SOD), glutathione peroxidase (GSH-Px) and catalase (CAT); alanine aminotransferase (ALT), aspartate aminotransferase (AST), gama glutamyltransferase (GGT), creatinine kinase (CK), CK-MB isoenzyme, urea, creatinine and uric acid. Positive changes caused by dietary seaweeds on somatic index and histological changes in the liver, heart, kidney, brain, spleen and eye of the experimental animals are shown. The comparative in vivo cardiovascular protective effects of red and green tropical seaweeds in mammals fed on a rich lipogenic or sometimes called Western diet (24% fat and 1% cholesterol) are elaborated as a case. The potential anti-infective, antiviral and tissue healing properties of seaweeds are also incorporated.

---

*Tel: +03 89472168; +60127843190. E-mail: mohamed.suhaila@gmail.com.

1. INTRODUCTION

Seaweeds are low in calories, rich in soluble dietary fiber, proteins, mineral, vitamin, antioxidants, phytochemical and polyunsaturated fatty acids. Presently, they are widely used as gelling agent and stabilisers in the food and pharmaceutical industries. However, recent researches have revealed their therapeutic potential for various pathological conditions. Most researches were done in vitro or on animals, but the small numbers of reports on humans trials and epidemiological studies supported these findings.

2. CHEMICAL COMPOSITION AND PHENOLIC CONTENT OF SEAWEEDS

The brown and green seaweeds usually have higher phenolic content as compared to the red seaweeds [1, 2]. A significant correlation exists between antioxidant activity and phenolic content of seaweeds [1, 3, 4]. Many algal species contain polyphloroglucinol phenolics (phlorotannins) [5, 6]. The phenolic content in the seaweed extracts has a higher correlation with reducing power ($R^2 = 0.96$) than the radical-scavenging activity ($R^2 = 0.56$).

Table 1. Example of the chemical constituents of the 3 types of seaweeds [7]

Chemical constituents Type of seaweed DW: dry weight basis	*Euchema cottonii* (Red)	*Caulerpa lentillifera* (Green)	*Sargassum polycystum* (Brown)
Soluble fiber (g/100g)	18.25 ± 0.09^a	17.21 ± 0.09^a	5.57 ± 0.03^b
Insoluble fiber (g/100g)	6.80 ± 0.00^c	15.78 ± 0.12^b	34.10 ± 0.03^a
Total dietary fiber (g/100g)	25.05 ± 0.01^c	32.99 ± 0.21^b	39.67 ± 0.06^a
Polyphenols (mg PGE/g extract)	22.50 ± 0.28^b	52.85 ± 0.12^a	55.16 ± 0.74^a
β-carotene (mg/100g)	1.9 ± 0.00^b	6.8 ± 0.00^a	7.5 ± 0.00^a
Vitamin C (mg/100g)	35.3 ± 0.00^a	34.7 ± 0.00^a	34.5 ± 0.00^a
α-tocopherol (mg/100g)	5.85 ± 0.03^c	8.41 ± 0.01^b	11.29 ± 0.06^a
Selenium (mg/100g)	0.59 ± 0.00^c	1.07 ± 0.00^b	1.14 ± 0.03^a
Na (mg/100g DW)	1771.84 ± 0.01^b	8917.46 ± 0.00^a	1362.13 ± 0.00^c
K (mg/100g DW)	$13{,}155.19 \pm 1.14^a$	1142.68 ± 0.00^c	8371.23 ± 0.01^b
Ca (mg/100g DW)	329.69 ± 0.33^c	1874.74 ± 0.20^b	3792.06 ± 0.51^a
Mg (mg/100g DW)	271.33 ± 0.20^c	1028.62 ± 0.58^a	487.81 ± 0.24^b
Fe (mg/100g DW)	2.61 ± 0.00^c	21.37 ± 0.00^b	68.21 ± 0.03^a
Zn (mg/100g DW)	4.30 ± 0.02^a	3.51 ± 0.00^b	2.15 ± 0.00^c
Cu (mg/100g DW)	0.03 ± 0.00^b	0.11 ± 0.00^a	0.03 ± 0.00^b
Se (mg/100g DW)	0.59 ± 0.00^c	1.07 ± 0.00^b	1.14 ± 0.03^a
I (μg/g DW)	9.42 ± 0.12^a	4.78 ± 0.59^c	7.66 ± 0.10^b
Na/K ratio	0.14	7.8	0.16

Values are mean ± S.E.M, n=3.
Values in the same rows with different superscript are significantly different ($P<0.05$).
Total polyphenols is expressed as mg PGE (pholoroglucinol equivalents)/g extract.

2.1. Dietary Fibers

The soluble dietary fibres contents in seaweeds are higher than terrestrial plants (up to 55% dry weight) [8]. Soluble fibers are good for gut health through its water holding, fecal volume increase and digestive passage time decreasing effects [7], which ultimately help prevent colon cancer [9]. The main soluble fibres include alginate from brown seaweeds, carrageenan and agar from red seaweeds; while minor polysaccharides such as fucoidans, xylans and ulvans are present in brown, red and green seaweeds correspondingly [10]. Other bioactive polysccharides from seaweed include fucan, galactan sulphate, xylomannan sulphate, sodium alginate, porphyran and alginic acid. The therapeutic applications of fucoidans with notable anticoagulant and anti-thrombic properties will be discussed later [11]. The high soluble dietary fibre contents of *Eucheuma cottonii, Caulerpa lentilifera, Sargassum polycystum* (up to 40 % dry weight) and Nori, help lower blood cholesterol levels and prevent metabolic syndrome [7, 12]. High dietary fiber consumption can reduce chronic degenerative disease risks such as diabetes, heart disease and cancers.

The high molecular weight sulphated polysaccharides, from the cell walls of brown seaweed [13], such as fucans and fucoidans have antioxidant, antiproliferative, antitumor, antiviral, antiflammatory, anticoagulant, antipeptic and antiadhesive pharmacological activities [14, 15]. The sulphated polysaccharides from red seaweed (galactan sulphate, carrageenan and xylomannan sulphate) have antiviral properties [16]. The soluble fibres help manage and prevent overweight and obesity [17]. Alginic acid help reduce hypertension [18]. The polysaccharide porphyran, from red Porphyra spp. have immunoregulatory, antioxidative and anticancer properties [19-21].

2.2. Protein and Amino Acid Profiles

Green and red seaweeds generally have higher protein contents (10-47% of dry weight) than brown seaweeds (5-24% dry weight) [7, 10, 22]. Most seaweed has all the essential amino acids at a quantity near the FAO/WHO recommended values [7, 8]. Seaweed such as *S. vulgare* is high in methionine (1.7%), which are not abundant in other seaweeds [23]. Many seaweeds such as *Ulva spp.* contains high amounts of acidic amino acid which behave as flavour enhancers [24]. The protein chemical scores of tropical seaweeds are between 20-67% [7], some of which may be comparable to certain animal proteins. Biliproteins, from red seaweeds are fluorescent and phycobiliproteins has antioxidant properties which are beneficial for the prevention and treatment of degenerative diseases [25].

2.3. Lipids and Fatty Acids

The seaweeds lipid content varies from 0.12% dry weight for *Jania rubens* to 6.73% for *L. papillosa* [22]. Low lipid levels are also found in *Codium fragile, Gracilaria chilensis, Macrocystis pyrifera* (0.7-1.5%), *Eucheuma cottonii, Caulerpa lentifera* and *Sargassum polycystum* (0.29-1.11% dry weight) [7, 26].

Red seaweeds (*Laurencia papillosa* and *Jania rubens*) and brown seaweeds (*Cystoseira corniculata, Poetino, pavonia*) contains 31.66-47.97% saturated fatty acids (SFA), 23.04-

32.06% monounsaturated fatty acids (MUFAs) and 11.58-18.08% polyunsaturated fatty acids (PUFAs) [22]. The % of polyunsaturated fatty acids in the tropical seaweed are high (*E. cottonii* - 52%; *C. lentillifera* - 17% and *S. Polycystum* - 20% [7]. *C.fragile* and *G.chilensis* contains linoleic, linolenic and oleic acid [26]. The seaweeds are a source of omega-3-fatty acids such as eicosapentanoic acid (EPA) [27]. EPA accounted for 24.98% of all fatty acids in *E.cottonii* [7]. Seaweeds have a low ω6/ω3 ratio (0.10 for *E. cottonii* and 1.06 for *C. lentillifera*) and may help reduce the ω6/ω3 ratio. To reduce coronary diseases, diabetes and osteoathritis, a dietary ω6/ω3 ratio of less than 10 is recommended by WHO [10]. Seaweeds can be used as an ingredient in slimming foods to help weight loss [7].

2.4. Minerals

Seaweeds are rich in macro minerals (Ca, Mg, Na, P and K) and trace elements (Zn, I and Mn) [7, 22]. Seaweeds have higher contents of calcium and phosphorus, than apples, oranges, carrots or potatoes [12]. The high calcium levels (5.24 g kg−1) in *Ulva rigida* [9, 28] varies from that of *Eucheuma cottonii* (0.3%), *Caulerpa lentillifera* (1.9%) and *Sargassum polycystum* (3.8%) [7]. The high macrominerals (Na, K, Ca and Mg) contents of *E. cottonii*, *C. lentillifera* and *S. polycystum* (12.01-15.53%) [7] are also different to that of the brown *Undaria pinnatifida* and red *Chondrus crispus* (8.1-17.9 mg/100g) [29]. The trace minerals (Fe, Zn, Cu, Se and I) (7.53-71.53 mg.100g-1) contents of *E. cottonii*, *C. lentillifera* and *S. polycystum* [7] also varies widely to that of *U. pinnatifida* and red *C. crispus* (5.1-15.2 mg/100g) [29].

Seaweeds have low Na/K ratios of 0.14-0.16 [7]; >1 [28] and <1.5 [29]. Low Na/K ratios consumption can reduce the incidence of hypertension. Dietary seaweeds help reduce the high Na/K ratio widespread in current diets [7]. Seaweed iodine content are in several chemical forms (I⁻, I2, IO2-), which can serve as an antioxidant, anti-goitre and anticancer agent [30]. *Laminaria* contains 1500-8000 ppm dry weight iodine [10].

2.5. Micronutrients and Vitamins

Edible seaweeds contain vitamin C and α-tocopherol [7], that help prevent LDL oxidation and thromboxin formation [10]. Some are rich in beta carotene (e.g. 197.9 µg/g dried *C. fragile* and 113.7 µg/g dried *G. chilensis*), while others are low (17.4 µg/g dried *M. pyrifera*) [26]. δ-Tocopherol and α-tocopherol were found in *C. fragile* while Y-tocotrienol and α-tocopherol were found in *G. chilensis* and *M. pyrifera*. Brown seaweeds contain high levels of carotenoids especially fucoxanthin, beta carotene and violaxanthin, while the green and red seaweeds only contain α-tocopherol [10].

3. IN VITRO AND IN VIVO ANTIOXIDANT PROPERTIES

The onset of various chronic degenerative diseases such as coronary heart disease (CHD), certain cancers, rheumatoid arthritis, diabetes, retinopathy of prematurity, certain inflammatory diseases, Alzeimer's disease [31], other neurological disorders and the ageing process have been associated with reactive oxygen species (ROS) and oxidative stress [32]. ROS can oxidize biomolecules leading to malfunction, cell death and tissue injury.

Seaweeds have developed a strong antioxidative defense to protect their structural components (polyunsaturated fatty acids) against the exposure to strong sunlight, oxygen and other oxidizing agents [2]. Seaweeds contain phloroglucinol phenolic (phlorotannins) [5] which are powerful ROS scavengers, metal chelators, enzyme modulators and prevent lipid peroxidation [33]. Polyphenols are reducing agents and together with other dietary reducing agents such as vitamin C, vitamin E and carotenoids, are protective against oxidative stress and associated pathologies [34].

The antioxidant activities of several seaweeds were published [1-3, 35-37]. In comparison to the well known, three interconnected ringed terrestrial flavonoids, seaweeds phlorotannins have up to eight interconnected rings, making them 10-100 times more potent and stable as antioxidant than other polyphenols. The *in vivo* half-life of the partially fat soluble phlorotannins is up to 12 hours, compared to 30-180 minutes for the water-soluble flavonoids [39]. The total polyphenol contents varied with species, but generally the green seaweeds have higher free radical-scavenging properties, followed by the brown seaweed and the red seaweeds [1, 40].

Seaweeds can reduce lipid oxidation which increases with disease and aging. Lipid peroxidation, was significantly reduced in the liver, heart and brain of aging mammal by porphyran and sulfated galactan from *Porphyra spp.* Seaweed consumption generally upregulate the endogenous antioxidant defense enzymes superoxide dismutase (SOD), glutathione peroxidise (GSH-Px) and sometimes catalase activities [41-43].

4. SEAWEEDS FOR CARDIOVASCULAR HEALTH

Wakame effectively reduced the cholesterol, systolic and diastolic blood pressure of hypertensive human patients in a random, case controlled study [44]. Several seaweed hydrolysates strongly inhibit angiotensin-1-converting enzyme (ACE) and possess anti-hypertensive properties. These include (i) peptides from *Undaria pinnafitida* hydrolysates [45] and (ii) phlorotannins from *Ecklonia stolonifera* which was comparable to the physiological vasodilative hormone bradykinin that participated in the rennin-angiotensin system [46].

4.1. Obesity, Dislipidemia and Low Density Lipoprotein Cholesterol Reduction

The seaweeds high soluble dietary fibre content help moderate appetite by delaying digestion rate and calorie absorption in animals [17]. However these effects are difficult to prove in humans. The main carotenoid in brown seaweed, fucoxanthin, has anti-obesity properties especially for obese patients with non-alcoholic fatty liver disease and elevated indices of chronic inflammation [47]. The mechanism by which fucoxanthin help reduce bodyweight include: (i) upregulating mitochondrial uncoupling protein 1 (UCP1) gene expression in white adipose tissue (WAT) [48], (ii) increasing resting energy expenditure, through uncoupling step during cellular metabolism [49], (iii) increase oxidation of fatty acids, (iv) increase heat production in WAT20. [50, 48]; (v) reduce insulin resistance [50] and (vi) reduce blood glucose and plasma insulin [51].

The seaweeds omega-3 fatty acids help lower triglycerides [52] and help normalize inflammation indices that positively correlate with central adiposity [53]. Prolyl endopeptidase inhibition properties are shown by several *Sargassum spp. (S. hemiphyllum, S. thenbergii, S. Singgildianum)* [54], while *Caulerpa racemosa* and *Spatoglossum schtoderi* inhibited α-amylase [55]. Lipase, 5-lipoxygenase and Phospholipase A2 inhibition are also shown by *Caulerpa spp.* [56]. Human muscle aldose reductases are inhibited by species such as *Ecklonia cava* and *Eisenia bicyclis* [57].

Carrageenan was clinically proven to reduce blood cholesterol and lipid in humans [58]. Various edible seaweeds (Wakame, Bladderwrack, Nori, Porphyran, *Solieria robusta, Iyengaria, Euchema cottonii, Sargassum polycystum, Caulerpa lentilliferra* and *Caulerpa racemosa*) helped reduce cholesterol absorption and metabolism, thus lowering plasma cholesterol level, triglyceride and low density lipoprotein (LDL) cholesterol levels in hyperlipidemic mammals [12, 43, 45, 59]. Seaweeds polysaccharides (e.g. from *Fucus vesiculosus*) not only reduce the undesirable triglyceride and LDL cholesterol levels but moderately increased the desirable high density lipoprotein (HDL) cholesterol levels [60].

The anti-obesity, anti-hyperlipaemic and *in vivo* antioxidant effects of the best antioxidative red, green and brown tropical seaweeds were compared in rats. After four months, the seaweeds (5 %) supplementation to high-cholesterol/high fat diet rats, significantly reduced plasma total cholesterol, (up to 20%), low density cholesterol (up to 50%) and triglyceride levels (up to 36%) and significantly increased high density cholesterol levels (up to 55%).

Diet	Body Weight Gain Mean ± SEM (g/16 weeks/rat)
Normal diet	121 ± 13.28^{cd}
Normal diet +*K. alvarezii*	148 ± 12.12^{bc}
Normal diet +*C. lentillifera*	118 ± 4.62^{cd}
Normal diet +*S. polycystum*	111 ± 6.93^{d}
Lipogenic diet	230 ± 12.70^{a}
Lipogenic diet +*K. alvarezii*	163 ± 9.81^{b}
Lipogenic diet +*C. lentillifera*	139 ± 9.24^{bcd}
Lipogenic diet +*S. polycystum*	132 ± 10.97^{bcd}

Seaweeds supplements to lipogenic diets resulted in rats having similar weight and size to rats on normal diets.

Amongst the seaweeds, the brown *S. polycystum* showed the best anti-obesity and blood glutathione peroxidase properties; the red seaweeds *E. cottonii* showed the best anti-hyperlipidaemic and *in vivo* anti-oxidation effects; whilst the green seaweeds *C. lentillifera* was most effective at reducing plasma total cholesterol. All seaweeds significantly reduced body weight gain, erythrocytes glutathione peroxidase and plasma lipid peroxidation of high-cholesterol/high fat diet rats towards the values of normal diet rats [43].

5. ORGAN PROTECTIVE EFFECTS OF DIETARY SEAWEEDS IN RATS FED RICH LIPOGENIC (1% CHOLESTEROL 24% FAT) DIET

Shown below are examples of the *in vivo* antioxidative, morphological and biochemical effects of red seaweeds (*Kappaphycus alvarezii*) on the organs of rats fed with rich lipogenic (sometimes called Western) diet. The protocols/ methodology used were as previously documented by the authors [43, 61].

5.1. Liver

Gross, Somatic Index, Microscopy and Lesion Scoring

Figure 1 A shows the liver of rats on rich lipogenic diet that were pale, soft, mottled, fatty and larger in size compared to normal rat liver. In comparison Fig 1B shows the livers of rats from lipogenic diet supplemented with *K. alvarezii* (also called *Euchema cottonii*) \which were less fatty compared to A. The livers of normal rats or normal rats supplemented with *K. alvarezii* were similar.

The rich lipogenic diet increased the liver weight, hepatosomatic index and hepatocytes damage (Table 2) and seaweeds supplementation significantly reduced these changes. The seaweeds did not affect the liver weight of normal rats.

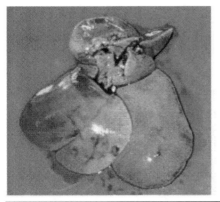
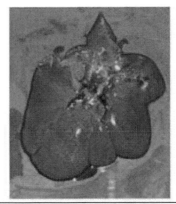

| A. Liver of rat on lipogenic diet. The enlarged liver appeared much paler than normal, soft, mottled and very fatty | B. Livers of rats on lipogenic diet+*K. alvarezii*. The livers were slightly pale but not so fatty |

Figure 1. The livers of rats on lipogenic diet (with and without 5% seaweed supplementation).

Table 2. Liver weight, hepatosomatic index (liver weight/terminal body weight x 100) and liver damage in rats after 16 weeks on lipogenic or normal diet

Group	Liver weight (g)	Hepatosomatic index	Liver (Necrotic hepatocytes and Kupffer cells)
normal	12.71 ± 0.75^c	3.21 ± 0.12^c	13.50 ± 0.57^f
normal+*K. alvarezii*	13.54 ± 0.92^{bc}	3.18 ± 0.13^c	14.67 ± 0.44^{ef}
lipogenic diet	23.46 ± 2.33^a	4.44 ± 0.18^a	27.06 ± 1.08^a
lipogenic diet+*K. alvarezii*	14.58 ± 1.22^{bc}	3.38 ± 0.16^{bc}	16.28 ± 0.93^e

Values are means ± S.E.M, N= 10 rats per group.
[a-c]: Values in the same column (between groups) with different superscript/s are significantly different ($P<0.05$).

Figure 2 shows the HandE stained photomicrograph of liver of rats from all groups at necropsy. The microscopic appearance of the liver of rats from the normal, normal+*K. alvarezii* rats displayed a normal appearance and was devoid of significant cellular degeneration or infiltration. In contrast, the lipogenic diet caused hepatocytic degeneration, inflammation with or without cellular infiltration indicative of non-alcoholic fatty injury, and seaweeds supplementations reduced these damages.

The fatty liver is due to hypertriglyceride (TG) depots or hyperlipidaemia. Supplementation of *K. alvarezii* can partially suppress the development of rich diet-induced fatty liver or hepatic steatosis.

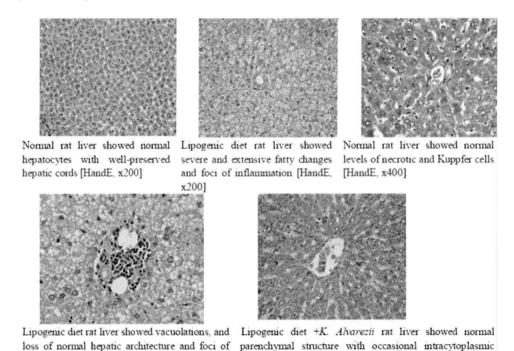

Figure 2. Histology of the rat livers from various groups.

Table 3. The Liver function biomarker levels in rat blood at the start and end of the study

Group	ALT (U/L) Week 0	Week 16	AST (U/L) Week 0	Week 16	GGT (U/L) Week 0	Week 16
normal	32.3 ± 4.7^{aB}	67.0 ± 3.6^{bcA}	70.8 ± 3.0^{aB}	115.3 ± 7.7^{bA}	1.7 ± 0.3^{aA}	1.8 ± 0.5^{abA}
normal+*K. alvarezii*	21.1 ± 4.7^{aC}	71.2 ± 6.0^{bcA}	84.2 ± 6.1^{aB}	90.7 ± 6.7^{bAB}	1.6 ± 0.4^{aA}	1.3 ± 0.2^{bA}
lipogenic diet	20.5 ± 3.6^{aD}	90.5 ± 3.0^{aA}	81.0 ± 5.9^{abC}	180.5 ± 7.1^{aA}	1.6 ± 0.2^{aB}	2.5 ± 0.3^{aAB}
lipogenic diet+*K. alvarezii*	21.6 ± 3.3^{aB}	69.9 ± 6.1^{bcA}	84.7 ± 9.6^{aA}	113.6 ± 16.5^{bA}	1.3 ± 0.3^{aA}	2.0 ± 0.4^{abA}

Values are means ± S.E.M., N=10 rats per group. [a-c]: Values in the same column (between groups) with different superscript/s are significantly different ($P<0.05$). [A-D]: Values in the same row (between time) with different superscript/s are significantly different ($P<0.05$).

Plasma alanine aminotransferase (ALT), aspartate aminotransferase (AST) and glutamyltransferase GGT detect liver injury (Table 3). These enzymes leak out of hepatocytes into the blood in large quantities when the liver is damaged. The long term injestion of the lipogenic diet caused significant elevations of blood AST, ALT and GGT concentrations and seaweeds supplementation reduced these increases. The AST enzyme is localized both in the cytoplasm and mitochondria and leaks more readily into the blood stream than ALT. The microscopic observations in all groups especially those of the cholesterol fed livers did not reveal evidence of cholestasis, which accounts for the rather lack of a huge increase in serum GGT.

Seaweeds like *Halimeda monile, Porphyra spp. Ecklonia stolonifera Okamura, Ulva reticulata, Ecklonia cava, Colpomenia sinuosa, Sargassum hemiphyllum, Myagropsis myagriodes, Sargassum henslowianum, Sargassum siliquastrum, Undaria pinnatifida, Hizikia fusiformis* and *Laminaria japonica* have hepatoprotective properties [20, 62, 63]. The bioactive compounds were (i) porphyran [20]; (ii) free phenolic acids [62] and (iii) polysaccharides [64]. The seaweeds bioactive compounds help balance (i) the hepatic enzymes expression [62], (ii) the rate of synthesis and release of triglycerides [64], (iii) oxidative stress on liver cholesterol metabolism [64] and (iv) the excretion of lipophilic toxins (e.g. dioxin) [65].

5.2 Heart

Although the rich lipogenic diet did not cause significant changes to the heart weight and cardiosomatic index, it caused vacuolations, fat droplets presence, extensive myonecrosis with fibroblastic proliferation and chronic inflammation to the myocardium indicating pathological significance (Table 4). During hyperlipidaemia, many cardiomyocytes died due to ischaemia. Being a post-mitotic cell, replacement of dead cells by new ones (hyperplasia) is not possible and is compensated by hypertrophied myocytes (slight) to maintain the normal function and replacement by fibrous tissue. The lipogenic diet caused areas of cardiomyocytic loss which were replaced by fibrous tissue which is probably lighter (in weight).

Table 4. Final heart weight (g), cardiosomatic index (organ weight/body weight x 100) and quantitative histological tissue lesions (vacuolation and necrosis) of rat hearts (Mean±S.E.M.)

Group	Heart weight	Cardiosomatic index	(Vacuolation and necrosis)
normal	1.45 ± 0.09^a	0.37±0.03^{ab}	0.33 ± 0.11^e
normal+*K. alvarezii*	1.42±0.06^a	0.34±0.01^{ab}	0.57 ± 0.15^{de}
lipogenic diet	1.66±0.15^a	0.32±0.05^b	2.81 ± 0.21^a
lipogenic diet+*K. alvarezii*	1.38±0.09^a	0.33±0.03^b	1.05 ± 0.20^{cd}

Values are means ± S.E.M, N= 10 animals per group.
^{a,b}: Values in the same column (between groups) with different superscript/s are significantly different ($P<0.05$).

The seaweeds supplementations to the lipogenic diet, significantly ($P<0.05$) but only partially reduced the vacuolation, fibrosis, inflammation and necrosis to the heart. The dietary seaweeds did not cause significant changes to the hearts of rats on normal diets (Figure 3).

Rich lipogenic diet caused interstitial fibrosis in the hearts of the rats whereby collagen are formed replacing the necrotic myocardium, which is the common response to all forms of heart injury, and is one of the major determinants of morbidity and mortality from CVD. It enhances myocardial stiffness because collagen-I is a very rigid protein, generating arrhythmias because fibrosis creates myocardial electrical heterogeneity and hampers systolic ejection by rendering the myocardium heterogeneous. Myocardial fibrosis is probably one of the major biological determinants of fatal episodes in cardiac fibrosis, including congestive heart failure, severe arrhythmias and sudden cardiac death.

Lipogenic diet rat heart showed homogenously pinkish myocardial fibers that lacked nuclei and hypertrophied with fibroblastic infiltration between muscle fibers [HandE, x100]

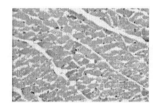

Lipogenic diet rat heart depicted numerous intracytoplasmic "lipid" vacuolation interspersed within the myocardial fibers [HandE, x400]

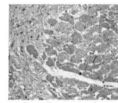

Lipogenic diet rat heart showed fibrosis, inflammation, intracytoplasmic vacuolation and necrosis [HandE, x400].

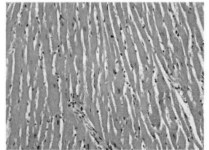

Normal rat heart showing size and orientation of the myocardial fibers [HandE, x100]

Lipogenic diet + *K. Alvarezii* rat heart resembled the normal rat heart [HandE, x100]

Figure 3. The microscopic appearances of the myocardium of rats from the experimental groups.

With *K. alvarezii* supplementation to the rich lipogenic diet, the rat heart showed moderate prevention to these undesirable changes. On the other hand, normal rats supplemented with seaweeds showed normal myocardial fibers with notable absence of cardiac muscle hypertrophy.

5.2.1. Cardiac Marker Enzymes

The activity of cardiac marker enzymes such as creatinine kinase (CK), creatinine kinase-MB (CK-MB) isoenzyme, ALT and AST are presented in Tables 3 and 5 respectively. Rich lipogenic diet caused marked elevations in the activities of these enzymes in rats plasma and supplementation with *K. alvarezii* significantly ($P<0.05$) reduced these enzyme towards the normal levels. Increased serum CK, CK-MB, ALT and AST are released from the heart into the blood during myocardial damage due to myofibrils degeneration and myocyte necrosis. The amount of enzymes appearing in serum is proportional to the number of necrotic cells.

Fucoidan blocks selectins, thus producing better recovery of left ventricular function, coronary blood flow and myocardial oxygen consumption after cold ischemia [66]. Seaweeds are prophylactic as a dietary supplement for the heart and coronary atherosclerosis.

Table 5. Creatinine kinase (CK) and Creatinine Kinase-MB (CK-MB) isoenzyme plasma of rats at the beginning and end of the experimental period

Group	CK (U/L) Week 0	Week 16	CK-MB (U/L) Week 0	Week 16
normal	112.0±12.0^{aB}	190.5±6.6^{bcA}	151.1±36.6^{aA}	191.2±19.0^{cA}
normal+*K. alvarezii*	120.0±23.6^{aA}	180.0±31.9^{bcA}	157.0±23.5^{aB}	199.0±8.7^{cB}
lipogenic diet	141.2±15.8^{aC}	398.3±58.6^{aA}	115.4±22.1^{aC}	388.9±45.5^{aA}
lipogenic diet+*K. alvarezii*	109.2±26.6^{aB}	226.2±39.5^{bcA}	113.4±18.4^{aC}	222.3±26.9^{bcAB}

Values are means ± S.E.M., normal=10 rats per group. $^{a-c}$: Values in the same column (between groups) with different superscript/s are significantly different ($P<0.05$). $^{A-B}$: Values in the same row (between time) with different superscript/s are significantly different ($P<0.05$).

5.3 Kidney

Table 6. Final kidney weight, nephrosomatic index (% kidney weight/terminal body weight), abnormal glomeruli and abnormal necrotic tubule of rats

Group	Kidney weight	Nephrosomatic index	Abnormal glomeruli	Abnormal Necrotic tubule
al	2.53±0.14^{bc}	0.64±0.03^{ab}	0.29 ± 0.12^{b}	0.86 ± 0.17^{e}
normal+*K. alvarezii*	2.86±0.15^{b}	0.67±0.02^{a}	0.62 ± 0.16^{b}	1.52 ± 0.19^{cd}
lipogenic diet	3.31±0.27^{a}	0.63±0.01^{abc}	1.43 ± 0.15^{a}	3.14 ± 0.17^{a}
lipogenic diet+*K. alvarezii*	2.51±0.15^{bc}	0.59±0.03^{bc}	0.57 ± 0.13^{b}	1.86 ± 0.20^{bc}

Values are means ± S.E.M, N= 10 rats per group.
$^{a-c}$: Values in the same column (between groups) with different superscript/s are significantly different ($P<0.05$).

The rich lipogenic diet caused significantly increased ($P<0.05$) kidney weight and renal lesions comprising of tubulonephropathy (abnomal glomeruli and necrotic tubules) (Table 6). Seaweeds supplementation especially with *K. alvarezii* group decreased these injuries. However 5% dietary *K. alvarezii* caused slight increase in the number abnormal necrotic tubules in the normal diet rats.

The rich lipogenic diet caused glomerulus loss, glomerular and tubular degeneration, dilatation of the Bowman's space with glomerular atrophy or complete obstruction of the Bowman's space and significant increase in kidney weight. The kidney weight increase could have resulted from pathological changes in the kidneys such as inflammation, increase in cellular density, ballooning of the tubular epithelium, oedema, fatty degeneration and changes in the Bowman's space (Figure 4). Likewise, the expansion and disappearance of Bowman's space may be due to accumulation of exudates or hypercellularity of the mesangium, respectively. Here, tubular epithelium dilatation along with glomerulopathy denoted renal injury. Hypertriglyceridaemia and severe hypercholesterolaemia are associated with renal lipid deposition and impaired rennin secretion in apoliprotein E deficient mammals, which explains the vacuolated tubules caused by the lipogenic diet. Addition of seaweeds to the lipogenic diet prevented all these renal injuries, and made the kidneys similar to rats on normal diets.

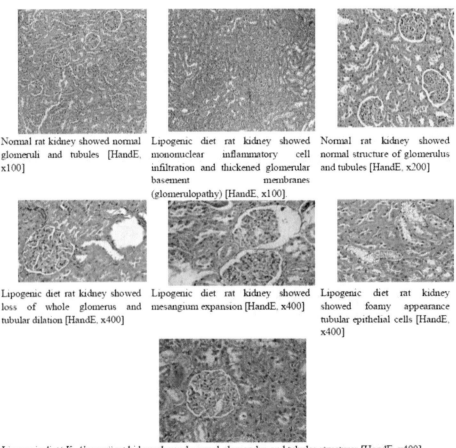

Figure 4. Representative microscopic appearance of the kidney of various diet rats at necropsy.

Table 7. Urea (mmol/L), Creatinine and Uric acid in rats plasma at the end of the study

Group	Urea (mmol/L) Week 0	Week 16	Creatinine (μmol/L) Week 0	Week 16	Uric acid (μmol/L) Week 0	Week 16
normal	7.7 ± 0.7$^{a\,A}$	6.8 ± 0.3$^{a\,AB}$	46.9 ± 2.4$^{a\,B}$	60.0 ± 1.4$^{b\,A}$	15.6 ± 1.0$^{a\,AB}$	34.4 ± 2.5$^{b\,A}$
normal+*K. alvarezii*	6.6 ± 0.4$^{a\,AB}$	6.4 ± 0.3$^{a\,AB}$	46.0 ± 2.8$^{a\,B}$	60.5 ± 2.1$^{b\,A}$	13.5 ± 0.6$^{a\,B}$	36.2 ± 7.6$^{ab\,A}$
lipogenic diet	6.4 ± 0.5$^{a\,A}$	5.4 ± 0.3$^{c\,AB}$	43.6 ± 5.4$^{a\,C}$	74.8 ± 4.7$^{a\,A}$	15.6 ± 0.4$^{a\,B}$	47.2 ± 4.8$^{a\,A}$
lipogenic diet+*K. alvarezii*	6.6 ± 0.8$^{a\,A}$	5.5 ± 0.3$^{c\,AB}$	49.3 ± 7.2$^{a\,A}$	61.5 ± 2.3$^{b\,A}$	14.2 ± 1.4$^{a\,B}$	33.2 ± 7.2$^{bc\,A}$

Values are means ± S.E.M., N=10 rats per group.
$^{a-e}$: Values in the same column (between groups) with different superscript/s are significantly different ($P<0.05$).
$^{A-C}$: Values in the same row (between time) with different superscript/s are significantly different ($P<0.05$).

The rich lipogenic diet caused significant (p<0.05) decrease in plasma urea and increases in plasma creatinine and uric acid (Table 7). The seaweeds supplements, prevented or reduced these kidney biomarker changes significantly (Table 7). Reduced plasma urea may also be caused by lack of eating activity especially that of protein. Diets high in cholesterol may impede protein absorption and give rise to low plasma urea. Large changes in renal function caused relatively small changes in the serum urea levels in the early stages of renal disease but as the condition progressed, relatively small changes in function caused large serum urea increase. Here, the lipogenic diet rats were degenerative to the kidneys.

In some seaweed, the antioxidative fucoidans help normalize the kidneys respiratory enzyme activities under hyperoxaluria, thus protect against mitochondrial damage [67] and chronic renal failure [68].

5.4. Brain

The rich lipogenic diet increased brain weight but reduced the encephalosomatic index and active neurons (Table 8). Lipogenic diets increased brain weight gain resulting in the lower recorded encephalosomatic index. The lower active neuron counts caused by the the lipogenic diet (Figure 5) may either be due to neuron loss, caused by increased lipid peroxidation or bigger % of fat in the brain. Again the seaweeds supplements prevented or reduced these changes (Table 8).

Table 8. Brain weight, encephalosomatic index (organ weight/terminal body weight x 100) and Quantitative histology of rat brains at the end of the experimental period

Group	Brain weight	Encephalosomatic index	Active neurons
normal	1.54±0.08^b	0.39±0.02^a	21.76 ± 0.71^a
normal+*K. alvarezii*	1.80±0.13^a	0.43±0.03^a	20.86 ± 0.35^a
lipogenic diet	1.81±0.09^a	0.35±0.01^b	14.48 ± 0.49^c
lipogenic diet+*K. alvarezii*	1.78±0.05^a	0.42±0.01^a	17.14 ± 0.54^b

Values are means ± S.E.M., normal= 10 animals per group. $^{a-b}$: Values in the same column (between groups) with different superscript/s are significantly different ($P<0.05$).

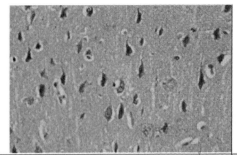 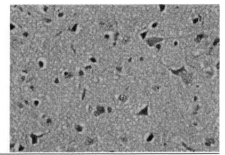

| Normal rat brain showing evenly distributed viable neurons [HandE, x400] | Lipogenic diet rat brain showed reduced viable neurons [HandE, x400] |

Figure 5. The histology of normal and lipogenic diet rat brain.

5.4.1. Nerves and Brain

A human study, on seaweeds extract supplement for three months, produced a significant improvement in well-being, sleep amelioration and skin tonicity [69]. The sulfated polysaccharide fucoidan blocked the generation of reactive oxygen species and inhibited cellular and neurotoxic effects of beta-amyloid in rat cholinergic basal forebrain neurons [70]. Fucoidan was also a potent neuroprotector against cerebral ischemia and Alzheimer's disease [70-72], as well as in dopaminergic neurons *in vivo* and *in vitro* [73].

Cholinesterase inhibitors increase acetylcholine levels in the synapse, help enhance cholinergic activities in the brain, prevent beta amyloidal plaques formation and neuronal death due to inflammation and are used for treating Alzheimer's disease (AD) dementia [74]. *Sargassum sagamianum* sargaquinoic acid [75], phlorotannins (dieckol and phlorofurofucoeckol) and sterols from *Ecklonia stolonifera* [76] and *Gracilaria gracilis* and *Sargassum* compounds [77] were inhibitory against various cholinesterases.

5.5. Spleen

The rich lipogenic diet caused a 30.8% higher increase in spleen weight but insignificant changes to the splenosomatic index (Table 9) or the spleen morphology (periarteriolar lymphatic sheath, splenic cord and sinuses). Nevertheless, K. alvarezii supplementation to lipogenic diet prevented this change in spleen size. The spleen is not involved in loading or overloading of tissues by the lipogenic diet.

Table 9. Spleen weight and splenosomatic index (% organ weight/terminal body weight) of rats at the end of the study

Group	Spleen weight	Splenosomatic index
normal	0.65±0.04[bc]	0.16±0.01[ab]
normal+*K. alvarezii*	0.58±0.04[c]	0.14±0.01[b]
lipogenic diet	0.85±0.18[a]	0.16±0.02[ab]
lipogenic diet+*K. alvarezii*	0.59±0.03[c]	0.14±0.01[b]

Values are means ± S.E.M., N= 10 animals per group.[a-c]: Values in the same column (between group) with different superscript/s are significantly different ($P<0.05$).

K. alvarezii consumption did not cause any damage to the livers, hearts, kidneys and spleens of normal rats yet reduced or prevented the injuries caused by the lipogenic diet. The brains, eyes and lungs were not significantly affected by the rich lipogenic diet.

5.6. Lungs

The seaweed sulfated oligosaccharide inhibited TGF-β1-induced pulmonary fibrosis *in vitro*, suppressed pulmonary fibrosis in the lungs, *in vivo* and decreased lung collagen contents [78]. Fucoidan inhibited lung metastasis *in vivo* [79].

6. LIPID OXIDATION LEVELS IN ORGAN

The rich lipogenic diet consumption for 16 weeks significantly increased malondialdehyde (MDA) levels, a marker for lipid peroxidation in the liver, heart, kidney and brain tissues (Table 10). In contrast, the lipogenic diet rats supplemented with seaweeds had significantly reduced MDA levels in the liver, heart, kidney and brain.

Liver is the main organ involved in lipid metabolism and thus prone to potential oxidative damage in conditions of hypercholesterolemia. Brain is also susceptible to free radical damage since the membrane lipids are very rich in polyunsaturated fatty acids (PUFA). Increased superoxide anion production in hypercholesterolemia conditions contribute to the atherosclerotic process and increased tissue MDA levels. Dietary seaweeds reduced the susceptibility to peroxidative damage under the challenge of oxidative stress such as lipogenic diet. Rats fed on standard (basal) diet supplemented with seaweeds showed no significant difference in MDA levels in all organs as compared to normal control diet.

7. SOD, GSH-PX AND CATALASE ACTIVITIES IN THE ORGANS

The rich lipogenic diet significantly reduced SOD and GSH-Px activities in most organs especially the liver, heart and kidney by up to 60% compared to normal control diet. Seaweeds supplements to the lipogenic diet prevented or reduced these decreases to values similar to that of normal control rats. High-fat diet diminishes the antioxidant defense system. The decrease in the SOD and GSH-Px activity is partly due to the oxidative inactivation of the enzyme caused by excessive reactive oxygen species, the α-hydroxyl ethyl radical or other free radical generation. SOD is the first line of defense against oxidative insults. GSH-Px is responsible for most of the decomposition of lipid peroxide in cells and may thus protect the cell from the deleterious effects of peroxides. The increased levels of superoxide anion could inactivate GSH-Px, which lead to an enhancement of H_2O_2 level which, in turn, would inactivate SOD, resulting in elevated MDA levels and low antioxidant enzyme activities found in these organs. A good negative correlation exist between the MDA concentration and the SOD activity (r=-0.78) in the liver of the rats.

Table 10. Effects of seaweed supplementation on plasma lipid concentrations, malondialdehyde (MDA) levels, superoxide dismutase (SOD), glutathione peroxidase (GSH-Px) and catalase (CAT) activities in various organs of in normal and hypercholesterolemic rats at the end of the experimental period of 16 weeks

Group	Normal (N)	N+*K. alvarezii*	lipogenic diet (LD)	LD+*K. alvarezii*
TC (mmol/L)	1.48 ± 0.03^a	1.46 ± 0.12^a	2.71 ± 0.07^b	2.39 ± 0.07^{bc}
HDL-C (mmol/L)	1.08 ± 0.10^{ab}	1.04 ± 0.09^{abc}	0.80 ± 0.04^c	1.24 ± 0.08^a
LDL-C (mmol/L)	0.24 ± 0.02^a	0.27 ± 0.05^a	1.50 ± 0.09^b	0.76 ± 0.08^c
TG (mmol/L)	0.86 ± 0.10^a	0.89 ± 0.09^a	1.69 ± 0.15^b	1.08 ± 0.25^a
AI (LDL-C/HDL-C)	0.23 ± 0.03^a	0.26 ± 0.03^a	1.88 ± 0.14^b	0.55 ± 0.05^c
Liver				
MDA(nmol/mg protein)	3.44 ± 0.26^b	3.39 ± 0.14^b	5.69 ± 1.20^a	3.69 ± 0.17^b
SOD(U/min/mg protein)	6.76 ± 0.87^a	5.16 ± 0.86^{ab}	3.32 ± 0.56^b	6.55 ± 1.03^a
GSH-Px(U/mg protein)	2.91 ± 0.36^a	2.13 ± 0.31^{abc}	1.67 ± 0.21^c	2.56 ± 0.27^{ab}
CAT(k/sec/mg protein)	0.23 ± 0.03^a	0.24 ± 0.02^a	0.23 ± 0.04^a	0.25 ± 0.01^a
Heart				
MDA(nmol/mg protein)	7.30 ± 0.26^{bc}	5.01 ± 0.21^{bc}	8.67 ± 0.43^a	7.04 ± 0.50^{bc}
SOD(U/min/mg protein)	9.79 ± 1.88^a	7.20 ± 0.67^{abc}	6.51 ± 0.95^{bc}	8.76 ± 0.69^{ab}
GSH-Px(U/mg protein)	8.62 ± 0.49^a	7.88 ± 0.70^{ab}	5.74 ± 0.41^{cd}	8.13 ± 0.29^{ab}
CAT(k/sec/mg protein)	0.50 ± 0.05^a	0.49 ± 0.02^a	0.45 ± 0.02^a	0.50 ± 0.02^a
Kidney				
MDA(nmol/mg protein)	4.79 ± 0.26^b	5.01 ± 0.32^b	6.06 ± 0.46^a	4.20 ± 0.13^b
SOD(U/min/mg protein)	11.09 ± 0.59^a	11.47 ± 0.29^a	8.15 ± 2.21^{ab}	9.60 ± 2.14^{ab}
GSH-Px(U/mg protein)	6.38 ± 0.84^{ab}	8.50 ± 1.07^a	2.34 ± 0.54^{cd}	4.78 ± 0.81^b
CAT(k/sec/mg protein)	0.31 ± 0.02^a	0.30 ± 0.03^a	0.33 ± 0.04^a	0.28 ± 0.01^a
Brain				
MDA(nmol/mg protein)	12.73 ± 0.81^b	11.16 ± 0.16^b	14.61 ± 1.61^a	12.83 ± 1.00^{ab}
SOD(U/min/mg protein)	10.38 ± 2.27^a	6.84 ± 0.92^a	13.62 ± 4.51^a	8.71 ± 1.88^a
GSH-Px(U/mg protein)	5.94 ± 0.89^a	7.69 ± 1.27^a	5.19 ± 1.16^a	8.69 ± 1.71^a
CAT(k/sec/mg protein)	1.15 ± 0.06^{abc}	0.95 ± 0.06^c	1.26 ± 0.13^a	1.21 ± 0.07^{ab}
Lung				
MDA(nmol/mg protein)	5.14 ± 0.20^a	5.31 ± 0.30^a	4.90 ± 0.10^a	4.86 ± 0.09^a
SOD(U/min/mg protein)	2.13 ± 0.13^a	2.46 ± 0.88^a	2.92 ± 0.44^a	2.01 ± 0.09^a
GSH-Px(U/mg protein)	1.52 ± 0.31^a	1.56 ± 0.23^a	1.88 ± 0.78^a	1.07 ± 0.21^a
CAT(k/sec/mg protein)	0.44 ± 0.04^a	0.47 ± 0.03^a	0.51 ± 0.06^a	0.41 ± 0.03^a
Spleen				
MDA(nmol/mg protein)	4.09 ± 0.11^a	3.70 ± 0.11^a	3.98 ± 0.08^a	3.85 ± 0.12^a
SOD(U/min/mg protein)	5.02 ± 1.37^a	3.85 ± 0.27^a	2.66 ± 0.70^b	4.54 ± 1.58^a
GSH-Px(U/mg protein)	2.62 ± 0.40^a	2.44 ± 0.37^a	3.00 ± 0.71^a	2.96 ± 0.80^a
CAT(k/sec/mg protein)	0.52 ± 0.06^b	0.67 ± 0.06^{ab}	0.70 ± 0.04^{ab}	0.66 ± 0.07^{ab}

Values are means ± SEM, n=10 animals per group.
[a-d] Values in the same row with different superscript letters are significantly different ($P<0.05$).
normal, control rats provided with standard (basal) diet.

Rats fed on standard (basal) diet supplemented with seaweeds showed no significant difference in SOD and GSH-Px activities in all organs as compared to normal control diet. The SOD and GSH-Px activities in the brain, lung and spleen were less affected by the lipogenic diet. The lipogenic diet did not affect catalase activities because GSH-Px is usually the rate determining enzyme.

8. TISSUE HEALING PROPERTIES OF SEAWEEDS

8.1. Dermal Wound Healing

The highly viscous bioabsorbable guluronic and mannuronic acid polysaccharide (alginate) from seaweeds can inhibit scar formation. They present a physical barrier to the invading fibroblast to enhance wound healing. The highly viscous bioabsorbable polysaccharide can be applied in tissue engineering and clinical purpose, particularly with the addition of collagen [80, 81]. The water soluble seaweed based dressing protects the healing granulation tissue against disturbances, are painless and are beneficial for venous, neuropathic and ischaemic ulcers.

Wound healing and tissue repair involves fibroblast and epithelial cell development, TGF-β secretion and growth factor dependant pathways. All these processes are stimulated by fucoidan [82, 83]. Seaweeds fucoidan-chitosan hydrogels efficiently shrink and repaired dermal burns [11]. *E. cottonii* ethanolic extract significantly accelerated wound healing and hair growth in rats [84]. In a controlled, randomized single-blinded human study, a silver-loaded seaweed-based cellulosic fiber greatly enhanced epidermal skin physiology, barrier function (transepidermal water loss), stratum corneum hydration and surface pH in atopic dermatitis and eczema patients [85].

9. RETARD BLOOD SUGAR INCREASES

Most seaweeds are high in soluble dietary fibres (carrageenan, agar, alginates), that help retard digestion and glucose absorption. Seaweed consumption help reduce diabetes risk in men (A Korean national Health and Nutrition Survey) [86]. Many seaweeds (e.g. the pacific edible brown seaweed *Petalonia binghamiae* extract; the red seaweed *Grateloupia elliptica* bromophenols extracts; the red seaweed *Hypnea musciformis* extract; *Eklonia Stolonifera;* the brown seaweeds, sea tangle; *Pelvetia babingtonii; Ascophyllum nodosum* extracts) have antidiabetic properties [87-94].

The seaweed extracts antidiabetic mechanisms include (i) producing insulin-like and insulin sensitizing action in adipocytes, of diabetic mammals, (ii) stimulating the differentiation of 3T3-L1 preadipocytes (iii) stimulating basal glucose uptake into 3T3-L1 adipocytes, (iv) increased transcriptional activities of peroxisome proliferator-activated receptor gamma (PPARγ) which improve insulin sensitivity and glucose metabolism [95, 96], (v) actively inhibiting intestinal α-glucosidase, and (vi) antioxidant activities. Many contains polyphenols that strongly inhibits α-glucosidase activities, that help reduce postprandial hyperglycemia [89, 91, 97] and also responsible for the stimulatory activities on glucose uptake [98, 99].

10. ANTICOAGULANT PROPERTIES

The seaweeds anticoagulant behavior is related to the sulfated polysaccharides and depends on having preferably two sulfate groups and a glycosidic linkage on the pyranose ring [100], which enable them to interact with the basic groups in the protein of probably more than one protease. The anticoagulant activity is chiefly by antithrombin and/or heparin cofactor II mediated thrombin inhibition. An anticoagulative polysaccharide was produced from an enzymatic hydrolysate of *Ecklonia cava* [101].

11. ENDOCRINE MODULATING PROPERTIES

11.1. Thyroid

Seaweeds are natural sources of iodine and iodine deficiency or excess can be harmful to the reproductive function and to the growth of the fetus and newborn [102]. Regular dietary seaweeds increased the thyroid stimulating hormone (TSH) in healthy postmenopausal women [103] and slightly decreased serum free thyroxine FT4 [104].

12. PHYTOESTROGENIC PROPERTIES

Seaweeds can effectively increase menstrual cycle length, stimulate ovulation and lower the estrogen/progesterone ratio in pre-menopausal women [105]. Seaweed (Alaria) positively altered estrogen and phytoestrogen metabolism [103] for women with fertility problems and/or of high risk estrogen-dependent disease.

13. IMMUNE REGULATION/INFLAMMATION/ALLERGY

A decreased prevalence of allergic rhinitis in Japanese females was associated with a high dietary intake of seaweed, calcium, magnesium and phosphorus [106]. The sulfolipids and fucoidans from seaweeds are anti-inflammatory [107]. Fucoidan has anti-coagulant, anti-thrombic and anti-inflammatory effects [11, 38]. The brown seaweed *Ecklonia cava* activated anti-inflammatory responses and suppressed the proinflammatory responses [108]. A clinical study (phase I and II) on fucoidan based complex, dose-dependently decreased pain and stiffness, in confirmed osteoarthritis patients [109]. Other seaweeds that have immunomodulatory properties are the brown seaweeds *Ishige okamurae* [110], *Hizikia fusiformis, Meristotheca papulosa* [111], *Porphyra yezoensis* [112], *Sargassum thunbergii* [79] and *Ecklonia cava* [108]. The B-1,3;1,6-D-glucan (translam) from *Laminaria cichorioides,* fucoidan from *Fucus evanescens* [113, 114], laminarans [115] and porphyrans [117] are amongst the bioactive compounds.

The immune-modulatory mechanisms by the bioactive seaweeds compounds are numerous. The methoxylated fatty acid from brown seaweed *Ishige okamurae* inhibited Phospholipase A2 and inflammation *in vivo* [110]. The *Hizikia fusiformis, Meristotheca papulosa* [111] and *Porphyra yezoensis* extracts stimulated B cells and macrophages *in vivo* [112]. The *Hizikia fusiformis* also showed immune modulatory effects on human lymphocytes [118]. The *Sargassum thunbergii* fucoidan increased peritoneal macrophages, total cells and macrophages in the lung [79] and enhanced the phagocytosis and chemiluminescence of macrophages [79]. The fucoidan and B-1,3;1,6-D-glucan increased *in vitro* adhesion of the intact cells, re-established neutrophil functions [113] and improved the number and functional activities of immunocompetent cells in humoral immune reactions [114]. The laminarans were immunomodulatory on anterior kidney leukocytes [115]. Porphyran increased primary antibody response (IgM), enhanced macrophages and depressed Th-2 type immune system without affecting Th-1 type immune system [117].

14. ANTI-CANCER, ANTI-TUMOR AND ANTIPROLIFERATIVE EFFECTS

A reduced risk of human breast cancer was linked to regular seaweeds consumption [119]. Dietary seaweeds inhibit the development of both benign and cancer neoplasis [30]. Regular seaweeds consumption decreased rectal [166], breast [119], in addition to benign and neoplasis cancer risks [120, 121]. Various seaweeds successfully reduced mammary tumours [122]. Seaweeds such as *Champia Feldmannii*, Diaz-Pifferer and *Undaria Pinnafitida* possess antitumour activities *in vitro* and *in vivo* [123]. Additionally *Sargassum thunbergii* and *Dictyopteris divaricata* extracts were antiproliferative against the HL-60 (human promyelocytic leukemia cell line) [124]. Furthermore, seaweed extracts such as *L. Japonica*, *P. tenera* and *Gelidium amansii*, have been shown to dose-dependently inhibit human gastric (AGS) and HT-29 colon cancer cells growth [125] and mammary tumours [122].

The anticancer compounds from seaweeds include water-soluble polysaccharides, such as laminarans, sulphated polysaccharides and fucoidans [126, 127]. Fucoidans effectively suppressed bile duct cancer [128] and human breast cancer [122, 129]. *Ascophyllum nodosum* fucans were antiproliferative on a non-small cell human bronchopulmonary carcinoma cell line [130]. The polysaccharide porphyran from *Porphyra sp.* possess antitumor properties [20, 21] and caused apoptosis in AGS gastric cancer cell lines [19].

Seaweeds are rich in iodine and potent against breast cancer, *in vitro* [122] and constant administration of iodine to mammals consequently causes strong and persistent reduction of breast tumors incidence [121]. *L. japonica* water extract produced apoptosis in several human breast cancer cell lines and was associated to SOD inhibition by iodine [122]. Selenium operates synergistically with iodine to influence hormone homeostasis and iodine availability. Additionally, iodine is derivatised into anti-proliferative iodilipids in the thyroid; which help manage the proliferation of other tissues, including the breast. Furthermore, arachidonic acid and its metabolites, cyclooxygenase (COX) and LOX, upregulate aromatase gene expression in the breast fatty tissues and increases intramammary estrogen. When arachidonic acid is iodinated, the iodolactones formed inhibit the EGF (epidermal growth factor) receptor, which consequently inhibited arachidonic acid metabolism.

Seaweeds sulfur compounds possess anti-proliferative properties [107]. Excess sulfation of fucoidans enhances their anti-angiogenic and antitumor activities [131]. Fucoidan itself is effective against bile duct cancer [128], human breast cancer [126] and reduced chemotherapy toxicities in colorectal cancer patients [132]. Fucoxanthin too has antiproliferative effect on human leukemia and colon cancer cell line [133]. Additionally, seaweeds (Laminariales sp. and *P. tenera*) polyphenols were anticarcinogenic and antiproliferative in fibroblasts [134].

The seaweeds compounds showed anticancer effects by inhibiting (i) cancer induction [135] (ii) hyaluronidase activity [136] (iii) inflammation [107] (iv) cell proliferation, [107] (v) mammary gland integrity disregulation, [120] (vi) oxidation, [30] (vii) cell survival, [19, 126] through the expression of transforming growth factor (TGF-β),[122] (viii) tumour initiation, [137] (ix) angiogenesis (e.g. by fucoidan), that inhibited VEGF165 (vascular endothelial growth factor 165) binding to its cell surface receptors [127, 129, 138] (x) tumor cells invasion, [127] (xi) immune responses suppression, [79], (xii) SOD activity [122] and (xiii) estrogen production.

15. ANTI-INFECTIVE

15.1. Antiviral Activity

The dextrin sulfate and fucoidan from seaweeds strongly subdued diverse strains of human immunodeficiency virus type 1 HIV-1 [139], herpes simplex virus-1 and 2 (HSV-1, HSV-2), human cytomegalovirus HCMV [140] and respiratory synctial virus-RSV [141]. In human T-lymphotropic virus type-1-associated neurological disease patients, fucoidan effectively reduced proviral load [142].

Carrageenan was also effective against HCMV, HIV, dengue virus, papilloma virus [143-146] and diverse strains of HSV [147]. The brown seaweed sulphated polysaccharides were helpful against HSV, RSV, Vaccinia virus *(Undaria pinnatifida* [148]; *Stypopodium zonale* [149]; *Hydroclathrus clathratus* [150]; *Sargassum wightii* [151] and herpatic virus by *Stoechospermum marginatum)*.

The seaweed antiviral biocompounds identified were (i) sulfoglycolipids, (ii) carrageenans (such as λ-carrageenan and partially cyclized μ/ι-carrageenan), from *Gigartina skottsbergii* [152] (iii) fucoidan, (iv) sesquiterpene hydroquinones, (v) xylomannan sulphate (from *Nothogenia fastigiata)* (vi) sulfated glucurono-galactan, (vii) Galactan sufate (from *Schizymenia binderi),* (viii) sulphated fucans and (vi) DL-Galactan (from *Gymnogongrus torulosus)* against HSV-1 and HSV-2, HIV, RSV, HCMV, Dengue virus, Pseudorabies virus and Influenza A and B virus, [16].

The sulfated polysaccharides were anti-viral on HSV-1, HSV-2, HCMV, HIV-1, RSV, influenza virus and bovine viral diarrhea virus. The antiviral mechanisms studied include (i) competitive inhibition of virus adsorption to cells [153], (ii) synergistically interacting with the target cell to block virus entry [139, 16], (iii) inhibiting retrovirus transcriptase *in vitro* (*Schizymenia pacifica* extracts) [154], (iv) inhibiting HIV replication *in vitro* (sulfated glucurono-galactan from *Schizymenia dubyi*) [155] and (v) modulating coagulation, thrombosis, immune and inflammatory process (by sulfated polysaccharide, like fucans) [156].

15.2. Antimicrobial (Bacterial and Fungal)

Many seaweeds (*e.g. D.humifusa, Acrosiphonia orientalis, Gelidella acerosa, Haligra species, Solieria robusta* and *Hypnea musciformis)* have good antimicrobial properties [157,

158]. The phaeophytes usually have the strongest effects, followed by rhodophytes and chlorophytes [159]. Most seaweeds extract are more effective against gram positive bacteria [160], but a few are also effective against gram negative *E.coli* [161]. The antibacterial compounds identified include (i) fatty acids [59], (ii) lipophilic compounds, (iii) phenolic compounds [162], (iv) lectin [157], (iv) secondary metabolites mainly acetogenins, terpenes, alkaloids and polyphenolics, especially the halogenated compounds [163], (v) isoprenoid metabolites [164] and (vi) hydrogen peroxide [165]. Good antifungal properties have also been shown by the brown seaweeds [160].

CONCLUSION

Many present ailments and disorders are related to the current lifestyle and eating habits and recent research indicates that seaweeds may be a good potential ingredient for use as functional food to help combat these degenerative conditions.

REFERENCES

[1] Matanjun, P., Mohamed, S., Mustapha, N.M., Muhammad, K. and Cheng, H.W. (2008).Antioxidant activities and phenolics content of eight species of seaweeds from north Borneo. *Journal of Applied Phycology*, 20(4), 367-373.

[2] Jiménez-Escrig A, Jiménez-Jiménez I, Pulido R, Saura-Calixto F (2001) Antioxidant activity of fresh and processed edible seaweeds. *J. Sci. Food Agric.* 81:530-534.

[3] Nagai T, Yukimoto T (2003) Preparation and functional properties of beverages from sea algae. *Food Chem.* 81:327-332.

[4] Duan X-J, Zhang X-M, Li X-M, Wang B-G (2006) Evaluation of antioxidant property of extract and fractions obtained from a red alga, Polysiphonia urceolata. *Food Chem.* 95:37-43.

[5] Pavia H, Aberg P (1996) Spatial variation in polyphenolic content of Ascophyllum nodosum (Fucales, Phaeophyta). *Hydrobiol.* 326/327:199-203.

[6] Nakamura T, Nagayama K, Uchida K, Tanaka R (1996) Antioxidant activity of phlorotannins from the brown alga Eisenia bicyclis. *Fish Sci.* 62:923-926.

[7] Matanjun, P., Mohamed, S., Mustapha, N.M. and Muhammad, K. (2009). Nutrient content of tropical edible seaweeds, Eucheuma cottonii, Caulerpa lentillifera and Sargassum polycystum. *Journal of Applied Phycology*, 21, 1-6.

[8] Wong, K.H.P, and Cheung, C.K. (2000). Nutritional evaluation of some subtropical red and green seaweeds Part I Ð proximate composition, amino acid profiles and some physico-chemical properties. *Food Chemistry*, 71, 475-482.

[9] MacArtain, P., Gill, C.I.R., Brooks, M., Campbell, R. and Rowland, I.R. (2007). Nutritional value of edible seaweeds. *Nutrition Review*, 65, 535–543.

[10] Burtin, P. (2003). Nutritional value of seaweeds. *Electronic Journal of Environment Agriculture Food Chemistry*, 2, 498–503.

[11] Sezer, A.D., Erdal, C., Fatih, H.P., Zeki, O., Ahmet, L. B., and Jülide, A. (2008). Preparation of Fucoidan-Chitosan Hydrogel and Its Application as Burn Healing Accelerator on Rabbits. *Biological and Pharmaceutical Bulletin*, 31(12), 2326—2333.

[12] Bocanegra, A., Bastida, S., Benedí, J., Ródenas, S. and Sánchez-Muniz, F. J. (2009). Characteristics and nutritional and cardiovascular-health properties of seaweeds. *Journal of Medicinal Food*, 12 (2), 236-258.

[13] Queiroz, K.C.S., Medeiros, V.P., Queiroz, L.S., Abreu, L.R.D., Rocha, H.A.O., Ferreira, C.V., Juca, M.B., Aoyama, H. and Leite, E.L. (2008). Inhibition of reverse transcriptase activity of HIV by polysaccharides of brown algae. *Biomedicine and Pharmacotherapy*, 62, 303-307.

[14] Cumashi, A., Ushakova, N.A., Preobrazhenskaya, M.E., D'Incecco, A., Piccoli, A., Totani, L. (2007). A comparative study of the anti-inflammatory, anticoagulant, antiangiogenic and antiadhesive activities of nine different fucoidans from brown seaweeds. *Glycobiology*, 17, 541–542.

[15] Azevedo, T.C.G., Bezerra, M.E., Santos, M.D., Souza, L.A., Marques, C.T., Benevides, N.M.B. and Leite, E.L. (2009). Heparinoids algal and their anticoagulant, hemorrhagic activities and platelet aggregation. *Biomedicine and Pharmacotherapy*, 63, 477–83.

[16] Pujol, C.A., Estevez, J.M., Carlucci, M.J., Ciancia, M., Cerezo, A.S. and Damonte, E.B. (2002). Novel DL-galactan hybrids from the red seaweed Gymnogongrus torulosus are potent inhibitors of herpes simplex virus and dengue virus. *Antiviral Chemistry and Chemotherapy*, 13(2), 83-89.

[17] Paxman, J.R., Richardson, J.C., Dettmar, P.W. and Corfe, B.M. (2008). Daily ingestion of alginate reduces energy intake in free living subjects. *Appetite*, 51, 713-719.

[18] Ikeda, K., Kitamura, A., Machida, H., Watanabe, M., Negishi, H., Hiraoka, J. and Nakano, T. (2003). Effect of Undaria pinnatifida (Wakame) on the development of cerebrovascular disease in stroke-prone spontaneously hypertensive rats. *Clinical and Experimental Pharmacology and Physiology*, 30, 44-48.

[19] Kwon, M.J. and Nam, T.J. (2006). Porphyran induces apoptosis related signal pathway in AGS gastric cancer cell lines. *Life Science*, 79, 1956–1962.

[20] Guo, T.T., Xu, H.L., Zhang, L.X. (2007). In vivo protective effect of Porphyra yezoensis polysaccharide against carbon tetrachloride induced hepatotoxicity in mice. *Regulatory toxicology and Pharmacology*, 49, 101–106.

[21] Zhao, T., Zhang, Q., Qi, H. Zhang, H. Niu, X. and Li Z. (2006). Degradation of porphyran from Porphyra haitanensis and the antioxidant activities of the degraded porphyrans with different molecular weight. *International Journal of Biological Macromolecules*, 38(1), 45-50.

[22] Polat, S. and Ozogul, Y. (2009). Fatty acid, mineral and proximate composition of some seaweeds from the northeastern mediterranean coast. *Italian Journal of Food Science*, 21(3), 317-324.

[23] Barbarino, E. and Lourenco, S.O. (2005). An evaluation of methods for extraction and quantification of protein from marine macro- and microalgae. *Journal of AppliedPhycology*, 17(5), 447-460.

[24] Fleurence, J. (1999). Seaweed proteins, biochemical, nutritional aspects and potential uses. *Trends in Food Science Technology*, 10, 25-28.

[25] Gonzalez, R., Rodriguez S., Romay, C., Ancheta, O., Gonzalez, A., Armesta, J., Remirez, D. and Merino, N. (1999). Anti-inflammatory activity of phycocyanin extract in acetic acid-induced colitis in rats. *Pharmacology Research*, 39(1), 55-59.

[26] Ortiz, J., Edgar, U., Paz, R., Nalda. R., Vilma, Q. and Catherine, L. (2009). Functional and nutritional value of the Chilean seaweeds Codium fragile, Gracilaria chilensis and Macrocystis pyrifera. *European Journal of Lipid Science Technology*, 111(4), 320 – 327.

[27] Sanchez-Machado, D.I., Lopez-Cervantes, J., Lopez-Hernandez, J. and Paseiro-Losada, P. (2004). Fatty acids, total lipid, protein and ash contents of processed edible seaweeds. *Food Chemistry*, 85, 439–444.

[28] Taboada, C., Rosendo, M. and Isabel, M. (2009). Composition, nutritional aspects and effect on serum parameters of marine algae Ulva rigida. *Journal of Science Food and Agriculture*, 90(3), 445 – 449.

[29] Ruperez, P. (2002). Mineral content of edible marine seaweeds. *Food Chemistry*, 79, 23–26.

[30] Eskin, B.A., Grotkowski, C.E. and Connolly, C.P. (1995). Different tissue responses for iodine and iodide in rat thyroid and mammary glands. *Biological Trace Element Research*, 49, 9–19.

[31] Chauhan V. and Chauhan A (2006) Oxidative stress in Alzheimer's disease. *Pathophysiol.* 13:195-208.

[32] Temple N.J. (2000) Antioxidants and disease: more questions than answers. *Nutr. Res.* 20: 449-459.

[33] Rodrigo R, and Bosco C (2006) Oxidative stress and protective effects of polyphenols: Comparative studies in human and rodent kidney. A review. *Comparative Biochem. Physiol. Part C* 142:317-327.

[34] Tapiero H, Tew KD, Nguyen Ba G, Mathé G (2002) Polyphenols: do they play a role in the prevention of human pathologies? Review. *Biomed. Pharmacother.* 56:200-207.

[35] Matsukawa R, Dubinsky Z, Kishimoto E, Masakki K, Masuda Y, Takeauchi, T, et al. (1997). A comparison of screening methods for antioxidant activity in seaweeds. *J. Appl. Phycol.* 9:29-35.

[36] Anggadiredja J., Andyani R, and Muawanah H. (1997) Antioxidant activity of Sargassum polycystum (Phaeophyta) and Laurencia obtusa (Rhodophyta) from Seribu Islands. *J. Appl. Phycol.* 9:477-479.

[37] Yan X, Nagata T, Fan X (1998) Antioxidant activities in some common seaweeds. *Plant Foods Human Nutrtion*, 52(3) 253-262.

[38] Rupérez, P., Ahrazem, O. and Leal, J.A. (2002). Potential antioxidant capacity of sulfated polysaccharides from the edible marine brown seaweed Fucus vesiculosus. *Journal of Agriculture and Food Chemistry*, 50, 840–845.

[39] Kang, K., Park, Y., Hwang, H. J., Kim, S. H., Lee, J. G., Shin, H.-C. (2003). Antioxidative Properties Of Brown Algae Polyphenolics And Their Perspectives As Chemopreventive Agents Against Vascular Risk Factors. *Archives of Pharmacal Research*, 26, 286-293.

[40] Chandini, S. K., Ganesan, P. and Bhaskar, N. (2008). In vitro antioxidant activities of three selected brown seaweeds of India. *Food Chemistry*, 107, 707-713.

[41] Zhang, Q., Ning, L., Xiguang, L., Zengqin, Z., Zhien, L. and Zuhong, Xu. (2004). The structure of a sulfated galactan from Porphyra haitanensis and its in vivo antioxidant activity. *Carbohydrate Research*, 339, 105–111.

[42] Yuan, Y. V. and Walsh, N.A. (2006). Antioxidant and antiproliferative activities of extracts from a variety of edible seaweeds. *Food and Chemistry Toxicology*, 44, 1144-1150.

[43] Matanjun, P., Mohamed, S., Kharidah, M. and Noordin, M.M. (2010). Comparison of Cardiovascular Protective Effects of Tropical Seaweeds, Eucheuma cottonii, Caulerpa lentillifera and Sargassum polycystum, on High-Cholesterol/High-Fat Diet in Rats, *Journal of Medicinal Food*, 13(4), 792–800.

[44] Hata, Y., Nakajima, K., Uchida, J.-I., Hidaka, H., Nakano, T. (2001). Clinical effects of brown seaweed, Undaria pinnatifida (wakame), on blood pressure in hypertensive subjects. *Journal of clinical Biochemistry and Nutrition*, 30, 43-53.

[45] Sato, M., Takashi, Oba., Toshiyasu, Y., Toshiki, N., Takashi, K., Katsura, F., Akio, K. and Takahisa, N. (2002). Antihypertensive Effects of Hydrolysates of Wakame (Undaria pinnatifida) and Their Angiotensin-I-Converting Enzyme Inhibitory Activity. *Annals of Nutrition and Metabolism*, 46, 259-267.

[46] Jung, H.A., Sook, K.H., Hyeung R. K. and Jae S. C. (2006). Angiotensin-converting enzyme I inhibitory activity of phlorotannins from Ecklonia stolonifera. *Fish Science*, 72, 1292–1299.

[47] Abidov, M., Ramazanov, Z., Seifulla, R. and Grachev, S. (2010). The effects of Xanthigen in the weight management of obese premenopausal women with non-alcoholic fatty liver disease and normal liver fat. *Diabetes, Obesity and Metabolism*, 12, 72-81.

[48] Maeda, H., Hosokawa, M., Sashima, T., Funayama, K. and Miyashita, K. (2005). Fucoxanthin from edible seaweed, Undaria pinnatifida, shows antiobesity effect through UCP1 expression in white adipose tissue. *Biochemical and Biophysical Research Communications*, 332, 392-397.

[49] Miyashita, K. (2006). Seaweed carotenoid, fucoxanthin, with highly bioactive and nutritional activities. *Journal of Marine Bioscience and Biotechnology*, 1, 48-58.

[50] Kazuo, M. (2009). The carotenoid fucoxanthin from brown seaweed affects obesity. *Lipid Technology*, 21, 186-190.

[51] Maeda, H., Tsukui, T., Sashima, T., Hosokawa, M. and Miyashita, K. (2008). Seaweed carotenoid, fucoxanthin, as multi-functional nutrient. *Asia Pacific Journal of Clinical Nutrition*, 17, 196-199.

[52] Skulas-Ray, A.C., West, S.G., Davidson, M.H. and Kris-Etherton, P.M. (2008). Omega-3 fatty acid concentrates in the treatment of moderate hypertriglyceridemia. *Expert Opinion on Pharmacotherapy*, 9(7), 1237-1248.

[53] Von, S.H., Wallentin, L., Lennmarken, C. and Larsson, J.(1992). Lipoprotein metabolism following gastroplasty in obese women. *Scandanavian Journal of Clinical and Laboratory Investigation*, 52(4), 269-274.

[54] Lee, H.J., Kim, J.H., Lee, C.H., Kim, J.S., San, T. K., Lee, K.B., Song, K.S., Byung W. C. and Bong H. L. (1999). Inhibitory activities of sea weeds on prolyl endopeptidase, tyrosinase and coagulation. *Korean Journal of Pharmacognosy*, 30(3), 231-237.

[55] Teixeira, V.L., Fabíola, D. R., Peter, J. H., Maria, A. C. K. and Renato, C. P. (2007). α-Amylase inhibitors from Brazilian seaweeds and their hypoglycemic potential. *Fitoterapia*, 78, 35–36.

[56] Bitou, N., Ninomiya, M., Tsujita, T. and Okuda, H. (1999). Screening of lipase inhibitors from marine algae. *Lipids*, 34 (5), 441.

[57] Iwahori, Y., Enomoto, S., Okada, Y., Tanaka, J. and Okuyama, T. (1999). Naturally occurring substances for prevention of complications of diabetes. IV. Screening of seavegetables for inhibitory effect on aldose reductase. *Nature medicine*, 53(3), 138-140.

[58] Panlasigui, L.N., Baello, O.Q., Dimatangal, J.M., Dumelod, B.D. (2003). Blood cholesterol and lipid-lowering effects of carrageenan on human volunteers. *Asia Pacific Journal of clinical Nutrition*, 12 (2), 209-214.

[59] Ara, J., Sultana, V., Qasim, R., Ehteshamul-Haque, S. and Ahmad, V. U. (2005). Biological Activity of Spatoglossum asperum, A Brown Alga. *Phytotherapy Research,* 19, 618–623.

[60] Vizquez-Freire, M. J., Lamela, M. and Calleja, J.M. (1996). Hypolipidaemic Activity of a Polysaccharide Extract from Fucus vesiculosus L. *Phytotherapy Research.* 10, 647-650.

[61] Jaffri J.M., Mohamed, S., Ahmad, I.N., Mustapha, N.M., Manap, Y.A., and Rohimi. N. (2011) Effects of Catechin-rich Oil Palm Leaf Extract on normal and hypertensive rats' kidney and liver. *Food Chem*. http://dx.doi.org/10.1016/j.foodchem.2011.03.050.

[62] Mancini, F.J., Novoa, A.V., González, A.E.B., de Andrade-Wartha, E.R.S., de O e Silva, A.M., Pinto, J.R. and Portari Mancini, D.A. (2009). Free phenolic acids from the seaweed Halimeda monile with antioxidant effect protecting against liver injury Zeitschrift fur Naturforschung. Section C *Journal of Bioscience*, 64(9-10), 657-663.

[63] Kim, Y.C., Ren, B.A., Na, Y.Y., Taek, J.N. and Jae, S.C. (2005). Hepatoprotective constituents of the edible brown alga Ecklonia stolonifera on tacrine-induced cytotoxicity in Hep G2 cells. *Archives of Pharmacal Research*, 28(12), 1376-1380.

[64] Arumugam, S., Rao, H., Raghavendran, B., Srinivasan, P. and Devakic, T. (2008). Anti-peroxidative and anti-hyperlipidemic nature of Ulva lactuca crude polysaccharide on d-Galactosamine induced hepatitis in rats. *Food Chemistry and Toxicology*, 46(10), 3262-3267.

[65] Morita, K. and Nakano, T. (2002). Seaweed accelerates the excretion of dioxin stored in rats. *Journal of Agriculture and Food Chemistry*, 50, 910–917.

[66] Takuya, Miura., David, P. N., Marc L. S., Toshiharu, S., Gregor, Z., Paul, R. H., Ellis, J. N. and John, E. M.J. (1996). Blockade of selectin-mediated leukocyte adhesion improves postischemic function in lamb hearts. *The Annals of Thoracic Surgery*, 62(5), 1295-1300.

[67] Veena, C. K., Anthony, J., Sreenivasan, P. P., Nachiappa, G. R. and Palaninathan, V. (2008). Mitochondrial dysfunction in an animal model of hyperoxaluria, A prophylactic approach with fucoidan. *European Journal of Pharmacology*, 579, 330–336.

[68] Zhang, Q., Li, Z., Xu, Z., Niu, X. and Zhang, H. (2003). Effects of fucoidan on chronic renal failure in rats. *Planta medica,* 69, 537–541.

[69] Gaigi, S., Elati, J., Ben Osman, A., Beji, C. (1996). Seaweed algae and obesity, A trial study | [Etude experimentale de l'effet des algues marines dans le traitement de l'obesite]. *Tunisie Medicale,* 74 (5), 241-243.

[70] Jhamandas, J.H., Wie, M.B., Harris, K., MacTavish, D. and Kar, S. (2005). Fucoidan inhibits cellular and neurotoxic effects of beta-amyloid (A beta) in rat cholinergic basal forebrain neurons. *European Journal of Neuroscience*, 21, 2649–2659.

[71] Uhm, C.S., Kim, K.B., Lim, J.H., Pee, D.H., Kim, Y.H., Kim, H., Eun, B.L. and Tockgo, Y.C. (2003). Effective treatment with fucoidan for perinatal hypoxic–ischemic encephalopathy in rats. *Neuroscience Letters*, 353, 21–24.

[72] Alarcon, R., Fuenzalida, C., Santibanez, M. and Von Bernhardi, R. (2005). Expression of scavenger receptors in glial cells. Comparing the adhesion of astrocytes and microglia from neonatal rats to surface-bound beta-amyloid. *Journal of Biological Chemistry*, 280, 30406–30415.

[73] Luo, D., Quanbin, Z., Haomin, W., Yanqiu, C., Zuoli, S., Jian, Y., Yan, Z., Jun, Journal of ., Fen, Y., Xuan, W. and Xiaomin, W. (2009). Fucoidan protects against dopaminergic neuron death in vivo and in vitro. *European Journal of Pharmacology*, 617, 33–40.

[74] Giacobini, E. (2004b). Cholinesterases, New roles in brain function and in Alzheimer's disease. *Neurochemistry Research*, 28(3–4), 515–522.

[75] Choi, B. W., Rhu, G., Park, S. H., Kim, E. S., Shin, J., Roh, S. S., Shin, H. C. and Lee, B. H. (2007). Anticholinesterase Activity of Plastoquinones from Sargassum sagamianum, Lead Compounds for Alzheimer's Disease Therapy. *Phytotherapy Research*, 21, 423–426.

[76] Yoon, N.Y., Hae, Y. C., Hyeung, R. K. and Jae, S. C. (2008). Acetyl- and butyryl-cholinesterase inhibitory activities of sterols and phlorotannins from Ecklonia stolonifera. *Fish Science*. 74, 200–207.

[77] Natarajan, S., Shanmugiahthevar, K.P. and Kasi, P.D. (2009). Cholinesterase inhibitors from Sargassum and Gracilaria gracilis, seaweeds inhabiting South Indian coastal areas (Hare Island, Gulf of Mannar). *Natural Product Research*, 23(4), 355-69.

[78] Han, D.J. and Hua, S.G. (2009). A novel sulfated oligosaccharide, inhibits pulmonary fibrosis by targeting TGF-β1 both in vitro and in vivo. *Acta Pharmacologica Sinica*, 30, 973–979.

[79] Itoh, H., Noda, H., Amano, H. and Ito, H. (1995). Immunological analysis of inhibition of lung metastases by fucoidan (GIV-A) prepared from brown seaweed Sargassum thunbergii. *Anticancer Research*. 15(5B), 1937-1947.

[80] Gong, T.F., Fang, C.Y., Chen, W. and Wang, P.N. (2007). Physiochemical properties of alginate in tissue engineering research and its clinical application. *Journal of Clinical Rehabilitative Tissue Engineering Research*, 11(18), 3613-3616.

[81] Dae, H.P., Won, S.C., Sean, H.Y., Jung, S.S. and Chul, H.S. (2007). A developmental study of artificial skin using the alginate dermal substrate. *Key Engineering Materials*, 342-343, 125-128.

[82] Shakespeare, P. (2001). Burn wound healing and skin substitutes. *Burns*, 27, 517—522.

[83] Matou, S., Helley, D., Chabut, D., Bros, A. and Fischer A.M. (2002). Effect of fucoidan on fibroblast growth factor-2-induced angiogenesis in vitro. *Thrombosis Research*, 106, 213–221.

[84] Fard, S.G., Shamsabadi, F.T., Emadi, M., Meng, G.Y., Muhammad, K., Mohamed, S. (2011). Ethanolic extract of Eucheuma cottonii promotes in vivo hair growth and wound healing. *Journal of Animal Veterinary Advances*, 10 (5), 601-605.

[85] Fluhr, J.W., Breternitz, M., Kowatzki, D., Bauer, A., Bossert, J., Elsner, P., Hipler, U.-C. (2010). Silver-loaded seaweed-based cellulosic fiber improves epidermal skin physiology in atopic dermatitis, Safety assessment, mode of action and controlled, randomized single-blinded exploratory in vivo study. *Experimental Dermatology*, 19 (8), e9-e15.

[86] Lee, H.J., Kim, H.C., Vitek, L., Nam, M.C. (2010). Algae consumption and risk of type 2 diabetes, Korean National Health and Nutrition Examination Survey in 2005. *Journal of Nutrition Science and Vitaminology*, 56 (1), 13-18.

[87] Kang, S., Young, J.J., Hee,C. K., Soo, Y. C., Joon, H. H., Ilson, W., Moo, H. K., Hye, S. S., Hyung, B. J. And Se, J. K. (2008). Petalonia Improves Glucose Homeostasis In Streptozotocin-Induced Diabetic Mice. *Biochemical And Biophysical Research Communications*, 373, 265–269.

[88] Kim, K.Y., Nama, K.A., Kurihara, H. and Kim, S.M. (2008). Potent a-glucosidase inhibitors purified from the red alga Grateloupia elliptica. *Phytochemistry*, 69, 2820–2825.

[89] Anandakumar, S., Balamurugan, M., Rajadurai, M. and Vani, B. (2008). Antihyperglycemic and antioxidant effects of red algae Hypnea musciformis in alloxan-induced diabetic rats. *Biomedicine*, 28(1), 34-38.

[90] Iwai, K. (2008) .Antidiabetic and Antioxidant Effects of Polyphenols in Brown Alga Ecklonia stolonifera in Genetically Diabetic KK-Ay Mice. *Plant Foods for Human Nutrition*, 63, 163–169.

[91] Lee, K., Young, S.C. and Jung, S. S. (2004). Sea Tangle Supplementation Lowers Blood Glucose and Supports Antioxidant Systems in Streptozotocin-Induced Diabetic Rats. *Journal of Medicinal Food*, 7(2), 130-135.

[92] Ohta, T., Sasaki, S., Oohori, T., Yoshikawa, S. and Kurihara, H. (2002). α- Glucosidase inhibitory activity of a 70% methanol extract from Ezoishige (Pelvetia babingtonii de Toni) and its effect on the elevation of blood glucose level in rats. *Bioscience, Biotechnology and Biochemistry*, 66, 1552–1554.

[93] Zhang, J., Christa, T., Jingkai, S., Can, W., Gabrielle, S. G., Dorothy, D., Colin J. B., Mingsan, M. and Stephen, E.H. (2007). Antidiabetic properties of polysaccharide- and polyphenolic-enriched fractions from the brown seaweed Ascophyllum nodosum. *Can. Journal of Physiology and Pharmacology*, 85(11), 1116–1123 (2007).

[94] Wu Z, Rosen, E.D., Burn, R., Hauser, S., Adelmant, G., Troy, A., Mckeon, C., Darlington, G.J. and Spiegelman, B.M. (1999). Cross-regulation of C/EBP alpha and PPAR gamma controls the transcriptional pathway of adipogenesis and insulin sensitivity. *Molecular Cell*, 3, 151–158.

[95] Rieusset, J., Touri, F., Michalik, L., Escher, P., Desvergne, B., Niesor, E. and Wahli, W. (2002). A new selective peroxisome proliferator-activated receptor gamma antagonist with antiobesity and antidiabetic activity. *Molecular Endocrinology*, 16, 2628–2644.

[96] Baron, A. D., (1998). Postprandial hyperglycemia and a-glucosidase inhibitors. *Diabetes Research and clinical Practice*, 40, S51–S55.

[97] Kurihara, H., Mitani, T., Kawabata, J. and Takahashi, K. (1999). Inhibitory potencies of bromophenols from Rhodomelaceae algae against a-glucosidase activity. *Fish Science*, 65, 300–303.

[98] Xu, N., Fan, X., Yan, X., Li, X., Niu, R. and Tseng, C.K. (2003). Antibacterial bromophenols from the marine red algae Rhodomela confervoides. *Phytochemistry*, 62, 1221–1226.

[99] Ciancia, M., Quintana, I., Cerezo, A.S. (2010). Overview of anticoagulant activity of sulfated polysaccharides from seaweeds in relation to their structures, focusing on those of green seaweeds. *Current Medicinal Chemistry*, 17 (23), 2503-2529.

[100] Athukorala, Y., Jung, W. K., Vasanthan, T. and Jeon, Y. J. (2006). An anticoagulative polysaccharide from an enzymatic hydrolysate of Ecklonia cava. *Carbohydrate Polymers*, 66, 184–191.

[101] Dunn, J.T. and Delange, F. (2001). Damaged reproduction, The most important consequence of iodine deficiency. *Journal of Clinical Endocrinology and Metabolism*, 86, 2360-2363

[102] Teas, J., Thomas, G. H., James, R. H., Adrian, A. F., Daniel W. S. and Mindy, S. K. (2009). May Dietary Seaweed Modifies Estrogen and Phytoestrogen Metabolism in Healthy Postmenopausal Women. *Journal of Nutrition*, 139(5), 939-944.

[103] Kiyoshi, M., Tomoyasu, T. and Masahiko, K. (2008). Suppression of thyroid function during ingestion if seaweed "Kombu" (Laminaria japonica) in normal Japanese adults. *Endocrine Journal*, 55(6), 1103-1108.

[104] Skibola, C.F., Curry, J., Gansevoort, C., Conley, A. and Smith, M.T. (2005). Brown Kelp Modulates Endocrine Hormones in Female Sprague-Dawley Rats and in Human Luteinized Granulosa Cells. *Journal of Nutrition*,135, 296–300.

[105] Miyake, Y., Sasaki, S., Ohya, Y., Miyamoto, S., Matsunaga, I., Yoshida, T., Hirota, Y., Oda, H. (2006). Dietary Intake of Seaweed and Minerals and Prevalence of Allergic Rhinitis in Japanese Pregnant Females, Baseline Data From the Osaka Maternal and Child Health Study. *Annals of Epidemiology*, 16 (8), 614-621.

[106] Berge, J. P., Debiton, E., Dumay, J., Durand, P. and Barthomeuf, C. (2002). In vitro anti-inflammatory and anti-proliferative activity of sulfolipids from the red alga Porphyridium cruentum. *Journal of Agriculture and Food Chemistry*, 50, 6227–6232.

[107] Ginnae, Ahn., Insun, H., Eunjin, P., Jinhe, K., You-Jin, J., Jehee, Lee., Jae W. P. and Youngheun, J. (2008). Immunomodulatory Effects of an Enzymatic Extract from Ecklonia cava on Murine Splenocytes. *Marine Biotechnology*, 10, 3.

[108] Myers, S.P., O'Connor, J., Fitton, J.H., Brooks, L., Rolfe, M., Connellan, P., Wohlmuth, H., Morris, C. (2009). A combined phase I and II open label study on the effects of a seaweed extract nutrient complex on osteoarthritis. *Biologics: Targets and Therapy*, 4, 33-44.

[109] Cho, J., Gyawali, Y. P., Ahn, S., Khan, M. N. A., Kong, I. and Hong, Y. (2008). A methoxylated fatty acid isolated from the brown seaweed Ishige okamurae inhibits bacterial phospholipase A2. *Phytotherapy Research*, 22(8), 1070 – 1074.

[110] Liu, J. N., Yoshida, Y., Wang, M. Q., Okai, Y. and Yamashita, U. (1997). B cell stimulating activity of seaweed extracts. *International Journal of Immunopharmacology*, 19(3), 135-142.

[111] Yoshizawa, Y., Ametani, A., Tsunehiro, J., Nomura, K., Ito, M., Fukui, F. and Kaminogawa, S. (1995). Macrophage stimulation activity of polysaccharide fraction from a marine alga (Porphyra yezoensis), structure-function relationships and improved solubility. *Bioscience, Biotechnology and Biochemistry*, 59, 1933–1937.

[112] Zaporozhets, T.S. (2003). Mechanisms of neutrophil activation by sea hydrobiont biopolymers. *Antibiotiki i Khimioterapiya*, 48(9), 3-7.
[113] Chertkov, K. S., Davydova, S. A., Nesterova, T. A., Zviagintseva, T. N. and Eliakova, L. A. (1999). Efficiency of polysaccharide translam for early treatment of acute radiation illness. *Radiatsionnaya Biologiya Radioekologiya*, 39, 572–577.
[114] Dalmo, R.A., Bogwald, J., Ingebrigtsen, K. and Seljelid, R. (1996). The immunomodulatory effect of laminaran [β-(1,3)-D-glucan] on Atlantic salmon, Salmo salar L., anterior kidney leucocytes after intraperitoneal, peroral and peranal administration. *Journal of Fish Diseases*, 19, 449–457.
[115] Bhatia, S., Sharma, A., Sharma, K., Kavale, M., Chaugule, B. B., Dhalwal, K., Namdeo, A. G. and Mahadik, K. R. (2008). Novel algal polysaccharides from marine source, Porphyran, Pharmacognosy Reviews, 2(4), 271-276.
[116] Okai, Y., Okai, K. H., Ishizaka, S. and Yamashita, U. (1997). Enhancing effect of polysaccharides from an edible brown alga, Hizikia fusiforme (Hiziki), on release of tumor necrosis factor-α from macrophages of endotoxin-nonresponder C3H/*HeJournal of mice. Nutrition Cancer.* 27, 74–79.
[117] Setchell, K.D., Borriello, S.P., Hulme, P., Kirk, D.N. and Axelson, M. (1984). Nonsteroidal estrogens of dietary origin, possible roles in hormone dependent disease. *American Journal of Clinical Nutrition*, 40, 569–578.
[118] Aceves, C., Anguiano, B. and Delgado, G. (2005). Is iodine a gatekeeper of the integrity of the mammary gland? *Journal of Mammary Gland Biology and Neoplasia*, 10, 189-196.
[119] Garcia-Solis, P., Alfaro, Y., Anguiano, B. (2005). Inhibition of normal-methyl-normal-nitrosourea-induced mammary carcinogenesis by molecular iodine (I2) but not by iodide (I2) treatment evidence that I2 prevents cancer promotion. *Molecular and Cellular Endocrinology*, 236, 49–57.
[120] Funahashi, H., Imai, T., Mase, T., Sekiya, M., Yokoi, K., Hayashi, H., Shibata, A., Hayashi, T., Nishikawa, M., Suda, N., Hibi, Y., Mizuno, Y., Tsukamura, K., Hayakawa, A., Tanuma, S. (2001). Seaweed prevents breast cancer? *Japanese Journal of Cancer Research,* 92(5), 483-487.
[121] Maruyama, H., Tamauchi, H., Hashimoto and M., Nakano T. (2003). Antitumor activity and immune response of Mekabu fucoidan extracted from Sporophyll of Undaria pinnatifida. *In vivo*, 17 (3), 245-249.
[122] Kim, K.N., Ham, Y.M., Moon, J.Y., Kim, M.J., Kim, D.S, Lee, W.J., Lee, N.H. and Hyun, C.G. (2009). In vitro cytotoxic activity of Sargassum thunbergii and Dictyopteris divaricata (Jeju seaweeds) on the HL-60 tumour cell line. *International Journal of Pharmacology,* 5(5), 298-306.
[123] Cho, E. J., Rhee, S. H. and Park, K. Y. (1997). Antimutagenic and cancer cell growth inhibitory effects of seaweeds. *Journal of Food Science Nutrition*, 2, 348–353.
[124] Yamasaki, M. Y., Yamasaki, M., Tachibana, H., and Yamada, K. (2009). Fucoidan Induces Apoptosis through Activation of Caspase-8 on Human Breast Cancer MCF-7 Cells. *Journal of Agriculture and Food Chemistry*, 57, 8677–8682.
[125] Ye, J., Li, Y., Teruya, K., Katakura, Y., Ichikawa, A., Eto, H., Hosoi, M., Hosoi, M., Nishimoto, S. and Shirahata, S. (2005). Enzyme-digested Fucoidan Extracts Derived from Seaweed Mozuku of Cladosiphon novae-caledoniaekylin Inhibit Invasion and Angiogenesis of Tumor Cells. *Cytotechnology*, 47, 1-3.

[126] Fukahori, S., Yano, H., Akiba, J., Ogasawara, S., Momosaki, S., Sanada, S., Kuratomi, K., Ishizaki, Y., Moriya, F., Yagi, M. and Kojiro, M.. Fucoidan. (2008). A major component of brown seaweed, prohibits the growth of human cancer cell lines in vitro. *Molecular medicine Report*, 1(4), 537-542.

[127] Yumi, Y., Masao, Y., Hirofumi, T. and Koji, Y. (2009). Fucoidan Induces Apoptosis through Activation of Caspase-8 on Human Breast Cancer MCF-7 Cells. *Journal of Agriculture and Food Chemistry*, 57, 8677–8682.

[128] Riou, D., Colliec-Jouault, S., Du Sel, D.P., Bosch S.,Tomasoni, C., Sinquin, C., Durand, P., and Roussakis, C. (1996). Antitumour and antiproliferative effects of a fucan extracted from Ascophyllum nodosum against non-small cell bronchopulmonary carcinoma line. *Anticancer Research*, 16, 1213-218.

[129] Koyanagi, S., Tanigawa, N., Nakagawa, H., Soeda, S,. and Shimeno, H. (2003). Oversulfation of fucoidan enhances its anti-angiogenic and antitumor activities. *Biochemical Pharmacology*, 65, 173-179.

[130] Ikeguchi, M., Yamamoto, M., Arai, Y., Maeta, Y., Ashida, K., Katano, K., Miki, Y., Kimura, T. (2011). Fucoidan reduces the toxicities of chemotherapy for patients with unresectable advanced or recurrent colorectal cancer. *Oncology Letters*, 2 (2), 319-322.

[131] Hosokawa, M., Kudo, M., Hayato, M., Kohno, H., Tanaka, T. and Miyashita, K. (2004). Fucoxanthin induces apoptosis and enhances the antiproliferative effect of the PPARg ligand, troglitazone, on colon cancer cells. *Biochimica Biophysica Acta*, 1675, 113–119.

[132] Okai, K., Higashi-Okai, S.I., Nakamura, Y., Yano, Y., and Otani, S. (1994). Suppressive effects of the extracts of Japanese edible seaweeds on mutagen-induced umu C gene expression in Salmonella typhimurium (TA 1535/pSK 1002) and tumour promoter-dependent ornithine decarboxylase induction in BALB/c 3T3 fibroblast cells. *Cancer Letters*, 87, 25–32.

[133] Reddy, B.S., Sharma, C. and Mathews, L. (1984). Effect of Japanese seaweed (Laminaria angustata) extracts on the mutagenicity of 7,12-dimethylbenz[a]anthracene, a breast carcinogen and of 3,2'-dimethyl-4-aminobiphenyl, a colon and breast carcinogen. *Mutation Research*, 127, 113–118.

[134] Shibata, T., Fujimoto, K., Nagayama, K., Yamaguchi, Y. and Nakamura, T. (2002). Inhibitory activity of brown algal phlorotannins against hyaluronidase. *International Journal of Food Science and Technology*, 37, 703–709.

[135] Lee, E. and Sung, M.K. (2003). Chemoprevention of azoxymethane-induced rat colon carcinogenesis by seatangle, a fiber-rich seaweed. *Plant Foods for Human Nutrition*, 58, 1–8.

[136] Satoru, K., Noboru, T., Hiroo, N., Shinji, S. and Hiroshi, S. (2003). Oversulfation of fucoidan enhances its anti-angiogenic and antitumor activities. *Biochemical Pharmacology*, 65, 173-179.

[137] Mc Clure, M.O., Whitby, D., Patience, C., Gooderham, N.J., Bradshaw, A., Cheingsong-Popov, R., Weber, J.N., Davies, D.S., Cook, G.M.W., Keynes, R.J. and Weiss, R.A. (1991). Dextrin sulphate and fucoidan are potent inhibitors of HIV infection in vitro. *Antiviral Chemistry and Chemotherapy*, 2, 149–156.

[138] Feldman, S.C., Reynaldi, S., Stortz, C.A., Cerezo, A.S. and Damont, E.B. (1999). Antiviral properties of fucoidan fractions from Leathesia difformis. *Phytomedicine*, 6, 335–340.

[139] Malhotra, R.,Ward, M., Bright, H., Priest, R., Foster, M.R., Hurle, M., Blair, E. and Bird, M. (2003). Isolation and characterisation of potential respiratory syncytial virus receptor(s) on epithelial cells. *Microbes and Infection*, 5, 123–133.

[140] Araya, N., Takahashi, K., Sato, T., Nakamura, T., Sawa, C., Hasegawa, D. ando, H., Yamano, Y. 2011Fucoidan therapy decreases the proviral load in patients with human T-lymphotropic virus type-1-associated neurological disease. *Antiviral Therapy*. 16 (1), 89-98.

[141] Gonzalez, M.E., Alarcon, B. and Carrasco, L. (1987). Polysaccharides as antiviral agents, antiviral activity of carrageenan. *Antimicrobial Agents Chemotherapy*, 31, 1388-1393.

[142] Talarico, L.B., Pujol, C.A., Zibetti, R.G., Faria, P.C., Noseda, M.D., Duarte, M.E. and Damonte, E.B. (2005). The antiviral activity of sulfated polysaccharides against dengue virus is dependent on virus serotype and host cell. *Antiviral Research*,66, 103-110.

[143] Buck, C. B., Thompson, C. D., Roberts, J. N., Muller, M., Lowy, D. R. and Schiller, J. T. (2006). Carrageenan is a potent inhibitor of papillomavirus infection. *PLoS Pathogens,* 2, e69.

[144] Baba, M., Snoeck, R., Pauwels, R. and De, C. E. (1988). Sulfated polysaccharides are potent and selective inhibitors of various enveloped viruses, including herpes simplex virus, cytomegalovirus, vesicular stomatitis virus and human immunodeficiency virus. *Antimicrobial Agents and Chemotherapy*, 32, 1742-1745.

[145] C´aceres, P. J., Carlucci, M. J., Damonte, E. B., Matsuhiro, B., Z´ũniga, E. A. (2000). Carrageenans from Chilean samples of Stenogramme interrupta (Phyllophoraceae), Structural analysis and biological activity. *Phytochemistry,* 53, 81–86.

[146] Cooper, R., Charles, D., Kate, E., Fitton, J. H., John, G. and Ken, T. (2002). GFS, a preparation of Tasmanian Undaria pinnatifida is associated with healing and inhibition of reactivation of Herpes. *BMC Complementary and Alternative medicine*, 2, 11.

[147] Soares, A.R., Juliana, L. A., Thiago, M. L. S., Carlos, F. L. F., Renato, C. P., Izabel, C.D. P. P. F. and Valéria, L. T. (2007) In vitro Antiviral Effect of Meroditerpenes Isolated from the Brazilian Seaweed Stypopodium zonale (Dictyotales). *Planta medica*, 73(11), 1221-1224.

[148] Wang, H., Engchoon, V. O. and Put O. A. J. (2008). Antiviral activities of extracts from Hong Kong seaweeds. *Journal of Zhejiang University- Science* B, 9(12), 969-976.

[149] Premanathan, M., Kathiresan, K., Chandra, K. and Bajpai, S.K. (1994). In vitro anti-vaccinia virus activity of some marine plants. *Indian Journal of medical Research*, 99, 236-238.

[150] Matsuhiro, B., Ana, F., C., Elsa, B. D., Adriana, A., K., Marı´a, C. M., Enrique, G. M., Carlos A. P. and Elisa, A. Z. (2005). Structural analysis and antiviral activity of a sulfated galactan from the red seaweed Schizymenia binderi (Gigartinales, Rhodophyta). *Carbohydrate Research*, 340, 2392–2402.

[151] Duarte, M.E.R., Cauduro, J.P., Noseda, D.G., Noseda, M.D., Goncalves, A.G., Pujol, C.A., Damonte, E.B. and Cerezo, A.S. (2004). The structure of the agaran sulfate from Acanthophora spicifera (Rhodomelaceae, Ceramiales) and its antiviral activity. Relation between structure and antiviral activity in agarans. *Carbohydrate Research*, 339, 335–347.

[152] Nakashima, H., Kido, Y., Kobayashi, N., Motoki, Y., Neushul, M. and Yamamoto, normal (1987). Antiretroviral activity in a marine red alga; reverse transcriptase

inhibition by an aqueous extract of Schizymenia pacifica. *Journal of Cancer Research and Clinical Oncology*, 113, 413–416.

[153] [Bourgougnon, N., Lahaye, M., Quemener, B., Chermann, J. C., Rimbert, M., Cormaci, M., Furnari, G. and Kornprobst, J. M. (1996). Annual variation in composition andin vitro anti-HIV-1 activity of the sulfated glucuronogalactan fromSchizymenia dubyi (Rhodophyta, Gigartinales). *Journal of Applied Phycology*, 8, 155–161.

[154] Boisson-Vidal, C., Haroun, F., Ellouali, M., Blondin, C., Fischer, A. M., De Agostini, A. and Jozefonvicz, J. (1995). Biological activities of polysaccharide from marine algae. *Drugs of the Future*, 20, 1237–1249.

[155] Cordeiro, R.A., Gomes, V.M., Carvalho, A.F.U. and Melo, V.M.M. (2006). Effect of proteins from the red seaweed Hypnea musciformis (Wulfen) lamouroux on the growth of human pathogen yeasts. *Brazilian Archives of Biology and Technology*, 49(6), 915-921.

[156] Khanzada, A.K., Shaikh, W., Kazi, T.G., Kabir, S. and Soofia, S. (2007).Antifungal activity, elemental analysis and determination of total protein of seaweed, Solieria robusta (Greville) Kylin from the coast of Karachi. *Pakistan Journal of Botany*, 39(3), 931-937.

[157] Nightingale, V.D.R., Balasubramanian, R. and Anantharaman, P. (2009). Antifungal activity of seaweeds from Vellar and Uppanar estuaries. *Ecology Environment and Conservation,* 15(3), 465-468.

[158] Pesando, D. and Caram, B. (1984). Screening of marine algae from the french Mediterranean coast for antibacterial and antifungal activity. *Botany*, 27, 3281-3386.

[159] Kamenarska, Z., Julia, S., Hristo, N., Kamen, S., Iva, T., Stefka, D.K. and Simeon, P. (2009). Antibacterial, antiviral and cytotoxic activities of some red and brown seaweeds from the Black Sea. *Botanica Marina*, 52(1), 80–86.

[160] Devi, K.P., Natarajan, S., Periyanaina. K. and Shanmugaiahthevar K. P. (2008). Bioprotective properties of seaweeds, In vitro evaluation of antioxidant activity and antimicrobial activity against food borne bacteria in relation to polyphenolic content. *BMC Complementary and Alternative medicine*, 8, 38.

[161] Watson, S.B. and Cruz-Rivera, E. (2003). Algal chemical ecology, an introduction to the special issue. *Phycologia,* 42, 319–323.

[162] Paul, V.J. and Puglisi, M.P. (2004). Chemical mediation of interactions among marine organisms. *Nat. Prod. Report,* 21, 189-209.

[163] Weinberger, F. and Friedlander, M. (2000). Response of gracilaria conferta (rhodophyta) to oligoagars results in defense against agar-degrading epiphytes. *Journal of Phycology*, 36, 1079–1086.

[164] Kato, I., Tominaga, S., Matsuura, A., Yoshii, Y., Shirai, M., Kobayashi, S. (1990). A comparative case-control study of colorectal cancer and adenoma. *Japanese Journal of Cancer Research*, 81 (11), 1101-1108.

In: Seaweed
Editor: Vitor H. Pomin

ISBN 978-1-61470-878-0
© 2012 Nova Science Publishers, Inc.

Chapter 5

SEAWEEDS, FOOD, AND INDUSTRIAL PRODUCTS AND NUTRITION

Maha Ahmed Mohamed Abdallah[*]
National Institute of Oceanography and Fisheries, Egypt

ABSTRACT

Algae, as processed and unprocessed food, have a commercial value of several billion dollars annually. Approximately 500 species are used as food or food products for humans and about 160 species are valuable commercially.

Commercially available varieties of marine macroalgae are commonly referred to as ''seaweeds''. Seaweed is suitable for human and animal feed, as well as for fertilizer, fungicides, herbicides, and Phycocolloids (as Chlorophyta) are commonly used as food due to high contents of vitamins and minerals, Phaeophyta are typical suppliers of alginic acid. Rhodophyta are responsible to produce agar and carrageenan. However, there is a worldwide interest for macroalgal components, since they are used in medicine and in pharmacology for their antimicrobial, antiviral, antitumor, anticoagulant and fibrinolytic properties. Seaweeds contain high amounts of carbohydrates, protein and minerals. Because of their low fat contents and their proteins and carbohydrates, which cannot be entirely digested by human intestinal enzymes, they contribute few calories to the diet. The protein content of seaweed varieties varies greatly and demonstrates a dependence on such factors as season and environmental growth conditions. Thus edible marine seaweeds may be an important source of minerals, since some of these trace elements are lacking or very minor in land vegetables. About 221 seaweeds are utilized commercially worldwide of which 65% are used as human food. Most recently seaweeds have been utilized in Japan as raw materials in the manufacture of many seaweed food products, such as jam, cheese, wine, tea, soup and noodles. Currently, human consumption of green algae (5%), brown algae (66%) and red algae (33%) is high in Asia. However demand for seaweed as food has now also extended to other parts of the world.

Seaweeds have a wide range in mineral content, not found in edible land plants, is related to factors such as seaweed phylum, geographical origin and seasonal environmental and physiological variations. The mineral fraction of some seaweed even accounts for up to 40% of dry matter.

---
[*] E-mail: mahaahmed72007@yahoo.com.

Algal products are used in the preparation or manufacture of many nonfood products. The agars, carrageenans, and alginates, collectively termed hycocolloids, are a major source of industrially important algal products. Agar and agaroses are used in medical and biological sciences for culture media and for gel electrophoresis. Agars are also used in many other products, including ion-exchange and affinity chromatography, pharmaceutical products, and fruit fly foods. Carrageenans are used as binders and thickeners in a wide variety of pastes, lotions, and water-based paints. Alginates are used to bind textile printing dyes, to stabilize paper products during production, to coat the surfaces of welding rods, to serve as binders and thickeners in numerous pharmaceutical products, and to act as binders in animal feed products.

1. INTRODUCTION

Seaweeds are floating submerged plants of shallow marine meadow, having salt tolerance, because the osmolarity of cytoplasm is adjusted to match the osmolarity of the seawater, so that desiccation of the plant does not occur. Seaweeds (with 200,000 species) belonging to class Thallophyta of the plant kingdom, occur in aquatic ecosystems all over the world. Ecologically, seaweeds are worldwide in distribution, occurring on the land surface, on all types of soil excepting sandy deserts, on permanent ice, and snow fields. However, the major centers of distribution are in the waters. These are the primary and major organic producers. Besides occurring in the form of microscopic planktons, these occur as microscopic and macroscopic forms along the sea shores. In the Ocean, the littoral habitat is occupied by few plants other than seaweed. Hence, these are known to occur practically in every habitable environment on earth. The greatest diversity is seen on rocky seashores and coral reefs. Most of the seaweed is photosynthetic. They are the major primary producers of organic compounds and play a key role in food chains. However, there are a few parasitic forms, which are not able to photosynthesize but are classified as seaweed because of their close resemblance to photosynthetic forms. The reproductive structures of seaweed lack sterile cells and do not form embryos [1].

The importance of seaweeds for human consumption is well known in many countries, especially in Asian countries, such as Malaysia, Indonesia, Korea, Australia, Japan and Singapore where algae are used in the form of salads, soups, jellies and in vinegar dishes. The species used as food includes *Caulerpa* sp., *Codium* sp., *Hydroclathum* sp., *Sargassum* sp., *Porphyra* sp. and *Laurencia* sp. and *Enteromorpha, Gracilaria, Sargassum, Padina, Dictyota* are used as a feed for cattle and poultry. In many coastal environments, seaweeds are a source of diatomaceous earth and hydrocolloids, such as alginic acid, carrageenan, and agar.

Seaweeds are extremely important not only ecologically but also phylogenetically. It is thought that all the major groups of animals and plants originated in the sea and that, even today, this is where one can find representatives of many ancient evolutionary lineages. Thus, if we are to understand the diversity and the phylogeny of the plant world, it is of fundamental importance, indeed essential, to investigate seaweed [2]. Lately, a great deal of interest has been centered on seaweed as potential candidates for bioremediation of polluted water bodies [3,4]. The groups of seaweed with such potential are Cyanobacteria (blue green seaweed), microseaweed (generally green) and macroseaweed. The term Cyanobacteria is often preferred since the members show their affinity with bacteria (cell wall composition and prokaryotic nature) and identify themselves as microbes. These seaweeds have great capacity to accumulate dissolved metals and hold an important position in *'green clean'* technology. The technology deals with metal ion uptake that has been suggested as taking place in two stages. The first stage is rapid and reversible reaction called physical adsorption or ion-exchange, which occurs at the cell surface. The subsequent stage is slower and called chemisorptions. Dead cells accumulate heavy metal ions to the same or greater extent than the living cells. This kind of adsorption is called biosorption [5]. During past 10 years marine organisms have provided a large number of new natural products. Interesting compounds have mainly been derived from microorganisms such as algae, sponges, ascidians, corals and bryozoans. The number of secondary metabolites from marine organisms is smaller, but rapidly increasing [6]. Many of the marine algae constitute a major part of the diet in Eastern countries which are known for the longevity of life span.

2. SEAWEEDS AS FOOD PRODUCTS

Algae are used as food delicacies or supplements in some maritime countries but have probably never become a staple item. Approximately 500 species are used as food or food products for humans, and about 160 species are valuable commercially [7]. Commercially available varieties of marine macroseaweed are commonly referred to as "seaweeds". Seaweed is suitable for human and animal feed, as well as for fertilizer, fungicides, herbicides, and Phycocolloids (as Chlorophyta) are commonly used as food due to high contents of many vitamins and minerals, and some are said to contain substantial amounts of protein. In Japan and China, Wales, Iceland, and the Canadian Maritimes, selected macroscopic marine algae are served as delicacies. Algae, both macroscopic and microscopic, are considered as potential inexpensive and effective dietary supplements for the malnourished. The salt tolerant green alga *Dunaliella* is known to produce large amount of [β-carotene [8]. Phaeophyta are typical suppliers of alginic acid. Rhodophyta are responsible to produce agar and carragenan [9,10]. Unfortunately, we cannot digest many of the complex carbohydrates in the plants, but this may be an advantage for those counting calories. Seaweeds can add variety and taste to bland foods and may be used to wrap such foods as rice. The harvesting and marketing of edible seaweed is a growing business. There is no need for planting, fertilizing, weeding, or tilling. The ocean takes care of everything, though oil and sewage pollutants can spoil the best of harvests. Seaweeds are harvested by hand, rinsed in water, and dried on lines or nets. Connoisseurs use seaweed in salads, soups, omelets, casseroles, and sandwiches. Figure (1) illustrated some kinds of seaweeds used as food in different places of the world.

- *Ulva* is not called sea lettuce for nothing. It can be eaten fresh in salads.
- *Limu,* including *limu 'ele 'ele* (the green *Enteromorpha prolifera*) and *limu manauea* (the red *Gracilaria monopifolia),* are beloved by the Hawaiians.
- *Purple laver* (a species of *Porphyra,* a red alga) prepared in various ways is eaten in some parts of the British Isles. Washed and boiled, it is formed into flat cakes, rolled in oatmeal, and fried; it is then called *laverbread*. Purple laver is also eaten as a hot vegetable or fried with bacon.

- *Irish moss (Chondrus,* a red alga and a source of carrageenan) is dried and used in preparing *blancmange* and other desserts in eastern Canada, New England, and parts of northern Europe.
- *Palmana,* another red alga, is dried and eaten, mostly by those living along the Atlantic coast of Canada and northern Europe. Called *dulse,* it is sometimes still used in making bread and several types of desserts. For those on a diet, *dulse* can also be chewed like tobacco (of course it's nicotine-free).
- Species of *Laminaria* and *Alaria,* a kelp called kombu, are dried and shredded, then prepared in various ways. They are even used to make tea and candy.
- *Undaria pinnatifida,* or *wakame,* is edible kelp best when fresh or cooked for a very short time. It is a good source of protein, iron, calcium, sodium and other minerals and vitamins. *Wakame* is leafy and mild in flavor, it turns green after soaking. It is common to Japanese waters. Traditionally added to miso soup, *wakame* is also good with other vegetables or in salads, stir-fry dishes, and rice dishes.
- *Porphyra,* a red alga, is used to make thin sheets *of nori,* widely used in soups, and for wrapping *sushi,* boiled rice stuffed with bits of raw fish, sea urchin roe, or other ingredients.
- *Sargassum fusiforme*, a brown seaweed, called (*Hijiki*) - Found primarily in the Far East, contains the most calcium of any of the sea vegetables, 1400mg/100gr dry weight (compared to milk with 100mg/100gr.) In its natural state it is very tough; after harvesting it is dried, steamed and dried some more. When cooked, it rehydrates and expands about five times its dry volume.

Figure 1. Different kinds of seaweeds used as a food.

However, there is a worldwide interest for macroalgal components, since they are used in medicine and in pharmacology for their antimicrobial, antiviral, antitumor, anticoagulant and fibrinolytic properties [11-14]. Furthermore, they are used for animal nutrition as feed or as fertilizers and soil conditioning agents. Fresh seaweeds have been used directly as foodstuff in the Asian countries for centuries and are considered under-exploited resources [15].

3. Nutritional Value of Seaweeds

About 221 seaweeds are utilized commercially worldwide of which 65% are used as human food [16]. Most recently seaweeds have been utilized in Japan as raw materials in the manufacture of many seaweed food products, such as jam, cheese, wine, tea, soup and noodles [17]. Currently, human consumption of green seaweeds (5%), brown seaweeds (66%) and red seaweeds (33%) is high in Asia [18]. However demand for seaweed as food has now also extended to other parts of the world [19].

People in the western societies use algal products indirectly in commercially prepared foods, especially puddings, pies, ice cream, etc. People in oriental and Pacific island societies consume seaweeds directly in soups, rice dishes, and meat dishes. Hawaiians were eating about 75 species of seaweeds in the 1800s, and during the 1980s more than 50 species were still being consumed [7]. The use of seaweeds for food is not limited to humans, however. Phytoplanktons are used for feeding a variety of animals, especially shrimp and shellfish.

The different species consumed represent a great nutritional value as source of minerals, protein, carbohydrates and vitamins [20]. Because of their low fat contents and their proteins and carbohydrates which cannot be entirely digested by human intestinal enzymes seaweeds contributed few calories to the diet [21].

3.1. Minerals

Seaweeds draw from the sea an incomparable wealth of mineral elements, macro-elements and trace elements. Seaweeds have a wide range in mineral content, not found in edible land plants, and are related to factors such as seaweed phylum, geographical origin and seasonal environmental and physiological variations [22]. The mineral fraction of some seaweed even accounts for up to 40% of dry matter [23]; however, in some cases the mineral content of the seaweeds is recorded even higher than that of land plants and animal products [24]. Evaluation of minerals in any edible seaweed is important from both the nutritional and the toxicological point of view, Table 1 shows concentration of minerals and trace elements of some marine seaweed.

Marine seaweeds constitute potential sources of dietary fiber that differ chemically and physiochemical from those land plants and thus may have other physiological effect on man [25]. The green seaweeds (*Enteromorpha compressa, Codium fragile, Caulerpa racemosa* and *Ulva lactuca*), Red seaweeds (, *Pterocladia capellaceand Hypnea musciformis*) and brown seaweeds (*Padina pavonia* and *Colpomenia Sinuosa*) has great potential for commercial exploitation because of its abundant and varied chemical composition, quality and concentration of basic nutriments for other living organism.

Table 1. Concentrations ± Standard Division of minerals and trace elements
(mg/100g dry weight) in some marine seaweeds

Mineral and trace elements		Ulva Lactuca	Pterocladia capillacea	Padina pavonia	Caulerpa racemosa	Codium fragile	Hupnea musciformis	Colpomenia sinuosa
Minerals	Na	925±16.1	1140±31.2	855±21.1	1441±30.1	2948±26.1	8355±87.3	1441±30.2
	K	650±86.6	1000±31.2	750±39.6	9700±93.6	1073±41.6	3088±35.2	540±18.1
	Ca	3216±69.6	5360±20.2	7504±9.3	6513±23.6	1503±19.3	10521±26.6	10312±77.2
	Mg	13500±110.4	15370±158.1	17069±126.8	204±15.4	1014±65.8	1081±22.9	968±22.6
Total Mineral		18292	26178	22870	17858	6538	23045	13261
trace elements	Fe	119.7±24.2	28.7±11.3	56.8±13.4	53.88±12.5	18.58±8.2	131.45±18.2	102.9±15.3
	Mn	2.26±1.4	4.41±1.3	2.47±1.2	4.03±1.4	1.155±1.2	9.69±2.4	8.163±1.3
	Cu	0.34±0.007	0.28±0.05	0.21±0.1	0.04±0.07	0.21±0.1	0.675±0.11	0.63±0.09
	Zn	3.19±1.6	1.78±0.13	1.81±1.02	1.52±1.1	0.625±0.17	2.685±0.78	2.14±0.14
Total trace elements		125.5	61.3	35.1	59.45	20.57	144.5	113.83
Total contents		18417	26239	22905	105217.45	6558.57	23189.5	13374.83

Seaweeds contain high amounts of carbohydrates, protein and minerals [20]. Because of their low fat contents and their proteins and carbohydrates, which cannot be entirely digested by human intestinal enzymes, they contribute few calories to the diet [21]. The protein content of seaweed varieties varies greatly and demonstrates a dependence on such factors as season and environmental growth conditions. Thus edible marine seaweeds may be an important source of minerals [17], since some of these trace elements are lacking or very minor in land vegetables. The mineral fraction of some seaweeds accounts for up to 36% of dry matter.

Among all organisms, brown algae are the ones that accumulate and stock iodine with the highest efficiency. Certain species of laminaria are capable of maintaining an iodide concentration 100,000 times higher to that of sea water. Algae emit volatile iodinated organic compounds (VOC) and molecular iodine. This phenomenon highlights the importance of brown algae in the iodine biogeochemical cycle and explains their impact on atmospheric chemistry. Seaweeds, eaten regularly, are the best natural food sources of biomolecular dietary iodine, no land plants are reliable sources of dietary iodine. For comparison, someone would have to eat about 40 lb. of fresh vegetables and/or fruits to get as much iodine as he would from 1 gram of our whole leaf kelp. The brown seaweeds have traditionally been used for treating thyroid goiter [26]. The link between iodine and the thyroid hormones was established soon afterwards. Despite the fact that this raw material source of iodine has been abandoned (except in China), *Fucus vesiculosus* is still registered in the European pharmacopoeia for its high iodine content. As concerns iodine, laminaria is the main source as it contains 1500 to 8000 ppm dry weight (fucals contain 500 to 1000 ppm dry weight of iodine).

Seaweeds are also one of the most important vegetable sources of calcium. Calcium content may be as high as 7 % of the dry weight in macroseaweed and up to 25 to 34 % in the chalky seaweed lithotamne. Seaweed consumption may thus be useful in the case of expectant mothers, adolescants and elderly that all exposed to a risk of calcium deficiency. Despite high contents, the linkage of certain minerals with anionic polysaccharides (alginate, agar or carrageenan) might limit their absorption. For instance, the strong affinity of divalent cations (particularly calcium) for carboxylic polysaccharides (alginates) probably limits their availability. In contrast, the weakness of the linkages between polysaccharides and iodine allows rapid release of this element [27]. Sea vegetables contain 10 to 20 times the minerals and vitamins of land vegetables. Gram for gram, they are higher in vitamins and minerals than any other class of food. The minerals are available in chelated, colloidal forms that make them especially available to the bodies of humans and animals, a concept known as "bioavailability."

3.2. Polysaccharides and Dietary Fibers

Seaweeds contain large amounts of polysaccharides, notably cell wall structural polysaccharides that are extruded by the hydrocolloid industry: alginate from brown seaweeds, carrageenans and agar from red seaweeds. Other minor polysaccharides are found in the cell wall: fucoidans (from brown seaweeds), xylans (from certain red and green seaweeds), ulvans in green seaweeds. Seaweeds also contain storage polysaccharides, notably laminarin (β-1,3-glucan) in brown seaweeds and floridean starch (amylopectin- like glucan) in red seaweeds. When faced with the human intestinal bacteria, most of these polysaccharides (agars, carrageenans, ulvans and fucoidans), are not digested by humans and therefore can be regarded as dietary fibers [28,29].

Marine algae appear to be good sources of fibers presenting great chemical, physico-chemical and rheological diversities that may be beneficial in nutrition. Water-soluble and water-insoluble fibers have been associated with different physiological effects. Many viscous soluble polysaccharides (pectins, guar gum, etc.) have been correlated with hypocholesterolemic and hypoglycemic effects, whereas water-insoluble polysaccharides (cellulose) are mainly associated with a decrease in digestive tract transit time [30]. Some seaweed polysaccharides are used by the food industry as texture modifiers because of their high viscosity and gelling properties. The dietary fiber which constitutes 25–75% of the dry weight of marine algae and represents their major component is primarily soluble fiber. Nowadays, dietary fiber from different sources is known to decrease the risk of coronary heart disease, mainly due to its characteristics of dispersibility in water (water-holding capacity), viscosity, binding ability, absorptive capacity, faecal bulking capacity and fermentability in the alimentary canal. Indigestible viscous seaweed polysaccharides such as alginates, carrageenans and funorans, which are capable of forming ionic colloids, have shown positive effects on serum lipid levels in rats. The capacity of seaweed polysaccharides to lower serum cholesterol levels seems to be due to their ability to disperse in water, retain cholesterol and related physiologically active compounds and inhibit lipid absorption in the gastrointestinal tract. Some edible marine algae (*Undaria pinnatifida*, *Hijiki fusiformis*, *Himanthalia elongata*, *Eisenia bicyclis*, *Ulva lactuca*, *Enteromorpha* spp, and *Porphyra tenera*) are particularly rich in dietary fibers with total dietary fiber contents varying between 32.7 and 74.6% (on a dry weight basis) of which 51.6 to 85.0% are water soluble [29].

Among polysaccharides, fucoïdans were particularly studied as they showed interesting biological activities (anti-thrombotic, anti-coagulant, anticancer, anti-proliferative, anti-viral, and anti-complementary agent, anti-inflammatory). These properties open up a wide field of potential therapeutic applications, some of which are already the subject of patents concerning notably the anti-coagulant and anti-thrombotic properties [31,32]. As for xylans and laminarans, they are completely and rapidly degraded by intestinal bacteria, alginates are only partly degraded and lead to a substantial production of short chain fatty acids.

3.3. Protein

The protein content of seaweeds differs according to species. Proteins represent 30 percent to 40 percent of the dry weight of red seaweeds. Green seaweeds also have a high protein content of about 20 percent of the dry weight; all types of seaweeds are sources of sufficient protein. the protein fraction of brown seaweeds is low (3-15% of the dry weight) compared with that of the green or red seaweeds (10-47% of the dry weight) Table (2). Except for the species Undaria pinnatifda (wakame) which has a protein level between 11 and 24% (dry weight), most brown seaweeds industrially exploited (Laminaria digitata, Ascophyllum nodosum, Fucus vesiculosus and Himanthalia elongata) have a protein content lower than 15% (by dry weight). In some green seaweeds such as the species belonging to the genus Ulva, the protein content can represent between 10 and 26% (dry weight) of the plant. For instance, the species Ulva pertusa, which is frequently consumed under the name of "aonori" by the Japanese people, has a high protein level between 20 and 26% (dry product) [35]. In some red seaweed, such as Palmaria palmata (dulse) and Porphyra tenera (nori), proteins can represent up to 35 and 47% of the dry matter, respectively. These levels are comparable to those found in high-protein vegetables such as soybeans (in which proteins represents 35 % of the dry mass).

The protein content of marine algae also depends on the seasonal period. An annual monitoring of protein level from Palmaria palmata (Dulse) collected on the French Atlantic

Table 2. Protein levels in some seaweeds used by the food industry

Seaweed species	Palmaria palmata	Pterocladia capillacea	Porhyra tenera	Humea muscoformis	Ulva Lactuca	Ulva Pertusa	Caulerpa racemosa	Codium fragile	Padina pavonia	Laminaria digitata	Ascophullum nodosum	Colpomen ia sinuosa
	\multicolumn{4}{}{Red algae}		Green algae				Brown algae			
Protein (in % of dry weight)	8-35 [33,34]	10-14 [40]	33-47 [35]	23-48 [41]	10-21 [36, 3, 40]	20-26 [38, 17]	18-42 [41]	12-24 [41]	6-12 [40]	8-15 [39]	3-15 [35, 14]	15-33 [41]

coast showed that the protein content of this algae can vary between 9 and 25% (dry weight) [13]. Red seaweeds are on par with the protein content of legumes, and spirulina (a type of blue-green algae) provides a whopping 62 percent protein, according to the University of Maryland Medical Center. The protein levels of Ulva spp. are in the range 15-20 % of the dry weight. Except for Undaria pinnatifida, which contains 11-24 % proteins, other brown seaweeds (Laminaria digitata, Ascophyllum nodosum, Fucus vesiculosus and Himanthalia elongata) have low protein content. Spirulin, freshwater microalgae, is well-known for its high protein content (70% of dry matter). Proteins which are the building blocks of living world also have a hidden capacity of modulating physiology by means of small peptides which are formed either during the processing or during *in vivo* digestion. Small peptides with biological function or physiological effects are called bioactive peptides. These can be obtained from different sources like microorganisms, plants, animals etc., [42]. Dairy products are important sources of bioactive peptides [43]. Active peptides are potential modulator of various regulatory processes in the living system. These are immunomodulatory peptides, cytomodulatory peptides, Angiotensin-I-Converting Enzyme (ACE)-inhibitory peptides, opoid peptides, mineral binding peptides.

Comparable results are observed with proteins from green and brown seaweeds. However, for these seaweeds, the high phenolic content might limit protein availability in vivo and thus moderate in vitro figures. This situation is probably not found for the green and red seaweeds, which possess low levels of phenols and higher protein content.

3.4. Amino Acid Composition

The amino acid composition of seaweeds has been frequently studied and compared to that of other foods. For most seaweed, aspartic and glutamic acids constitute together a large part of the amino acid fraction. Munda (1977) [44] reported that these two amino acids can represent between 22 and 44% of the total amino acids. The protein in algae contains all essential amino acids (EAA) and all EAA are available throughout the year although seasonal variations in their concentrations are known to occur [34]. For example, the proportion of EAA is 45–49% in Hizikia sp. and Eisenia bicyclis (Arame). In both these brown algae varieties, tryptophan is the first limiting EAA, followed by lysine [45]. In the green seaweeds, the level of aspartic and glutamic acids can represent up to 26 and 32% of the total amino acids of the species Ulva rigida and Ulva rotundata, respectively [30]. The amino acid composition of Ulva pertusa shows that valine, leucine and lysine are the main essential amino acids present. Histidine, which is an essential amino acid for children, is present at a similar level to leguminous and egg proteins. However, aspartic and glutamic acids levels seem to be lower in red seaweed species such as Palmaria palmata and Porphyra tenera (14 and 19% of the total amino acids, respectively) [46,31]. The EAA contents of some species (e.g., Porphyra sp.) can be compared with those of soy and egg protein [13,34]. In addition, high concentrations of arginine, aspartic acid and glutamic acid are found in many seaweed species [13]. The content of taurine in red algae varieties studied was especially noteworthy since this sulphonic acid is not often found in conventional European food proteins. Generally, the most abundant amino acid score in seaweed varieties investigated were aspartic acid and glutamic acid.

3.5. Lipids and Fatty Acids

Lipids represent only 1-5 % of algal dry matter and show an interesting polyunsaturated fatty acid composition particularly regarding with omega 3 and omega 6 acids which play a role in the prevention of cardio-vascular diseases, osteoarthritis and diabetes. Lipids in herbivores originate from both lipoidal and non-lipoidal precursors formed by, and assimilated from, relative low-lipid primary producers [47]. Actively growing and dividing algae, including those from polar regions, contain 10-20% of their dry weight as total lipid, which are mainly polar glycolipids located in the cells' thylakoid membranes, these glycolipids being rich in (n-3) polyunsaturated fatty acids (PUFA) [48,49]. Algae may have variable amounts of neutral lipids, mainly triacylglycerols (TAG), present as membrane-bound oil droplets in their cytoplasm, with nutrient limitation generally favouring neutral lipid storage [50,51]. Studies of fatty acids in seaweeds have investigated their seasonal variation [52] differences among different plant tissues [53], the effect of growth conditions [54], the compositional characterization of a genus [55] and applications to aquaculture [56]. The fatty acids of seaweeds generally have linear chains, an even number of carbon atoms, and one or more double bonds [57]. In particular, seaweeds can be a source of essential fatty acids such as eicosapentaenoic acid, C20:5ω3 [58]; ω3 fatty acids such as C20:5ω3 are thought to reduce the risk of heart disease, thrombosis and atherosclerosis [59]. It has also been reported that the fatty acids of certain seaweeds have antiviral activity. The green algae show interesting levels of alpha linolenic acid (ω3 C18:3). The red and brown algae are particularly rich in fatty acids with 20 carbon atoms: eicosapentanoïc acid (EPA, ω3 C20 :5) and arachidonic acid (AA , ω6 C20 :4) [60].

Spirulin provides an interesting source of gamma linolenic acid (GLA) (20 to 25 % of the total lipidic fraction), which is a precursor of prostaglandins, leucotriens and thromboxans involved in the modulation of immunological, inflammatory and cardio-vascular responses. Spirulin thus provides an interesting alternative to the other known sources of GLA : onagre oil, blackcurrant pips and borage oil [60,61]. For quantitative determination of fatty acids in foods by gas chromatography it is usual either for crude extracts of samples to be derivatized to form fatty acid methyl esters, or for one-step transesterification with methanolic HCl to be followed by extraction with an apolar solvent [62].

Besides fatty acids, unsaponifiable fraction of seaweeds was found to contain carotenoids (such as β-carotene, lutein and violaxanthin in red and green seaweeds, fucoxanthin in brown seaweeds), tocopherols, sterols (such as fucosterol in brown seaweeds), and terpenoids [63, 64]. Lipidic extracts of some edible seaweed showed antioxidant activity and a synergistic effect with the tocopherol [65].

3.6. Micronutrients

For the past years, seaweeds have been a popular food additive in almost all parts of the world. According to studies, seaweeds are rich in vitamins A1, B1, B2, B6, B12, C, E and K which can keep the body healthy and strong enough to fight many types of diseases. In combination with other food, the red and brown algae contribute to satisfying the recommended daily doses of vitamins especially β-carotene and vitamin B2.

3.6.1. Vitamin A

Red and brown seaweeds contain carotenoids, which can be converted to active vitamin A in the body. Each 100 g of brown seaweed provide 10 mg active vitamin A. This is about the amount of vitamin A provided by an ear of the yellow sweet corn.

3.6.2. Vitamin B12

Many seaweed species are rich in B-complex vitamins, and unlike most plants, some seaweeds, notably Dulse and Nori, are high in B-12. Therefore, plants such as grains, fruits, and vegetables generally do not contain vitamin B-12. Seaweed is an exception. However, the vitamin B-12 content in seaweed is lower than that of most animal foods. The sheets of dried seaweed used to wrap sushi are nori, which is also available dried for seasoning in soups and other dishes. Although many vegans regard nori as an excellent source of B-12, the National Health Association states it may not be sufficient. Vitamin B-12 which is particularly recommended in the treatment of the effects of ageing, of chronic fatigue syndrome (CFS) and anemia. Among the edible seaweeds, Spirulin is richest in vitamin B12. Daily ingestion of one game of Spirulin would be enough to meet the daily requirements in B12 [66].

3.6.3. Vitamin C

Seaweeds provide a worthwhile source of vitamin C [67]. The levels of Vitamin C average between 500 to 3000 mg/kg of dry matter for the green and brown algae (a level comparable with that of parsley, lackcurrant, and peppers), whereas the red algae contains vitamin C levels of around 100 to 800 mg/ kg. Vitamin C is of interest for many reasons: it strengthens the immune defence system, activates the intestinal absorption of iron, controls the formation of conjunctive tissue and the protidic matrix of bony tissue, and also acts in trapping free radicals and regenerates Vitamin E.

3.6.4. Vitamin E

Due to its antioxidant activity, vitamin E inhibits the oxidation of the low-density lipoproteins. It also plays an important part in the arachidonic acid chain by inhibiting the formation of prostaglandins and thromboxan. The brown seaweeds contain higher levels of vitamin E than green and red seaweeds. Among the brown algae, the highest levels are observed in the Fucaceae, *Ascophyllum* and *Fucus sp.*, which contain between 200 and 600 mg of tocopherols / kg of dry matter. Brown algae contain alpha, beta and gamma tocopherol while the green and red algae only contain the alpha tocopherol. It was shown that the gamma and alpha tocopherols increase the production of nitric oxide and nitric oxide synthase activity (cNOS) and also play an important role in the prevention of cardio-vascular disease [68].

3.6.5. Vitamin K

Seaweed is high in vitamin K, which regulates blood clotting and coagulation. Drugs.com cautions that excessive seaweed consumption while taking the anticoagulant drug warfarin may cause a change in the international normalized ratio, or INR---how long it takes blood to clot.

4. INDUSTRIAL SEAWEED PRODUCTS

Industrial utilization of seaweed is mostly centered on the extraction of phycocolloids (marine hydrocolloids), and, to a much lesser extent, certain fine biochemicals. Fermentation and pyrolysis and the use of seaweed as biofuels are not an option on an industrial scale at present, but are possible options for the future, particularly as conventional fossil fuels run out. Seaweed products are used in the preparation or manufacture of many nonfood products. The agars, carrageenans, and alginates, collectively termed hycocolloids, are a major source of industrially important algal products [69]. Carrageenans are obtained from different species of red algae and used as binders and thickeners in a wide variety of pastes, lotions, and water-based paints [70]. Agar and agaroses are used in medical and biological sciences for culture media and for gel electrophoresis. Agars are also used in many other products, including ion-exchange and affinity chromatography, pharmaceutical products, and fruit fly foods [71]. Alginates are obtained from species of brown algae and used to bind textile printing dyes, to stabilize paper products during production, to coat the surfaces of welding rods, to serve as binders and thickeners in numerous pharmaceutical products, and to act as binders in animal feed products [72].

Seaweeds are being used in cosmetics, and as organic fertilizers. There is a long history of coastal people using seaweeds, especially the large brown seaweeds, to fertilize nearby land. Wet seaweed is heavy so it was not usually carried very far inland, in France; brown seaweed is regularly collected by farmers and used on fields up to a kilometer inland. For example in a more tropical climate like the Philippines, large quantities of *Sargassum* have been collected, used wet locally, but also sun dried and transported to other areas. In Puerto Madryn (Argentina), large quantities of green seaweeds are cast ashore every summer and interfere with recreational uses of beaches. Part of this algal mass has been composted and then used in trials for growing tomato plants in various types of soil. In all cases, the addition of the compost increased water holding capacity and plant growth, so composting simultaneously solved environmental pollution problems and produced a useful organic fertilizer. They have the potential to be much more widely used as a source of long- and short-chained biochemicals with medicinal and industrial uses. Marine algae may also be used as energy-collectors and potentially useful substances may be extracted by fermentation and pyrolysis. Seaweed extracts appear in the oddest of places; we almost certainly have eaten some sort of seaweed extract in the last 24 hours as many processed foods such as chocolate milk, yoghurts, health drinks, and even the highest-quality German beers contain seaweed polysaccharides such as agars, carrageenans and alginates! Seaweed baths have been popular in Ireland and Britain since Edwardian times, and seaweed wraps and treatments have become more popular in the last few years. A recent innovation is the apparent incorporation of seaweed into a fiber, although some commentators have cast doubt on the presence of any seaweed in fabrics advertised as such.

Diatomite is a rock-like deposit (fossilized diatom frustules) on the floor of the sea formed from indestructible siliceous frustules of the past diatoms over many millions of years. These deposits are mined in several parts of the world to obtain the diatomaceous earth, which is put to several commercial uses. It is highly porous, insoluble with abrasive qualities and is chemically inert. It is fireproof and highly absorbent. It is used for clearing solvents and as filters for oils and other solutions, and it is used as a blocking agent in the manufacture of plastics, as a mild abrasive in toothpastes and household cleaners, as a filtration compound for a wide variety of liquids, and as a binder for paints and pastes.

Seaweeds can be used to reduce the nitrogen and phosphorus content of effluents from sewage treatments. Many types of seaweed have a preference to take up ammonium as the form of nitrogen for their growth and ammonium is the prevalent form of nitrogen in most domestic and agricultural wastewater. Another important feature of many types of seaweed is their ability to take up more phosphorus than they require for maximum growth. It would be preferable to use seaweeds that have some commercial value, but these do not necessarily have the ability to withstand the conditions encountered in the processing of the wastewater. There is a need for the seaweed to be able to tolerate a wide variation in salinity because of the dilution of salinity by the sewage or wastewater. Intertidal and estuarine species are the most tolerant, especially green seaweeds such as species of *Enteromorpha* and *Monostroma*.

The accumulation of heavy metals (such as copper, nickel, lead, zinc and cadmium) by seaweeds especially of the large brown seaweeds, varied according to their geographic source and sometimes to their proximity to industrial waste outlets. From these studies came the idea of using seaweeds as biological indicators of heavy metal pollution, either from natural sources or from activities such as mining or disposal of industrial wastes. This has been successfully implemented using brown seaweeds such as *Sargassum*, *Laminaria* and *Ecklonia*, and the green seaweeds *Ulva* and *Enteromorpha*. A further extension of this ability of some seaweed to take up heavy metals is to use them to remove heavy metals in cleaning up wastewater. While there have been many small-scale trials, it is difficult to find reports of actual implementation on a large scale. Milled, dried species of the brown seaweeds *Ecklonia*, *Macrocystis* and *Laminaria* were able to adsorb copper, zinc and cadmium ions from solution. In another laboratory-scale trial, *Ecklonia maxima*, *Lessonia flavicans* and *Durvillaea potatorum* adsorbed copper, nickel, lead, zinc and cadmium ions, though to varying extents depending on the seaweed type and metal ion concentration.

8. MEDICAL IMPORTANCE

Seaweed are rarely used directly as cures for diseases, but there are some examples of antibiotic activity, vermifuge activity, antitumor activity, goiter treatment antipyretics, cough remedies, wound healing compounds, thirst quenching remedies, and for treatment of gout, gall stones, hypertension, diarrhoea, constipation, dysentry, burns, ulcers, skin diseases, lung diseases, and semen discharge [73,74].

Algae have also been used against a wide range of ailments. The role of algae and algal products in medicine is an area of knowledge restricted to coastal communities. Curative powers from selected algae, particularly the tropical and subtropical marine forms, are reported since ancient times. Thus, the use of algae for cures or preventives against diverse medical problems is widespread in certain regions. The kelps (Laminariales) and sargassums (Fucales) are rich source of iodine, and are used in treatment of goitre. The algal product, agar, is also used against prolapsed stomach. In Europe and North America, many claims have been made for the effectiveness of seaweeds on human health. It has been suggested, amongst other things, that seaweeds have curative powers for tuberculosis, arthritis, colds and influenza, worm infestations, and may even improve one's attractiveness to the opposite sex.

Digenea (Ceramiales; Rhodophyta) produces an effective vermifugal agent (kainic acid). Recently, aqueous extracts from two red algae of the family Dumontiaceae have been found to inhibit the *herpes simplex* virus but no tests have been carried out on humans. Carrageenans have been patented as anti-viral agents. Many of the reported medicinal effects of marine algae have not been substantiated. *Corallina* is being used used in bone-replacement therapy [73].

A series of Cyanophyta were screened at National Cancer Institute, USA, for anti-human immunodeficiency virus (HIV-1) properties. Macroalgae are also known to have good proportions of provitamin A, ascorbic acid (see above), cobalamine, thiamine, riboflavin, nicotinic acid, pantothenic acid, lipoic acid, and choline. Two green algae- *Scenedesmus* and *Chlorella* - are known to be the richest sources of vitamins. However, algal extracts are important in the manufacture of many pharmaceutical products, acting as emulsifiers and binding agents for syrups, tablets, capsules, and ointments. In Japan, where *Porphyra* is an economically important alga, some research with medical applications has been completed. A diet of *Porphyra* (and, in some cases, other seaweeds) lowered blood cholesterol levels in rats [75], was effective against stomach ulcers [76], reduced mammary cancer in mice, and lowered intestinal cancer in rats [77]. The search for medically important natural products from algae is becoming easier as more and more algal species are brought into culture. The maintenance of algae may be a key factor in exploiting algae as a source of natural products.

Algae cause few human diseases. Perhaps the best known algal disease is protothecosis, a subcutaneous infection that creates lesions. Protothecosis results from infections in humans of the achlorotic green alga *Prototheca* [78,73], and a similar disease is caused by a chlorophyll-containing species [82].

9. ECONOMIC IMPORTANCE

Seaweeds, as processed and unprocessed food, have a commercial value of several billion dollars annually. The seaweed industry provides a wide variety of products that have an estimated total annual value of US$ 5.5-6 billion. Food products for human consumption contribute about US$ 5 billion of this. Substances that are extracted from seaweeds - hydrocolloids - account for a large part of the remaining billion dollars, while smaller, miscellaneous uses, such as fertilizers and animal feed additives, make up the rest. The industry uses 7.5-8 million tones of wet seaweed annually [7, 79]. Major commercial algal products at present are based on only 10-20 algal species, almost all of them macroalgae. Polysaccharide products derived from algae, referred to as hydrocolloids or phycocolloids are available in all continents. At present their combined market value is well over US$ 500 million. Red and brown seaweeds are used to produce hydrocolloids; alginate, agar and carrageenan, which are used as thickening and gelling agents. Today, approximately 1 million tones of wet seaweed are harvested and extracted to produce about 55 000 tones of hydrocolloids, valued at almost US$ 600 million. The economic importance of the algae is not always recognized [80]. Algae are used as food or in food products, as plant fertilizers, in cosmetics, in manufacturing processes, in biomedical research, and in many other products and processes. Also, algae form the base of the food chain for 71% of the world's surface because they are the only significant primary producers in the oceans, i.e., commercial fisheries and aquaculture depend, directly or indirectly, on algae.

Algae, especially via algal products, account for billions of dollars worth of food and industrial products each year. Commercial products are derived from field collections of native algae (especially seaweeds), from cultivated algae raised either in natural habitats or in man-made ponds and laboratories, and from fossilized deposits (e.g., crude oil, diatomite, limestone) [80, 81]. The red alga *Porphyra* is the most important alga that is eaten directly as food and is commonly used to make sushi. Worldwide, the retail value of the *Porphyra* harvest for 1986 was about $2 billion p.a. (approx 130 t dry weight) [82], and this amount is increasing annually. The seaweed that produces the largest amount by aquaculture is *Laminaria japonica* - similar to the Irish Seaweed *Laminaria digitata*. China produces around 3.8 million tonnes p.a. of *Laminaria japonica*. Japan produces about 10 billion sheets of *Porphyra* (nori) each year, or about 3×10^6 kg dry weight; this amounts to about 60% of the world's commercial harvest of *Porphyra*. The utilization of many types of seaweed is restricted biogeographically (Table 3).

"Farming in the sea" is termed mariculture. The predominant form of algal mariculture is the farming of seaweeds in harbors, bays, and estuaries [83]. Interestingly, agriculturally important plants and animals were domesticated before recorded history, and since the beginning of recorded history about 7000 years ago, no major, economically important agricultural organisms have been domesticated. Conversely, mariculture has developed in the past few centuries.

Since seaweeds have high vitamin and mineral contents, researchers found out those seaweeds are not only good as food additive; they are also potent ingredients for cosmetic products. Studies show that these products are effective and safe to use in almost all types of skin. In fact, seaweed beauty products are considered as one of the best in the cosmetic industry today. For the past years, many players in the cosmetic industry came up with different beauty products using seaweeds are primary ingredients.

The seaweed industry was an important source of employment for poorer coastal communities long before the abalone farming industry developed. Production peaked during the 1970s, but since then it has been rather erratic due to fluctuations in the global seaweed

Table 3. World production of seaweed (from Algo Rhytme, No 31. CEVA, Pluebian, France)

Country/Area	Seaweed (fresh weight million t)	World's production (%)
China	4.093	59
Korea	0.771	11
Japan	0.737	10
Philippines	0.404	6
Far East countries (total)	6.263	90
Norway	0.185	2.6
Chile	0.182	2.6
USA	0.116	1.6
France	0.079	1.1
European countries	0.302	4.3
Total	6.941	100

market. The introduction of kelp as abalone feed has created a new market and undoubtedly stimulated the local seaweed industry by increasing profitability for companies located near abalone farms. Calculated on a dry weight basis, the price obtained for fresh kelp as abalone feed is more than double that for dry kelp that is exported for alginate production. However, a proper comparison must take into account capital and operational costs, and is subject to price fluctuations in the international kelp market [84].

REFERENCES

[1] Kaur I. Bhatnagar AK. *Algae-dependent bioremediation of hazardous wastes*: Elsevier Science B.V. 2002.
[2] Round, FE. *The Biology of the Algae London*. Edward Arnold Ltd., 1965.
[3] Hassett JM; Jennett C.; Smith JE. *App.l Environ. Microbiol.*, 1981. 41, 1097-1106.
[4] Fernandez-Pinas F.; Mateo P.; Bonilla I. *Arch. Environ. Contam. Toxicol.,* 1991. 21, 425-431.
[5] Leusch, A.; Holan, ZR.; Volesky, B. *Journal of Chemistry Tech. Biotechnol.*, 1995. 62, 279-288.
[6] Wanger-Dobler, I.; Meiners, WBM.; Laatsch, M. Integrated approach to explore the potencial of marine micro-organisms for the production of bioactive metabolites. *Adv. Biochem. Eng. Biotechnol,* 2002. 74, 207-23.
[7] Abbott, L. The mystery of the cosmological constant. *Scientific American* (ISSN 0036-8733), 1988. 258, 106-113.
[8] Radmer RJ. Algal Diversity and Commercial Algal Products. *Bioscience,* 1996. 46, 263-270.
[9] Kingman, AR.; Moore, J. Isolation, purification and quantitation of several growth regulating substances in Ascophyllum nodosum (Phaeophyta*). Botanica Marina*, 1982. 25, 149-154.
[10] Featonby-Smith BC.; Van Staden, J . Identification and seasonal variation of endogenous cytokinins Ecklonia maxima (Osbeck) Papenf. *Botanica Marina* 1984. 27,527-531.
[11] Parker, A. Using elasticity/temperature relationships to characterize gelling carrageenans. *Hydrobiologia*,1993. 260/261,583-588.
[12] Honya, M.; Kinoshita, T.; Tashima, K; Nisizawa, K.; Noda, H. Modification of the M/G ratio of alginic acid from Laminaria japonica areshong cultured in deep seawater, *Bot. Mar.*, 1994. 37,463- 466.
[13] Fleurence, J. Seaweeds proteins: Biochemical, nutritional aspects and protein uses, *Trends in Food Sci.and Technol.*, 1999. 10, 25-28.
[14] Fleurence, J.; Chenard, E.; Luçon, M. Determination of the nutritional value of proteins obtained from Ulva armoricana, *J. Appl. Phycol.*, 1999.11, 231-239.
[15] Chapman, VJ.; Chapman, DJ. Sea vegetables (algae as food for man). In *Seaweeds and their uses.* London: Chapman and Hall.1980. 62-97.
[16] Zemke-White, LW.; Ohno, M. World seaweed utilization: end-of-century summary. *Journal of Applied Phycology*, 1999. 11, 369–376.

[17] Nisizawa, K.; Noda, H.; Kikuchi, R.; Watanabe, T. The main seaweeds in Japan. *Hydrobiologia*, 1987. 151/152, 5–29.

[18] Dawes, CJ. *Marine Botany*. John Wiley and Sons, Inc. New York, 1998.

[19] Mchugh, D.J. A guide to the seaweed industry. *FAO Fisheries Technical Paper No. 441*, Rome, 2003. p. 105.

[20] Rupe re'z P.; Saura-calixto F. Dietary .fibre and physicochemical properties of edible Spanish seaweeds. *Europe. J. of Food Res. and Tech.*, 2001. 212, 349–354.

[21] Lahaye, M.; Kaeffer, B. Seaweed dietary fibers: Structure, physicochemical and biological properties relevant to intestinal physiology. *Science des Aliments*, 1997. 17, 619–639.

[22] Mabeau, S.; Fleurence, J. Seaweed in food products: Biochemical and nutritional aspects. *Trends in Food Science and Technology*, 1993.4, 103–107.

[23] Ortega-Calvo, Mazuelos, Hermos 'n, Sa'iz-Jime'nez. Chemical composition of Spirulina and eucaryotic algae food products marketed in Spain. *Journal of Applied Phycology,* 1993. 5, 425–435.

[24] Ito, K.; Hori, K. Seaweed: chemical composition and potential uses. *Food Review International*, 1989. 5, 101–144.

[25] Dreher, M.L. Handbook of dietary fiber. *An applied approach*. Marcel Dekker, New York, 1987.

[26] Suzuki, H.; Higuchi, T.; Sawa, K.; Ohtaki, S.; Tolli, J. Endemic coast goitre in Hokkaido, Japan. *Acta Endocr*, 1965. 50, 161-176.

[27] Burtin, P. Nutritional value of seaweeds. *Electron. J. Environ. Agric. Food Chem.* [ISSN: 1579- 4377]. 2003. 2, 498-503.

[28] Lahaye, M. and Thibault, j.F. 1990. Chemical and physio-chemical properties of fibers from algal extraction by-products. *In Dietary Fibre: Chemical and Biological Aspects* (Southgate, D.A.T., Waldron,K., johnson, I.T. and Fenwick, G.R., eds), pp. 68-72, Royal Society of Chemistry, Cambridge, UK.

[29] Lahaye, M. Marine algae as sources of fibres: Determination of soluble and insoluble dietary fiber contents in some 'sea vegetables'. *Journal of the Science of Food and Agriculture*, 1991. 54, 587-594.

[30] Southgate, DAT. Dietary fiber and health. In Dietary fibre : *Chemical and Biological Aspects* (Southgate, D.A.T., Waldron, K., johnson, I.T. and Fenwick, G.R., eds), The Royal Society of Chemistry, Cambridge, UK, 1990.

[31] Charreau, B. Efficiency of fucans in protecting porcine endothelial cells against complement activation and lysis by human serum. *Transplantation Proceedings*, 1997. 29 , 889-890.

[32] Nasu, T. Fucoidin, a potent inhibitor of L-selectin function, reduces contact hypersensitivity reaction in mice. *Immunology Letters*, 1997. 59, 47-51.

[33] Morgan, KC.; Wright, JCL.; Simpson, FJ. `Review of chemical constituents of the red alga Palmaria palmata (Dulse)' in *Econ. Bot,* 1980. 34, 27-50.

[34] Galland-Irmouli, AV.; Fleurence, J.; Lamghari, R.; Luc_on, M.; Rouxel, C.; Barbaroux, O. Nutritional value of proteins from edible seaweed Palmaria palmata (Dulse). *Journal of Nutritional Biochemistry,*1999. 10, 353–359.

[35] Indegaard, M.; Minsaas, J. In Animal and Human Nutrition. Seaweed Resources in Europe. *Uses and Potential* (Guiry, M.D. and Blunden, G., eds), John Wiley andSons,1991.

[36] Castro-Gonza, Á.; les, MI.; PereÂ z-Gil Romo, F.; PereÂ z-Estrella, S.; Carillo-Dominguez, SD. `Chemical composition of the green alga Ulva lactuca' in Ciencas Marinas,1996. 22,205-213.

[37] Smith, DG.; Young, EG. `Amino acids of marine algae' in *J. Biochem*, 1954. 217,845-853.

[38] Arasaki, A.; Arasaki, T. Low calories, High Nutrition. *Vegetables from the Sea to Help you Look and Feel Better*, Japan Publications Inc. 1983.

[39] Augier, H.; Santimoine, M. `Contribution á l'ètude de la composition en azote total, en protè ines et en acides amineè s protèiques des diffèrentes parties du thalle de Laminaria digitata (Huds.) Lamour. dans le cadre de son exploitation industrielle et agricole' in *Bull. Soc. Phycol. de France*, 1978. 23, 19-28.

[40] Abdallah, M.A.M., Chemical composition, mineral content and heavy metals of some Marine Seaweeds from Alexandria Coast, Egypt: potential uses. *Egyptian Journal of Aquatic Research*, 2008. 34, 84-94.

[41] Abdallah, M.A.M., Evaluation of some Marine Seaweed as Food Supplements. *International Conference on "Biodiversity of the Aquatic Environment"* (BAEC), Lattakia, Syria, 2010.

[42] Jones, JW.; McFadden, HW.; Chandler, FW.; Kaplan, W.; Conner, KH., Green algal infection in a human. *J. Clin. Pathol.* 1983. 80, 102-107.

[43] Wanger-Dobler, I.; Meiners, WBM.; Laatsch, H., Integrated approach to explore the potencial of marine micro-organisms for the production of bioactive metabolites. *Adv. Biochem. Eng. Biotechnol.*, 2002. 74: 207-23.

[44] Munda, I.M., Differences in amino acid composition of estuarine and marine fucoids. *Aquatic Botany*, 1977. 3,273-280.

[45] Kolb, N.; Vallorani, L.; Stocchi, V., Chemical composition and evaluation of protein quality by amino acid score method of edible brown marine algae Arame (Eisenia bicyclis) and Hijiki (Hijikia fusiforme). *Acta Alimentaria*, 1999. 28, 213–222.

[46] Boussiba, S.; Richmond A.E., Isolation and characterization of phycocyanins from the bluegreen alga spirulina platensis. *Arch. Microbiol.*, 1979. 120, 155-159.

[47] Sargent, JR.; Falk-Petersen, S., The lipid biochemistry of calanoid copepods. *Hydrobiologica*, 1988. 168,101-114.

[48] Sargent, JR.; Eilertsen, HC.; Falk-Petersen, S.; Taasen, JP., Carbon assimilation and lipid production in phytoplankton in northern Norwegian fjords. *Mar Biol*, 1985. 85,109-116.

[49] Nichols, PD.; Palmisano, AC.; Rayner, MS.; Smith, GA.; White, DC., Changes in the lipid composition of Antarctic sea ice diatom communities during a spring bloom: an indication of community physiological status. *Antarct. Sci.*, 1989. 1,133-140.

[50] Shifrin, NS.; Chisholm, SW., Phytoplankton lipids: interspecific differences and effects of nitrate, silicate and light-dark cycles. *J. Phycology*, 1981. 17,374-384.

[51] Reitan, KI.; Rainuzzo JR.; Olsen ,Y., 1994. Effect of nutrient limitation on fatty acid and lipid content of marine microalgae. *J. Phycology* 30,972-979.

[52] Nelson, M.; Phleger, CF.; Nichols, PD., Seasonal lipid composition in macroalgae of the northeastern Pacific ocean. *Botanica Marina*, 2002. 45, 58–65.

[53] Khotimchenko, SV.; Kulikova, IV., Lipids of different parts of the lamina of Laminaria japonica Aresch. *Botanica Marina*, 2000. 43, 87–91.

[54] Dawes, CJ.; Kovach, C.; Friedlander, M., Exposure of Gracilaria to various environmental conditions II. The effect on fatty acid composition. Botanica Marina, 1993. 36, 289–296.

[55] Khotimchenko, SV., Fatty acid composition of green algae of the Genus Caulerpa. *Botanica Marina*, 1995. 38, 509–512.

[56] De Roeck-Holtzhauer, Y.; Claire, C.; Bresdin, F.; Amicel, L.; Derrien, A., Vitamin, free amino acid and fatty acid compositions of some marine planktonic microalgae used in aquaculture. *Botanica Marina*, 1993. 36, 321–325.

[57] Shameel, M., Phycochemical studies on fatty acids from certain seaweeds. *Botanica Marina*, 1990. 33, 429–432.

[58] Khotimchenko, SV.; Vaskovsky, VE.; Titlyanova, TV., Fatty acids of marine algae from the Pacific coast of north California. *Botanica Marina*, 2002. 45, 17–22.

[59] Mishra, VK.; Temelli, F.; Ooraikul Shacklock, PF.; Craigie, JS., Lipids of the red alga Palmaria palmate. *Botanica Marina*, 1993. 36, 169–174.

[60] Renaud, SM.; Thinh, LV.; Parry, LD., The gross chemical composition and fatty acid composition of 18 species of tropical Australian microalgae for possible use in mariculture. *Aquaculture*, 1999. 170,147-159.

[61] Fleurence, J.; Gutbier, G.; Mabeau, S.; Leray, C., Fatty acids from 11 marine macroalgae of the French Brittany coast. *Journal of Applied Phycology*, 1994. 6 , 527-532.

[62] Adrian, J.; Potus, J.; Poiffait, A.; Dauvillier, P., *Ana' lisis nutricional de los alimentos*. Zaragoza, Spain: Editorial Acribia SA. 2000.

[63] Piovetto, L.; Peiffer P., Determination of sterols and diterpenoids from brown algae (Cystoseiraceae). *Journal of Chromatography*, 1991. 588, 99-105.

[64] Haugan, JA.; Liaaenn-Jensen, A., Algal carotenoids 54.*Carotenoids of brown algae (Phaeophycae) . *Biochemical Systematics and Ecology*, 1994. 22 , 31-41.

[65] Le Tutour, B., Antioxidative activities of algal extracts, synergistic effect with vitamin E. *Phytochem.*, 1994. 29, 3759-3765.

[66] Watanabe, F.; Takenaka, S.; Katsura, H.; Zakir Hussain Masumder, SAM.; Abe, K.; Tamura, Y.; Nakano, Y., Dried green and purple lavers (Nori) contain substantial amounts of biologically active vitamin B12 but less of dietary iodine relative to other edible seaweeds. *J. Agric. Food Chem.* 1999. 47 , 2341-2343.

[67] Qasim, R.; Barkati, S., Ascorbic acid and dehydroascorbic acid contents of marine algal species from Karachi. Pakistan J. Sci. Ind. Res. 1985. 28, 129-133.

[68] Solibami, VJ.; Kamat, SY., Distribution of Tocopherol (Vitamin E) in Marine algae from Goa, West Coast of India. *Indian Journal of Marine Sciences*. 1985. 14, 228 – 229.

[69] Lewis, JG.; Stanley, NF.; Guist, GG., Commercial production and applications of algal hydrocolloids. *In "Algae and Human Affairs"* (C. A. Lembi and J. R. Waaland, eds.), Cambridge Univ. Press, New York. 1988.

[70] Stanley, N., Production, properties and uses of carrageenan. *In "Production and Utilization of Products from Commercial Seaweeds,"* 1987. pp. 97-147. FAO/UN Rome.

[71] Armisen, R.; Galatas, F., Production, properties and uses of agar. *In "Production and Utilization of Products from Commercial Seaweeds,"* 1987. pp. 1-49. FAO/UN, Rome.

[72] McHugh, D., Production, properties and uses of alginates. *In "Production and Utilization of Products from Commercial Seaweeds,"* 1987. pp. 50-96. FAO/UN, Rome.

[73] Stein, J. R., and Borden, C. A. 1984. Causative and beneficial algae in human disease conditions: *A review. Phycologica.* 23, 485-501.

[74] Cooper, S.; Battat, A.; Marsot, P.; Sylvestre, M., Production of antibacterial activities by two Bacillariophyceae grown in dialysis culture. *Can. J. Microbiol.*, 1983. 29, 338-341.

[75] Abe, S.; Kaneda, T., The effect of edible seaweeds on cholesterol metabolism in rats. *Proc. Int. Seaweed Symp*, 1972. 9, 562-565.

[76] Sakagami, Y.; Watanabe, T.; Hisamitsu, A.; Kamibayshi Honma, K.; Manabe, H., Anti-ulcer substances from marine algae. *In "Marine Algae in Pharmaceutical Science"* (H. A. Hoppe and T. Levring, eds.), de Gruyter, Berlin, 1982. 99-108.

[77] Yamamoto, I.; Maruyama, H., Effect of dietary seaweed preparations on 1,2-dimethyhydrazine- induced intestinal carcinogenesis in rats. *Cancer Lett,* 1985. 26, 241-251.

[78] Schwimmer, M.; Schwimmer, D., The Role of Algae and Plankton in Medicine. New York: Grune and Stratton. *Selye H* ,1955. Pp.85.

[79] Druehl. LD., Cultivated edible kelp. *In: CA Lembi and JR Waaland, eds, Algae and Human Affairs,* Cambridge University Press, New York, 1988. pp.119-134.

[80] Lembi, C. A.; Waaland, JR., eds. "*Algae and Human Affairs.*" Cambridge Univ. Press, New York.1988.

[81] George, RW., Products from fossil algae. *In "Algae and Human Affairs"* (C. A. Lembi and J. R. Waaland, eds.), 1988. pp. 305-33.

[82] Mumford, TFJr.; Miura, A., Porphyra as food: Cultivation and economics. *In "Algae and Human Affairs"* (C. A. Lembi and J. R. Waaland, eds.), pp. 87-117. Cambridge Univ. Press, New York.Univ. Press, New York.1988.

[83] Ohno, M.; Critchley, AT., (Editors) "*Seaweed Cultivation and Marine Ranching.*"Japan International Cooperation Agency, Yokosuka, Japan. 151 pp.3. Cambridge. 1993

[84] Troell, M.; Robertson-Andersson, D.; Anderson, RJ.; Bolton, JJ.; Maneveldt, G.; Halling, C.; Probyn, T., Abalone farming in South Africa: an overview with perspectives on kelp resources, balone feed, potential for on-farm seaweed production and socio-economic importance. *Aquaculture*, 2006. 257, 266.

Chapter 6

THE POTENTIAL HEALTH BENEFITS OF SEAWEED AND SEAWEED EXTRACTS

I. A. Brownlee, A. C. Fairclough, A. C. Hall and J. R. Paxman

Centre for Food Innovation, Sheffield Hallam University,
Sheffield, UK

ABSTRACT

Edible seaweeds have historically been consumed by coastal populations across the globe. Today, seaweed is still part of the habitual diet in many Asian countries. Seaweed consumption also appears to be growing in popularity in Western cultures, due both to the influx of Asian cuisine as well as notional health benefits associated with consumption. Isolates of seaweeds (particularly viscous polysaccharides) are used in an increasing number of food applications in order to improve product acceptability and extend shelf-life.

Epidemiological evidence suggests regular seaweed consumption may protect against a range of diseases of modernity. The addition of seaweed and seaweed isolates to foods has already shown potential to enhance satiety and reduce the postprandial absorption rates of glucose and lipids in acute human feeding studies, highlighting their potential use in the development of anti-obesity foods. As seaweeds and seaweed isolates have the potential to both benefit health and improve food acceptability, seaweeds and seaweed isolates offer exciting potential as ingredients in the development of new food products.

This review will outline the evidence from human and experimental studies that suggests consumption of seaweeds and seaweed isolates may impact on health (both positively and negatively). Finally, this review will highlight current gaps in knowledge in this area and what future strategies should be adopted for maximising seaweed's potential food uses.

1. INTRODUCTION

Biologically, seaweeds are classified as macroalgae, with subclassification as brown (*Phaeophyta*), red (*Rhodophyta*) or green algae (*Chlorophyta*). Some examples of these edible algae are outlined in Table 1. The nutritional properties of these seaweeds are discussed earlier in this edition and reviewed elsewhere [1,2]. In 1994/95 over 2,000,000 tonnes (dry weight) of seaweed was harvested [3]. Much of this may be consumed as whole seaweed products, while a large proportion is also used in the production of over 85,000 tonnes of viscous polysaccharides for various food and industrial applications [4].

Historically, seaweed is a readily available food source that has been consumed by coastal communities likely since the dawn of time [15,16]. Seaweed is consumed habitually in many countries in South-East Asia [17]. However, as a wholefood it is not considered a habitual component of the Western diet [2]. In the West, seaweed isolates (e.g. alginate from brown algae and agar or carrageenan from red algae) are typically used industrially [18]. Seaweed consumption has gained a measure of acceptance in some Westernised cultures such as Hawaii, California and Brazil, where there are large Japanese communities who have had a tangible influence on the local dietary practices [19,20]. Low consumer awareness regarding potential health benefits and a lack of previous experience of seaweed challenges its use in the daily diet [21].

Table 1. Examples of edible algae

Subclassificaton	Genus	Common Name
Brown algae (*Phaeophyta*)	*Alaria*	Kelp/ bladderlocks
	Himanthalia/ Bifurcaria	Sea spaghetti, fucales
	Laminaria	Kelp/ kombu/ kumbu/ sea tangle
	Saccharina	Sugar wrack
	Undaria	wakame
	Ascophyllum	Egg wrack
	Fucus	Bladder wrack, rockweed
	Sargassum	Mojaban/Indian brown seaweed
	Hizikia	Hijiki
	Sargassum	Sea holly
	Dictyotales	
	Eisenia	Arame
Red algae (*Rhodophyta*)	*Rhodymenia/ Palmaria*	Dulse
	Porphyra	Nori/ haidai/ kim/ gim
	Chondrus	Irish moss/ carrigeen
	Mastocarpus/ Gigartina	Stackhouse, Guiry
	Gracilaria	
	Asparagopsis	Limu Kohu
	Grateloupia	
Green algae (*Chlorophyta*)	*Ulvaria/ Enteromorpha*	Laver/sea lettuce/ sea grass/nori

Adapted from [2,5,6]. Further details from [7-14].

2. SEAWEED AS A WHOLE FOOD

Consumption of seaweed in Europe and North America is minimal at present [22]. While instruments within Japan and Korea have been developed to assess dietary intake of seaweed [23,24], consumption is so infrequent in most Western cultures that it is not considered within nationwide dietary intake assessment surveys. In the USA and Canada seaweed is cultivated in onshore tanks and the market for it is growing. In Ireland there is a renewed interest in seaweed that once formed part of the traditional diet. Recipe books promoting the use of 'sea vegetables' or 'marine vegetables' [25] in home cooking are becoming more popular. As consumer health and nutrition become more influential in the food industry, the use of seaweed as an ingredient is on the rise [25], and product development involving salads and wraps appears to be slowly evolving.

The rich mineral and trace element content of seaweed compared to terrestrial plant foods can however, impact negatively on its sensory characteristics [2]. As an ingredient of composite foods, it has been shown to be acceptable to consumers when baked into breads (*Ascophyllum nodosum* up to 5% w/w; Hall et al., 2010; Mahadevan and Fairclough, personal communication) and added to pasta (Wakame - *Undaria pinnatafida* - up to 10% w/w;[26]). Further from these applications, seaweed has been added to low fat meat products where it contributes to water retention and gel formation [27]. Collectively, these results suggest that seaweed may be successfully included as an ingredient in a number of food applications. As dried seaweed is high in dietary fibre, along with a range of other potentially bioactive components, this addition has the potential to enhance the nutritional quality of a product.

Habitual consumption of seaweed may offer a nutritionally rich addition to the diet. However, micronutrient intakes in excess of the RNI could be of concern to nutritionists, particularly where bioavailability is high.

3. WHOLE SEAWEED USE IN FOODS

Whole seaweeds have been incorporated into a range of foods including meat, and bakery products. Fairclough and Williams (personal communication) have recently successfully incorporated *Ascophyllum nodosum* into sausages. This usage has also previously been reported with *Laminaria japonica* (sea tangle) powder [28]. Previous authors have included *Himanthalia elongate* (sea spaghetti), *Undaria pinnatifida* (Wakame) and *Porphyra umbilicalis* (Nori) in frankfurter type products (gel/emulsion meat systems [29] and *H. elongate* in frankfurters [27,30,31]. More recently, *U. pinnatifida* and *H. elongate* have been incorporated into beef patties and restructured poultry steaks, respectively [32,33]. Recent work has also resulted in the production of an acceptable wholemeal bread enriched with *Ascophyllum nodosum* [34]. Locally, our research group has also added *Ascophyllum nodosum* to pizza bases, cheese and frozen meat products. Prabhasankar and colleagues have incorporated *Sargassum marginatum* (Indian brown seaweed) and *U. pinnatifida* into pasta [26,35]. The previously published literature described above reports mixed success in terms of acceptability of whole seaweed-enriched food products. There may also be issues involved in the large-scale processing of whole seaweed-enriched foods.

4. ANTIMICROBIAL PROPERTIES OF WHOLE SEAWEEDS

The incorporation of seaweed into foods has also been shown to have a preservative effect, particularly with regards to Gram-negative bacteria (Gupta *et al.*, 2010), reducing the need to add salt. The antimicrobial properties of seaweed extracts have been well documented over the years [36-39]. However, there would appear to be a lack of published information regarding the antimicrobial properties and preservative effects when seaweed as a 'whole food' is incorporated into a food matrix. Previous studies would suggest that the overall antimicrobial capacity of seaweeds appears to be linked to their antioxidant content [40].

Several studies have been undertaken at Sheffield Hallam University where seaweed (*Ascophyllum nodosum*, Seagreens®) as a dried whole-food has been incorporated into various products for example processed meat products and bread products.

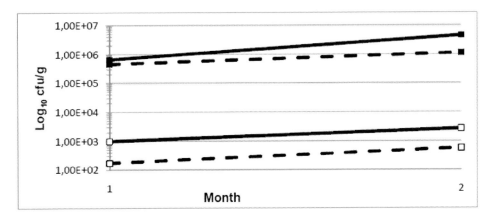

Figure 1. Changes in Total Viable Count (TVC ■) and coliform numbers (□) over shelf-life in frozen processed meat products containing seaweed (dashed line). Control product (no seaweed added) is shown by continuous line.

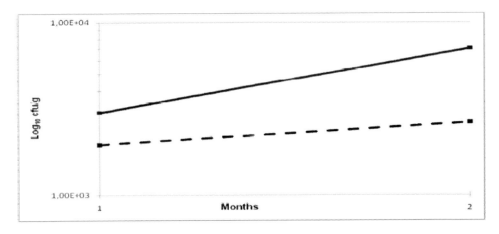

Figure 2. Changes in the population of lactic acid bacteria (LAB) in meat products containing Seagreens® (*Ascophyllum nodosum*). The dashed line represents the seaweed-enriched product, while the control product is denoted by a continuous line.

Figure 1 shows that the 3% w/w dried seaweed added to this processed meat product results in an overall 0.3 log$_{10}$ cfu/g reduction in the total viable count over the two month trial period; however, when looking at specific populations of micro-organisms, this antimicrobial activity is particularly effective against Gram negative micro-organisms such as coliforms (showing a 0.7 log$_{10}$ reduction cfu per gram of product) which is to be expected since Gram negative micro-organisms have a thinner cell wall than Gram positive micro-organisms making them more susceptible to antimicrobial agents. However, there is a significant reduction in the Gram positive lactic acid bacteria (LAB) population over shelf-life (0.5 log$_{10}$ - see Figure 2). Other data suggest similar reductions in yeast and mould populations (data not shown).

Interestingly when a methanolic extract of the seaweed is used in a typical antimicrobial susceptibility test then the trend is also mirrored with Gram positive organisms, on the whole, showing more susceptibility to the antimicrobial agent(s) contained within the seaweed; especially *Bacillus cereus* and *Staphylococcus aureus,* which show the greatest sensitivity to the extracted agent (see Table 2). *Listeria monocytogenes* also shows a noticeable susceptibility to the extract although to a lesser extent than the organisms named above.

Table 2. Antimicrobial effects of a methanolic extract from Seagreens® (*Ascophyllum nodosum*)

Micro-organism	Sensitivity
Bacillus cereus (NCTC 7464)	+++
Staphylococcus aureus (NCTC 12981)	+++
Listeria monocytogenes (NCTC 7973)	++
Bacillus subtilis (NCTC 10400)	+
Listeria innocua (NCTC 11288)	-
Enterococcus faecalis (NCTC 775)	-
E. coli (NCTC 12241)	±
E. coli 0157 * (NCTC 12900)	±
Pseudomonas fluorescens	-
Salmonella typhimurium (NCTC 12023)	+

+++ - zone of clearance ≥ 7.5mm, ++ - zone of clearance 2.5 - 7.5 mm, + - zone of clearance ≤ 2.5 mm, - - no discernible zone of clearance, ± - indeterminate zone of clearance.

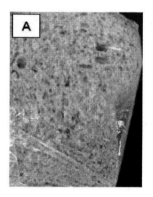 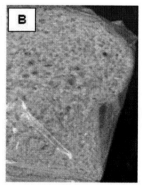 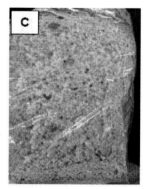

Figure 3. Mould growth at nine days post-production in wholemeal breads containing differing amounts of Seagreens®. Photograph A - control loaf with standard salt content. Photograph B - loaf containing a 50:50 mix of added salt and Seagreens®. Photograph C - loaf with Seagreens® instead of added salt.

When Seagreens® is incorporated into bakery products especially wholemeal bread as a replacement for salt (as sodium chloride) there is a suppression of mould growth for up to 9 days in preservative-free bread when compared to preservative-free control bread containing 5 g of salt (as sodium chloride) and no Seagreens®. Similar results have been recorded in all the other bread varieties baked at Sheffield Hallam University; with the exception of white bread where no significant suppression has been seen and mould growth occurs after 3 to 4 days. Figure 3 shows wholemeal breads containing different amounts of Seagreens® at 9 days.

5. COMPONENTS OF SEAWEEDS

As previously stated, the use of seaweeds in Western diets is predominantly limited to use in food additives or extracts [41]. In line with other natural foods like fruits, vegetables and grains, there has been marked interest within the scientific community to assess which fractions of seaweeds may be linked to the historically observed health benefits. For the purpose of this review, these factors are considered under two relatively crude banners: seaweed phytochemicals and seaweed polysaccharides. Seaweed proteins have also previously been assessed for their nutritional value [42-47] but will not be discussed within this chapter. As with other foods that have historically been consumed whole (e.g. fruits, vegetables and grain products), it must also be noted that isolation of such bioactive components may allow the development of food products and/or supplements with potential health benefits. This should not however preclude a drive to increase population-wide consumption of the original whole foods.

5.1. Seaweed Polysaccharides

Each seaweed subclassification differs in the type of dietary fibres they contain. Brown seaweeds for example contain the dietary fibres alginates, fucans and laminarans; red seaweeds contain galactans, agar and carageenans; whereas green seaweeds contain soluble ulvans and other insoluble fractions such as cellulose.As with plant polysaccharides, non-starch entities play a vital role in seaweed structure both at a microscopic and macroscopic level. The varying roles of these polysaccharides within the macroalgae structure should be considered when comparing these different types of polysaccharides.

In alginates, the presence and arrangement of carboxyl groups on spans of 3 or more guluronic acid residues can act to interact with hydrogen ions and divalent cations (particularly calcium) to cause gelation [48]. This allows gelation in specific formulations at room temperature. The presence and position of sulphate ester groups in carrageenans and other seaweed polysaccharides also appears to affect their gelation and ability to interact with other factors in composite foods [49,50]. The physicochemical variations in these polysaccharides allow for a wide variety of applications within the food industry. Polysaccharides of different viscosities that react differently under various conditions of temperature, pH and food chemistry are important tools in the arsenal of food formulators in order produce products with increased acceptability.

Seaweed polysaccharides are extensively used as thickening agents in sweet and savoury sauces and condiments [51-53]. A number of applications of seaweed polysaccharides are also utilised in order to stabilise food products against degradation, staling and heating or cooling/freezing. These applications also act to improve the consumer acceptability of such products, as well as extending the shelf-life. Further novel applications of seaweed polysaccharides in food manufacture are discussed elsewhere [54].

Seaweed polysaccharides are generally water-soluble and very hydrophilic. Their action as stabilizers within food oil-water emulsions is suggested to be a result of their ability to precipitate/adsorb onto oil droplets and sterically stabilize emulsions against flocculation and coalescence [55,56].

Previous studies have also suggested that seaweed polysaccharides may be used at fat replacers in a range of food applications. Where this is carried out, seaweed polysaccharides and other hydrocolloid thickening agents can be used to reduce or replace added fats within foods in order to produce an end-product with reduced total fat content, while still allowing for a product with improved moisture retention and consistency. This role is crucial in the development of low-fat products with high consumer acceptability. This use of seaweed polysaccharides has been shown to facilitate the production of low fat versions of meat-based, starch-based, fat-based and fruit/vegetable-based products [51,52,57-59].

High viscosity polysaccharides are likely to have detrimental effects both in terms of the manufacturing process and product acceptability. As with other types of viscous polysaccharides, low viscosity fractions of the indigestible carbohydrate material from seaweeds could be used to develop food products with higher fibre content.

5.2. Seaweed Phytochemicals

Seaweeds also contain a range of unique phytochemicals not present in terrestrial plants. As such, edible seaweeds may be the only relevant dietary source of some of these factors. A wide range of studies have described the high antioxidant capacity of a range of edible seaweeds [18,60-62]. This capacity is endowed by the presence of sulphated polysaccharides [63], polyphenolic compounds [64] and antioxidant enzymes [65]. Oxidative stress may play a key role in the development of cancers and cardiovascular disease [66]. Phytochemical-rich foods should clearly form part of a healthy balanced diet. However, the human body has a number of physiological, biochemical and enzymatic processes by which it can combat oxidative stress outside of dietary intake. The routes by which the wide variety of phenolic compounds enter the circulation is not well characterised, nor is the bioavailability and half-life/distribution of such factors in the human body. Previous intervention studies where dietary antioxidant intake has increased have not evidenced a parallel change in the total antioxidant capacity of the body [67,68]. While this casts doubt on the benefit of increasing polyphenolic consumption from the perspective of reducing oxidative stress, it must be noted that such compounds may have other physiological effects.

Previous studies in animal models and cell culture have suggested that seaweed phytochemicals have the potential to inhibit the progression of carcinoma formation [69,70]. *In vitro* studies have also suggested a potential for phenolic compounds from seaweeds to inhibit the action of digestive enzymes [71,72]. Such an action would be considered to have the potential to affect macronutrient uptake and act as a therapeutic agent to help combat metabolic diseases. Further from these findings, a recent study has suggested that a polyphenol-rich extract of *Ecklonia stolonifera* improved glycaemic control in a non-insulin dependent diabetic mouse model [73]. A similar effect was also noted in a chemically-induced diabetic mouse model [74]. Anti-inflammatory properties of a phlorotannin-rich extract from *Ecklonia cava* have also been demonstrated *in vitro* [75].

The above experimental evidence highlights some interesting ways in which these phytochemical compounds isolated from seaweed could benefit health. Certainly there has been great interest from the pharmaceutical industry in the high-throughput analysis of macroalgal compounds for the development of novel drugs [76-78]. The new, more stringent regulations on novel food ingredients within the EU is likely make the inclusion of specific seaweed phytochemicals as bioactives more problematic than whole seaweed or seaweed polysaccharides, both of which have been used historically within this context. This should not act as a barrier to research that characterises the physiological effects of this range of interesting compounds and potential therapeutic agents, however.

6. SEAWEEDS AND HEALTH

Previous studies on seaweed consumption in humans have centred in the areas of the world where reasonable amounts of seaweeds are habitually consumed (particularly South-East Asia). The evidence detailed below outlines a variety of potential benefits to health, with much of the research in whole seaweeds either focussed around their impact on metabolic disease (associated with increased phytochemical and fibre intake) or breast cancer (linked to increased iodine consumption). Both of these topics are reviewed in detail elsewhere [79,80].

6.1. Experimental Studies

Numerous researchers have studied the health benefits of seaweed incorporation in the diets of rats, particularly with reference to their effects on blood lipid profiles. Wong *et al.* (1999) examined the lipid changing effects of 4 types of seaweed (1 red, 1 green and 2 brown at 5% dry weight of feed) compared to a control group (cellulose) in 60 male Sprague Dawley rats. Comparisons were made between serum total cholesterol, HDL cholesterol, LDL cholesterol, triglycerides and hepatic cholesterol. The results suggested that the red algae *Hypnea charoides* had the greatest hypocholesterolaemic effect; however no significant reductions in cholesterol were seen between any of the seaweeds. On the contrary, *Colpomenia sinuosa* (a brown algae) induced a significant ($p<0.05$) increase in total serum cholesterol [81].

Carvalho *et al.* (2009) showed that the total cholesterol levels of rats fed a hypercholesterolaemic diet increased significantly ($p<0.05$) when supplemented with cellulose as opposed to the green seaweed *Ulva fasciata* (24% dry weight seaweed meal). The seaweed containing diet was able to keep the total cholesterol at levels similar to baseline, leading the authors to suggest the incorporation of seaweed into the diet might be important in the reduction of total cholesterol [41].

Bocanegra et al. (2009) conducted a study in groups of ten rats that were fed a diet containing a cholesterol raising agent with either a cellulose-wheat starch mix, Nori algae or Konbu algae (7% weight as freeze-dried material). Rats fed the Nori and cholesterol-raising diet had lower postprandial cholesterolaemia, and a more positive lipid profile with regards to HDL and LDL lipid fractions ($p<0.05$) when compared to the comparable Konbu diet [82].

These studies, among othersof similar design, hint towards the variability of the biological effects of different varieties of edible seaweeds. They also highlight the potential for cardiovascular health benefits in certain cases. As with most animal-based dietary interventions, the amounts of seaweed incorporated in the diet are extremely high and do not bear resemblance to the amounts eaten within the human diet.

6.2. Epidemiological Studies

The lack of a dietary intake assessment tool alongside the likely exceedingly low intake of seaweeds at a population level means that observational data linking seaweed intake to reduced disease risk have only been collected in South East Asian populations. The most recent, accessible data are summarised in Table 4.

Such data should be interpreted with caution, firstly as they do not necessarily represent a causal relationship between seaweed intake and health outcomes, but rather an association between the two factors. Also, different species of edible seaweeds appear to have different effects on disease risk. In a Korean case-controlled study, increasing frequency of *Porphyra* species consumption was associated with reduced risk of breast cancer, whereas *Undaria pinnatifida* consumption did not [15]. These results highlight the wide variability in the bioactive content of seaweed species. Even within a specific type of seaweed, previous research has suggested there are significant seasonal variations in nutritional content [83-85], which is likely to impact the biological effects of edible components. As outlined in the section below on the negative impact of seaweed intake, certain population groups may be at risk from global or national guidelines based on high seaweed consumption.

However, these data suggest that achievable daily intakes of seaweed (equating to approximately 30 g of fresh seaweed [15,24,86] or 2 g of dry seaweed [15] a day) appear to reduce disease risk compared to the lowest (close to zero intake) percentiles of seaweed consumption. Such data are also routinely extrapolated to represent lifelong patterns that reduce a disease risk and are therefore a rational basis on which to develop prudent lifestyle choices across the whole life-course.

6.3. Intervention Studies

Relatively few human intervention studies have assessed the impact on seaweed consumption on risk factors for future disease. One previous study [91] has assessed the impact of seaweed consumption over a number of weeks on markers of cardiovascular disease risk. The physiological effects of seaweed supplementation were investigated in terms of effect on a number of markers of health, including blood glucose levels and blood lipid profiles in males and females with type II diabetes mellitus and a BMI of $<35kg/m^2$. Dried brown seaweeds (sea tangle and sea mustard) incorporated into a pill were consumed 3 times a day for 4 weeks as a food supplement. Total daily consumption of seaweed was 48g. After random assignment to either the control group or the seaweed supplementation group, and the completion of the trial, fasting blood glucose levels (p<0.01) and 2 hour postprandial glucose levels (p<0.05) were significantly lower in the seaweed supplemented group. However, while serum concentrations of triglycerides decreased, and HDL cholesterol levels significantly increased (p<0.05), levels of total and LDL cholesterol were not affected by seaweed supplementation. Nutrient intake (% energy from macronutrients) was identical over the 4 weeks, but the study had a relatively small sample size (n= 20), and while there was a control group (n=11), the study did not have a cross over design.

Table 4. Summary of recent observational studies relating to dietary seaweed intake and health

Disease/health concern	Study design	Odds ratio (95% CI) of highest seaweed to lowest	Reference
Serum total cholesterol	Retrospective study in the Japanese population with data from 1980 and 1990 for > 7000 people. Data were adjusted for age, BMI and total energy intake.	Not reported. No significant effect of seaweed consumption	[87]
Type II diabetes and prediabetes	3,405 Korean individuals, aged 20 - 65 y. Retrospective study. Adjusted for diet and lifestyle.	0.66 (0.43-0.99) for men and 0.80 (0.51-1.24) for women	[24]
Osteoporosis	214 Japanese elderly participants. Prospective study assessing calcaneus stiffness changes over 5 years. No adjustment of data.	0.22 (0.07-67) in all individuals	[88]
Obesity	3760 Japanese women aged 18-20 y. Cross-sectional study assessing 3 different eating patterns.	0.57 (0.37-0.87) for BMI >25.0[a]	[89]
Cardiovascular disease mortality	40547 Japanese men and women aged 40-79 y. Prospective study over seven years of follow-up. Not adjusted.	0.73 (0.59 -0.90)[a]	[90]
Allergic rhinosinusitis	1002 pregnant Japanese women. Cross-sectional study. Data adjusted for lifestyle and risk factors.	0.51 (0.30–0.87)	[86]
Breast cancer occurrence	South Korean case-control study. 362 cases (30-65y) with controls matched for age and menopausal status. Data adjusted for multivitamin supplement use, number of children, breastfeeding, dietary factors, education, exercise, oral contraceptive use.	0.48 (0.27-0.86)	[15]

[a]Seaweed was included as part of a healthy/traditional Japanese eating pattern (i.e. high intakes of vegetables, mushrooms, seaweeds, potatoes, fish and shellfish, soy products, processed fish, fruit and salted vegetables) and was not assessed independently.

The amount of seaweed consumed within this intervention was very high compared to the amounts consumed in the observational studies above that appeared to have a biological effect. As such, they may not be sustainable within the diet of individuals on a long-term basis if such an amount were not consumed in pill-form. Nonetheless, these findings warrant the development of further participant-based interventions involving long-term seaweed consumption and cardiovascular health.

In a study with twelve healthy female volunteers of healthy BMI, inclusion of 3 g of Nori (in capsules) 15 minutes before eating significantly blunted the postprandial glycaemic rise elicited from consumption of a white bread meal (containing 50 g of starch) [92]. These results were not duplicated in a recent study where the impact of inclusion of *Ascophyllum nodosum* as an integral ingredient within bread, consumed within a composite meal, was compared to a standard seaweed-free meal [93].

Daily supplementation with seaweed (20 g of *Laminaria japonica* diluted in water or a beverage) was administered in combination with diet, exercise and behavioural therapy to female, Korean college students (19-24y) over 8 weeks. Pre-post test analysis showed there were significant improvements, consistent with recommendations for weight management, across a range of anthropometric measures. However, the lack of a control group prohibits the authors from attributing such effects to any particular aspect of the intervention. There were no significant changes in blood lipids during this time [94].

Dietary supplementation with seaweed (5 g of *Alaria esculenta* in capsule form) consumption (*Alaria esculenta* (L.)) did not signficantly affect serum oestradiol concentrations in a recent randomised, placebo-ontrolled crossover trial in fifteen healthy, postmenopausal females [95] . However, it was noted in this study that there was a signficant inverse correlation between seaweed dosage (expressed in terms of mg/kg body weight) and serum oestrodiol concentrations. The same intervention also elicited a significant increase in circulating levels of thyroid-stimulating hormone [96]. It was calculated that c.75mg/kg of body weight of seaweed would need to be ingested to have this oestodiol lowering effect, which would equate to approximately 4 - 5 g/day of dry seaweed consumption for females weighing between 55 and 75 kg. These preliminary results highlight the potential for seaweed as an important dietary factor in the prevention of breast cancer.

There is a growing body of evidence on the acute benefits of alginate consumption to health-related parameters. Alginates are widely researched due to their unusual gelling properties and relatively low viscosity, meaning that higher amounts can be incorporated in foods or beverages than other types of seaweed polysaccharide. Relevant food products could be used to deliver other types of seaweed polysaccharide in such studies. An example of this is the work of Panlasigui *et al.* (2004) in which carrageenan in a powdered form was incorporated into 4 food products common in the Philippines: a yeast bread, a corn pudding, fish balls and a gruel like product [97]. Following a two-week intervention with these products, participants had significant improvements to plasma concentrations of total and HDL cholesterol compared to the control group (no intervention).

Hoad et al. (2004) investigated the gastric emptying rates of a strong gelling (high-G) and a weaker gelling (low-G) alginate meal compared to a guar-based meal and a control (without added fibre). *In vitro* characterisation of the gelling properties of both alginate meals showed the formation of intragastric 'lumps' which were shown to be associated with feelings of fullness and a reduction in hunger in the strong-gelling alginate condition[98]. The authors purport that acid-gelling agents, such as alginates, may be useful for those aiming to adhere to a weight-reducing diet.

To that end, Paxman and colleagues (2008) demonstrated the capacity for a strong-gelling alginate formulation to restrict free-living energy intake compared to a commercially available control formulation. The 7% (134.8 kcal) reduction in reported daily energy intake was consistent with published guidelines for weight management [99]. Similar significant findings were reported previously in overweight and obese women [100]. Such effects may be explained by the potential for seaweed isolates, particularly alginate to enhance satiety [101].

The potentially satiating effects of seaweed isolates are by no means unaminously reported in the literature, however. Findings from a well-controlled intervention show an alginate and guar-based breakfast bar had no effect on energy and macronutrient intake when incorporated into the habitual diet over 5 days [102]. The breakfast bar was consumed daily for 5 days and food intake recorded on 3 randomly selected days, however the authors purport that poor intragastric gelation of the fibres may explain the lack of a treatment effect. Similarly, acute *ad libitum* food intake was reportedly unaltered after a meal replacer containing alginate (0.4% and 0.8%) compared to a meal replacer alone [103], though hunger was significantly reduced for several hours following the treatment.

Numerous authors have reported beneficial effects of seaweed isolates on postprandial glycaemia. 5.0 g of sodium alginate, added to food significantly attenuated the postprandial glycaemic response in type 2 diabetics by 31% compared to the control meal [104]. Wolf et al. (2002) incorporated 1.5 g of sodium alginate into a 100 g liquid glucose-based preload along with an acid-soluble calcium source to produce an acid-induced viscosity complex. The authors reported a non-significant fall in peak glycaemia and a significant reduction in incremental change from baseline AUC in healthy, non-diabetic adults following ingestion of the acid-induced viscosity complex compared to a soluble fibre-based control [105]. Williams and colleagues (2004) investigated the glycaemic response to a novel induced viscosity fibre (IVF) "crispy bar" (including 5.5 g guar gum and 1.6 g sodium alginate) compared to an alginate free bar in healthy adults. Postprandial glycaemia was significantly reduced at 15, 30, 45, and 120 minutes. The positive iAUC was significantly reduced by 33% following the IVF "crispy bar" compared to the control [106]. Previous work suggested that the existing positive correlation between AUC glycaemia and body fat percentage (control condition) could be attenuated when an ionic gelling sodium alginate preload was ingested prior to a test lunch. This finding suggests the enhanced postprandial glycaemic response at higher body fat percentages could be normalised in response to alginate ingestion [107].

Effects on lipid uptake are less well-reported. In subjects with ileostomies, alginate added to a meal increased the ileal output of fatty acids [108]. Similar to the previously reported findings for postprandial glycaemia, Paxman and colleagues suggested that the existing positive correlation between AUC cholesterolaemia and body fat percentage was also eliminated by ingestion of an ionic gelling sodium alginate preload [107].

6.4. Potential Negative Effects of Seaweed or Seaweed Isolate Consumption

In Japan and Korea seaweed (often added to soup) is ingested by lactating mothers who believe it to promote an adequate supply of breast milk. Iodine, found in high concentrations in seaweed, is transmissible from mother to infant through breast milk and this local practice has led to documented cases of neonatal iodine toxicity and consequent hypothroidism, with its associated negative clinical consequences [109]. Toxic blue-green algae species can grow on edible seaweed and have been noted in the literature to be the causative factor in a number of food poisoning occurrences [110].

Components of seaweed bind to adsorb heavy metals [111], meaning that seaweed is particularly prone to contamination from polluted water, and its consumption is a potential route of toxic heavy metals entering the body. However, a recent Korean study assessed that high seaweed consumption (8.5 g/day) would result in exposure of individuals to significantly less than 10% of the toxic quantities of arsenic, mercury, lead and cadmium [112]. Associated with this is the action of alginates and other seaweed polysaccharides in binding divalent cations. This has led to a concern over whether their inclusion in the diet could affect the bioavailability of calcium, iron and some trace elements (reviewed in [113]). However, it would be likely that these cations would be absorbed in the large intestine as dietary fibre breaks down. There is no evidence that alginate inclusion in the diet drives micronutrient deficiency in the toxicity studies previously carried out. Carrageenan intake has been linked to breast cancer progression by in vitro studies, which have become somewhat magnified in the safety literature [114].

CONCLUSION

Seaweed is a foodstuff that has been historically consumed around the globe but is only consumed in appreciable amounts in certain areas of the world today. Seaweed polysaccharides have been used within the food industry in a wide number of important applications aimed at improving the sensory properties and shelf-life of food products. Previous research would suggest that incorporation of whole seaweeds and seaweed polysaccharides into foods is generally acceptable to the consumer. Seaweed or seaweed-isolate enrichment may not only benefit the nutritional value of a food product, but may also benefit the product in terms of improving the shelf-life and in some cases actually improving the sensorial properties.

While chemical analysis would suggest a number of nutritional benefits of seaweed consumption, there is a need for a more evidence relating dietary intake to health. Acute intervention studies would suggest alginate consumption could have long-term benefits to parameters of cardiovascular health and in appetite regulation. As with whole seaweeds, there is a need for long-term dietary intervention studies in this area. Design of intervention studies is crucial to their success. As such, nutrition or health researchers should collaborate early on with food technologists/food industry in order to design and develop suitably appealing products with these ingredients.

REFERENCES

[1] Patarra RF, Paiva L, Neto AI, Lima E, Baptista J. *J. Appl. Phycol.* (2010):1.
[2] MacArtain P, Gill CIR, Brooks M, Campbell R, Rowland IR. *Nutr. Rev.* (2007);65:535.
[3] Zemke-White WL, Ohno M. *J. Appl. Phycol.* (1999);11:369.

[4] Bixler HJ, Porse H. *J. Appl. Phycol.* (2010):1.
[5] Gómez-Ordóñez E, Jiménez-Escrig A, Rupérez P. *Food Res. Int.* (2010);43:2289.
[6] Romarís-Hortas V, García-Sartal C, Barciela-Alonso MC, Moreda-Piñeiro A, Bermejo-Barrera P. *J. Agric. Food. Chem.* (2010);58:1986.
[7] Bittner L, Payri CE, Couloux A, Cruaud C, de Reviers B, Rousseau F. *Mol. Phylogenet. Evol.* (2008);49:211.
[8] Burreson BJ, Moore RE, Roller PP. *J. Agric. Food Chem.* (1976);24:856.
[9] [9] Denis C, Morançais M, Li M, Deniaud E, Gaudin P, Wielgosz-Collin G, et al. *Food Chem.* (2010);119:913.
[10] Kolb N, Vallorani L, Stocchi V. *Acta Aliment* (1999);28:213.
[11] Mattio L, Payri CE. *Bot. Rev.* (2011);77:31.
[12] Saunders GW. Philosophical Transactions of the Royal Society B: *Biological Sciences* (2005);360:1879.
[13] Serrão EA, Alice LA, Brawley SH. *J. Phycol.* (1999);35:382.
[14] Watanabe F, Takenaka S, Katsura H, Masumder SAMZH, Abe K, Tamura Y, et al. *J. Agric. Food Chem.* (1999);47:2341.
[15] Yang YJ, Nam S-, Kong G, Kim MK. *Br. J. Nutr.* (2010);103:1345.
[16] Teas J, Hebert JR, Fitton JH, Zimba PV. *Med. Hypotheses* (2004);62:507.
[17] Jiménez-Escrig A, Sánchez-Muniz FJ. *Nutr. Res.* (2000);20:585.
[18] Cofrades S, López-Lopez I, Bravo L, Ruiz-Capillas C, Bastida S, Larrea MT, et al. *Food Sci. Technol. Int.* (2010);16:361.
[19] Cardoso MA, Hamada GS, De Souza JMP, Tsugane S, Tokudome S. *Journal of Epidemiology* (1997);7:198.
[20] Yamori Y, Miura A, Taira K. *Asia Pac. J. Clin. Nutr.* (2001);10:144.
[21] Kadam SU, Prabhasankar P. *Food Res. Int.* (2010);43:1975.
[22] Rose M, Lewis J, Langford N, Baxter M, Origgi S, Barber M, et al. *Food and Chemical Toxicology* (2007);45:1263.
[23] Wakai K. *Journal of Epidemiology* (2009);19:1.
[24] Lee HJ, Kim HC, Vitek L, Nam MC. *J. Nutr. Sci. Vitaminol.* (2010);56:13.
[25] Lee B. (2008).
[26] Prabhasankar P, Ganesan P, Bhaskar N, Hirose A, Stephen N, Gowda LR, et al. *Food Chem.* (2009);115:501.
[27] López-López I, Bastida S, Ruiz-Capillas C, Bravo L, Larrea MT, Sánchez-Muniz F, et al. *Meat Sci.* (2009);83:492.
[28] Kim H-, Choi J-, Choi Y-, Han D-, Kim H-, Lee M-, et al. *Korean Journal for Food Science of Animal Resources* (2010);30:55.
[29] Cofrades S, López-López I, Solas MT, Bravo L, Jiménez-Colmenero F. *Meat Sci.* (2008);79:767.
[30] López-López I, Cofrades S, Ruiz-Capillas C, Jiménez-Colmenero F. *Meat Sci.* (2009);83:255.
[31] Jiménez-Colmenero F, Cofrades S, López-López I, Ruiz-Capillas C, Pintado T, Solas MT. *Meat Sci.* (2010);84:356.
[32] López-López I, Cofrades S, Yakan A, Solas MT, Jiménez-Colmenero F. *Food Res. Int.* (2010);43:1244.
[33] Cofrades S, López-López I, Ruiz-Capillas C, Triki M, Jiménez-Colmenero F. *Meat Sci.* (2011);87:373.

[34] Hall AC, Fairclough A, Mahadevan K, Paxman JR. P *Nutr. Soc.* (2010);69:E352.
[35] Prabhasankar P, Ganesan P, Bhaskar N. *Food Sci. Technol. Int.* (2009);15:471.
[36] Gupta S, Cox S, Rajauria G, Jaiswal AK, Abu-Ghannam N. *Food and Bioprocess Technology* (2011):1.
[37] Cox S, Abu-Ghannam N, Gupta S. *International Food Research Journal* (2010);17:205.
[38] Kumaran S, Deivasigamani B, Alagappan K, Sakthivel M, Karthikeyan R. *Asian Pacific Journal of Tropical Medicine* (2010);3:977.
[39] Vlachos V, Critchley AT, Von Holy A. *Bot. Mar.* (1999);42:165.
[40] Devi KP, Suganthy N, Kesika P, Pandian SK. *BMC Complementary and Alternative Medicine* (2008);8.
[41] Carvalho AFU, Portela MCC, Sousa MB, Martins FS, Rocha FC, Farias DF, et al. *Brazilian Journal of Biology* (2009);69:969.
[42] Dawczynski C, Schubert R, Jahreis G. *Food Chem.* (2007);103:891.
[43] Goñi I, Gudiel-Urbano M, Saura-Calixto F. *J. Sci. Food Agric.* (2002);82:1850.
[44] Marrion O, Fleurence J, Schwertz A, Guéant J-, Mamelouk L, Ksouri J, et al. *J. Appl. Phycol.* (2005);17:99.
[45] Ortiz J, Romero N, Robert P, Araya J, Lopez-Hernández J, Bozzo C, et al. *Food Chem.* (2006);99:98.
[46] Sánchez-Machado DI, López-Cervantes J, López-Hernández J, Paseiro-Losada P. *Food Chem.* (2004);85:439.
[47] Wong KH, Cheung PCK. *Food Chem.* (2000);71:475.
[48] Moe S, Draget KI, Skjak-Braek G, Smidsrod O. In: Stephen AM, editor. *Food polysaccharides and their applications* New York: Marcel Decker(1995). p. 245.
[49] Jiao G, Yu G, Zhang J, Ewart HS. *Marine Drugs* (2011);9:196.
[50] De Ruiter GA, Rudolph B. *Trends in Food Science and Technology* (1997);8:389.
[51] Mancini F, McHugh TH. *Die Nahrung* (2000);44:152.
[52] Wendin K, Aaby K, Edris A, Ellekjaer MR, Albin R, Bergenstahl B, et al. *Food Hydrocolloids* (1997);11:87.
[53] Gujral HS, Sharma P, Singh N, Sogi DS. *Journal of Food Science and Technology* (2001);38:314.
[54] Brownlee IA, Seal CJ, Wilcox M, Dettmar PW, Pearson JP. In: Rehm BHA, editor. *Alginates: Biology and applications* Berlin, Germany: Springer(2009). p. 221.
[55] Dickinson E. *Food Hydrocoll.* (2003);17:25.
[56] Garti N, Leser ME. *Polym. Adv. Technol.* (2001);12:123.
[57] Kumar M, Sharma BD, Kumar RR. *Asian-Australasian Journal of Animal Sciences* (2007);20:588.
[58] Kumar N, Sahoo J. *Journal of Food Science and Technology* (2006);43:410.
[59] Lin KW, Keeton JT. *Journal of Food Science* (1998);63:571.
[60] Yuan YV, Bone DE, Carrington MF. *Food Chem.* (2005);91:485.
[61] Wang T, Jónsdóttir R, Ólafsdóttir G. *Food Chem.* (2009);116:240.
[62] Gupta S, Cox S, Abu-Ghannam N. *LWT - Food Science and Technology* (2011);44:1266.
[63] Rupérez P, Ahrazem O, Leal JA. *J. Agric. Food Chem.* (2002);50:840.
[64] Keyrouz R, Abasq ML, Bourvellec CL, Blanc N, Audibert L, Argall E, et al. *Food Chem.* (2011);126:831.

[65] Heo S-, Park E-, Lee K-, Jeon Y-. *Bioresour. Technol.* (2005);96:1613.
[66] Valko M, Leibfritz D, Moncol J, Cronin MTD, Mazur M, Telser J. *International Journal of Biochemistry and Cell Biology* (2007);39:44.
[67] Briviba K, Bub A, Möseneder J, Schwerdtle T, Hartwig A, Kulling S, et al. *Nutr. Cancer* (2008);60:164.
[68] Frei B. *J. Nutr.* (2004);134:3196S.
[69] Ohigashi H, Sakai Y, Yamaguchi K, Umezaki I, Koshimizu K. *Biosci. Biotechnol. Biochem.* (1992);56:994.
[70] Yuan YV, Carrington MF, Walsh NA. *Food and Chemical Toxicology* (2005);43:1073.
[71] Apostolidis E, Lee CM. *J. Food Sci.* (2010);75:H97.
[72] Ben Rebah F, Smaoui S, Frikha F, Gargouri Y, Miled N. *Appl. Biochem. Biotechnol.* (2008);151:71.
[73] Iwai K. *Plant Foods for Human Nutrition* (2008);63:163.
[74] Zhang J, Tiller C, Shen J, Wang C, Girouard GS, Dennis D, et al. *Can. J. Physiol. Pharmacol.* (2007);85:1116.
[75] Shin H-, Hwang HJ, Kang KJ, Lee BH. *Arch. Pharm. Res.* (2006);29:165.
[76] Aneiros A, Garateix A. Journal of Chromatography B: *Analytical Technologies in the Biomedical and Life Sciences* (2004);803:41.
[77] Chakraborty C, Hsu C-, Wen Z-, Lin C-. *Current Topics in Medicinal Chemistry* (2009);9:1536.
[78] Dhargalkar VK, Verlecar XN. *Aquaculture* (2009);287:229.
[79] Bocanegra A, Bastida S, Benedí J, Ródenas S, Sánchez-Muniz FJ. *Journal of Medicinal Food* (2009);12:236.
[80] Teas J. *Nutr. Cancer* (1983);4:217.
[81] Wong KH, Sam SW, Cheung PCK, Ang Jr. PO. *Nutr. Res.* (1999);19:1519.
[82] Bocanegra A, Bastida S, Benedí J, Nus M, Sánchez-Montero JM, Sánchez-Muniz FJ. *Br. J. Nutr.* (2009);102:1728.
[83] Lamare MD, Wing SR. *N Z J Mar. Freshwat. Res.* (2001);35:335.
[84] Marinho-Soriano E, Fonseca PC, Carneiro MAA, Moreira WSC. *Bioresour. Technol.* (2006);97:2402.
[85] Martínez B, Rico JM. *J. Phycol.* (2002);38:1082.
[86] Miyake Y, Sasaki S, Ohya Y, Miyamoto S, Matsunaga I, Yoshida T, et al. *Ann. Epidemiol.* (2006);16:614.
[87] Kondo I, Funahashi K, Nakamura M, Ojima T, Yoshita K, Nakamura Y. *J. Epidemiol.* (2010);20 Suppl 3:S576.
[88] Nakayama Y, Sakauchi F, Mori M. *Sapporo Medical journal* (2008);76:33.
[89] Okubo H, Sasaki S, Murakami K, Kim MK, Takahashi Y, Hosoi Y, et al. *Int. J. Obes.* (2008);32:541.
[90] Shimazu T, Kuriyama S, Hozawa A, Ohmori K, Sato Y, Nakaya N, et al. *Int. J. Epidemiol.* (2007);36:600.
[91] Kim MS, Kim JY, Choi WH, Lee SS. *Nutr. Res. Practice* (2008);2:62.
[92] Goñi I, Valdivieso L, Garcia-Alonso A. *Nutr. Res.* (2000);20:1367.
[93] Hall AC, Fairclough AC, Mahadevan K, Paxman JR. *Submitted for publication* (2011).
[94] You JS, Sung MJ, Chang KJ. *Nutrition Research and Practice* (2009);3:307.
[95] Teas J, Hurley TG, Hebert JR, Franke AA, Sepkovic DW, Kurzer MS. *J. Nutr.* (2009);139:1779.

[96] Teas J, Braverman LE, Kurzer MS, Pino S, Hurley TG, Hebert JR. *Journal of Medicinal Food* (2007);10:90.
[97] Panlasigui LN, Baello OQ, Dimatangal JM, Dumelod BD. *Asia Pac. J. Clin. Nutr.* (2003);12:209.
[98] Hoad CL, Rayment P, Spiller RC, Marciani L, De Celis Alonso B, Traynor C, et al. *J. Nutr.* (2004);134:2293.
[99] Paxman JR, Richardson JC, Dettmar PW, Corfe BM. *Appetite* (2008);51:713.
[100] Pelkman CL, Navia JL, Miller AE, Pohle RJ. *Am. J. Clin. Nutr.* (2007);86:1595.
[101] Dettmar PW, Strugala V, Craig Richardson J. *Food Hydrocoll.* (2011);25:263.
[102] Mattes RD. *Physiology and Behavior* (2007);90:705.
[103] Appleton KM, Rogers PJ, Blundell JE. *Journal of Human Nutrition and Dietetics* (2004);17:425.
[104] Torsdottir I, Alpsten M, Holm G, Sandberg AS, Tolli J. *J. Nutr.* (1991);121.
[105] Wolf BW, Lai C-, Kipnes MS, Ataya DG, Wheeler KB, Zinker BA, et al. *Nutrition* (2002);18:621.
[106] Williams JA, Lai C-, Corwin H, Ma Y, Maki KC, Garleb KA, et al. *J. Nutr.* (2004);134:886.
[107] Paxman JR, Richardson JC, Dettmar PW, Corfe BM. *Nutr. Res.* (2008);28:501.
[108] Sandberg AS, Andersson H, Bosaeus I, Carlsson NG, Hasselblad K, Harrod M. *Am. J. Clin. Nutr.* (1994);60:751.
[109] Crawford, B.A., Cowell, C.T., Emder, P.J., Learoyd, D.L., Chua, E.L., Sinn, J. and Jack, M.M. *Med. J. Aust.* (2010);193:413-415.
[110] Marshall KLE, Vogt RL, Effler P. *West J. Med.* (1998);169:293.
[111] Bailey SE, Olin TJ, Bricka RM, Adrian DD. *Water Res.* (1999);33:2469.
[112] Hwang YO, Park SG, Park GY, Choi SM, Kim MY. *Food Additives and Contaminants: Part B Surveillance* (2010);3:7.
[113] Brownlee IA, Allen A, Pearson JP, Dettmar PW, Havler ME, Atherton MR, et al. *Critical Reviews in Food Science and Nutrition* (2005);45:497.
[114] Burges Watson D. *J. Appl. Phycol.* (2008);20:505.

Chapter 7

ANTIOXIDATIVE PROPERTIES OF SEAWEED COMPONENTS

R. Sowmya[1], N. M. Sachindra[1], M. Hosokawa[2] and K. Miyashita[2]

[1]Department of Meat, Fish and Poultry Technology,
Central Food Technological Research Institute
(Council of Scientific and Industrial Research), Mysore, India
[2]Faculty of Fisheries, Hokkaido University, Minato,
Hakodate, Japan

ABSTRACT

Seaweeds constitute one of the major components of diet in several Asian countries, particularly Japan, China and Korea. The consumption of seaweed as a part of diet has been shown to be one of the reasons for low incidence of breast and prostrate cancer in Japan and China compared to North America and Europe. Seaweeds are rich source of variety of nutrients and bioactive components. *In vivo* studies have demonstrated the anti-cancerous, anti-obesity, anti-inflammatory and anti-proliferative effects of seaweeds and their components. Reactive oxygen species (ROS) that include free radicals such as superoxide, hydroxyl and peroxyl are responsible for manifestation of oxidative stress related diseases like cancer and arteriosclerosis. ROS are the important initiators of lipid oxidation in biological membranes, which lead to many diseases. Thus, it is believed that the antioxidative components in seaweeds are responsible for their effects on disease protection. As the ROS are implicated in several diseases, antioxidants play an important role in preventing the interaction of reactive oxygen species with biological system. In addition to being implicated in diseases and aging, ROS also play an important role in chemical deterioration of food. The lipid oxidation initiated by ROS can lead to unacceptability of food products in addition to formation of harmful lipid oxidation products. Studies using model systems also reported that free radicals induce protein oxidation and prevention of protein oxidation by antioxidants has protective effects on lipid fractions. Antioxidants or ingredients having antioxidative properties are used extensively for improvement of food stability. With the focus being shifting towards finding alternatives for synthetic food ingredients, natural substances having antioxidative properties need to be further explored. Studies have been carried out on the antioxidative potential of different seaweeds mainly from the waters of China, Korea and Japan. Few studies have been reported on the antioxidative potential of extracts from Indian seaweeds. The presence of antioxidative substances in seaweeds is suggested to be an endogenous defence mechanism as a protection against oxidative stress due to extreme environmental conditions. The antioxidative components in seaweeds include chlorophyll and carotenoid pigments, vitamins like α-tocopherol and phenolic substances. The antioxidant activity of extracts from red algae, 'dulse' (*Palmaria palmate*) is associated with aqueous/alcohol soluble compounds characterized by phenolic functional groups with reducing activity. Antioxidant activity of red seaweed extracts correlates with their polyphenol content. Fucoxanthin, isolated from the brown seaweed Wakame exhibits various radical scavenging activities. The free radical scavenging activity of polysaccharide extracts from seaweeds, particularly the sulphated polysaccharides laminarin and beta-glucans, fucoidan, from seaweeds has been demonstrated. This chapter highlights the information available on the antioxidative potential of polyphenols and polysaccharides from different types of seaweeds and their potential health benefits.

1. INTRODUCTION

Marine macroalgae, which are commonly known as seaweeds, are one of nature's most biologically active resources, as they possess immense wealth of bioactive compounds as well as various biological activities. The most important use of seaweed is as food in many countries. It is also used as a fertilizer, as fuel, in cosmetics, etc [1]. Seaweeds are the only source of phycocolloides namely agar-agar, carrageenan and algin, which are extensively used in various industries such as food, confectionary, textiles, pharmaceuticals, dairy and paper industries mostly as gelling, stabilizing and thickening agents.

Seaweeds are abundantly available and valuable food source as they contain proteins, lipids, vitamins and minerals. They are an important source of dietary fiber, mainly soluble fiber which is considered important in prevention of constipation, colon cancer, cardiovascular disease and obesity [2-4]. Seaweeds are commonly classified into three main groups based on their pigmentation. Phaeophyta, or brown seaweeds, are predominantly brown due to the presence of the carotenoid fucoxanthin, and the primary polysaccharides present include alginates, laminarins, fucans and cellulose [5,6]. Chlorophyta, or green seaweeds, are dominated by chlorophyll a and b, with ulvan being the major polysaccharide component [7]. The principal pigments found in rhodophyta, or red seaweeds are phycoerythrin and phycocyanin and the primary polysaccharides are agars and carrageenans [1].

Seaweeds also contain vitamins and vitamin precursors, including α-tocopherol, β-carotene, niacin, riboflavin, thiamin and ascorbic acid [8,9]. It has also been reported that *Laurencia obtuse* (Rhodophyta) contains sesquiterpenes [10]. These compounds directly or indirectly contribute to the scavenging, inhibition or suppression of radical production. They also contain long chain polyunsaturated essential fatty acids from the omega -3 family such as eicosapentaenoic acid (EPA), which may reduce the risk of heart disease, thrombosis and atherosclerosis [11] and fatty acids of some seaweeds posses antiviral activity [12].

Different carotenoids, such as α- and β-carotene, lutein, zeaxanthin and fucoxanthin, among others have been identified in seaweeds. Fucoxanthin is the main carotenoid present in brown seaweeds, belonging to the group of xanthophylls and has exhibited a potent antioxidant activity [13]. In algal extracts, phospolipids, phenols and chlorophyll related compounds are also lipid-soluble antioxidative compounds having antioxygenic activity [14]. *Grateloupia elliptica,* one of the red algae, is rich in lipoxygenase inhibitors [15].

2. BIOFUNCTIONS OF SEAWEEDS

Seaweeds possess diverse biofunctions such as antibacterial activity [16], antioxidant potential [17], anti-inflammatory properties [18], anti-coagulant activity [19], anti-viral activity [20] and apoptotic activity [21]. An algal antioxidant-mediated mechanism was hypothesized as a contributing factor in the inhibition of mammary carcinogenesis by dietary kelp in the presence of enhanced antioxidant enzyme activity and reduced lipid peroxides in livers of treated rats [22]. Moreover, antioxidant and/or antimutagenic effects of dietary seaweeds have been observed in rodent model studies of colon and skin carcinogenesis, wherein treated animals exhibited suppression of tumor initiation [23,24].

Reactive oxygen species (ROS) that include free radicals such as superoxide, hydroxyl and peroxyl are responsible for manifestation of oxidative stress related diseases like cancer and arteriosclerosis. ROS are the important initiators of lipid oxidation in biological membranes, which lead to many diseases. Thus, it is believed that the antioxidative components in seaweeds are responsible for their effects on disease protection. Ingredients having antioxidative properties are used extensively for improvement of food stability. With the focus is being shifting towards finding alternatives for synthetic food ingredients, natural substances having antioxidative properties need to be further explored. Studies have been carried out on the antioxidative potential of different seaweeds mainly from the waters of China, Korea and Japan. The presence of antioxidative substances in seaweeds is suggested to be an endogenous defence mechanism as a protection against oxidative stress due to extreme environmental conditions.

3. ROS AND ITS EFFECT ON HUMAN HEALTH

Research in recent years has shown the implication of oxidative and free radical mediated reactions in degenerative processes related to aging and diseases like atherosclerosis, dementia and cancer [25-27]. ROS include superoxide anion (O_2^-), hydroxyl radical ($^{\cdot}OH$) which is the most reactive, hydrogen peroxide (H_2O_2), hypochlorite ion and nitric oxide (NO) which are highly toxic to cells, cause oxidative damage and target the membrane lipids, proteins and deoxyribonucleic acid and hence cause tissue injury [28,29]. They also cause qualitative decay of foods which leads to rancidity, toxicity and destruction of biomolecules important in physiologic metabolism. A radical (also called a "free radical") is a cluster of atoms one of which contains an unpaired electron in its outermost shell of electrons. This is an extremely unstable configuration, and radicals quickly react with other molecules or radicals to achieve the stable configuration of 4 pairs of electrons in their outermost shell.

When generation of the ROS overtakes the antioxidant defense of the cells, oxidative damage of the cellular macromolecules (lipids, proteins, and nucleic acids) occurs, leading finally to various pathological conditions. Cell membranes, DNA and proteins are the main target molecules of free radicals and singlet oxygen. Membranes are chiefly affected by peroxyl radicals that attack polyunsaturated phospholipids at the bisallylic methylenes thus compromising their integrity [30,31]. Radical reactions modify the purine and pyrimidine bases of DNA, whereas proteins are oxidized at the metal binding sites of amino acid residues leading, in the case of some amino acids, to loss of ammonia and the formation of carbonyl derivatives [32]. Lipid radical peroxidation yielding rancid smelling substances as end products is the main cause of the qualitative decay of foods rich in fatty acids [33].

DNA damage is known to be one of the most sensitive biological markers for evaluating oxidative stress representing the imbalance between free radical generation and efficiencies of the antioxidant system [34,35]. ROS-mediated lipid peroxidation, oxidation of proteins, and DNA damage are well-known outcomes of oxygen-derived free radicals, leading to cellular pathology and ultimately to cell death. Various necrotic factors, proteases, and ROS from damaged cells also attack the adjacent cells, resulting ultimately in tissue injury. Furthermore, tissue injury itself has been reported to cause severe oxidative stresses. Injury caused by ischemia reperfusion, heat, trauma, freezing, severe exercise, toxins, radiation or infection; leads to the generation of ROS, and development of various disease processes. Oxidative stress results in numerous diseases and disorders such as cancer, stroke, myocardial infarction, diabetes, septic and hemorrhagic shock Alzheimer's and Parkinson's diseases [36].

Oxidized LDL is known to alter endothelial cell functions, thereby modifying their capacity to prevent cell adhesion and their fibrinolytic properties. Recent studies have emphasized the role of oxidized LDLs (OxLDLs) in the pathogenesis of atherosclerosis, which is characterized by intimal thickening and lipid accumulation in the arteries [37]. Endothelial dysfunction or activation elicited by OxLDLs is the key step in the initiation of atherosclerosis. In response to stimulation by OxLDLs, endothelial cells reduce the release of NO, express adhesion molecules, and secrete chemoattractant and growth factors. Subsequently, OxLDLs are avidly ingested by macrophages, resulting in foam cell formation which eventually dumps cholesterol inside the arterial wall. OxLDLs are also involved in inducing smooth muscle cell (SMC) migration, proliferation, and transformation [38].

An inflammatory response is characterized by the attraction of large amounts of leukocytes (neutrophyles, monocytes, macrophages, and mast cells) to the inflamed area. These inflammatory cells are triggered by mediators of inflammation and generate superoxide anion and nitric oxide radicals. Thus, antioxidant and anti-inflammatory activities are very close to each other [39].

4. DEFENSE AGAINST ROS

The defense against ROS is achieved by two mechanisms, namely a) elimination of ROS by antioxidant enzymes (primary defence) and b) free-radical scavenging (secondary defence). Antioxidant enzymes are considered to be a primary defense that prevents biological macromolecules from oxidative damage. Superoxide dismutase (SOD), catalase, and peroxidases like glutathione peroxidase (GSH-PX) and heme peroxidase are the important antioxidant enzymes. SOD catalyses the breakdown of O_2^- to O_2 and H_2O_2 and has thereby been implicated as an essential defense against the potential toxicity of reactive oxygen [29]. Predominant forms of SOD are the copper-containing enzyme and the zinc-containing enzyme, located in the cytosol and manganese containing SOD in the mitochondrial matrix [40]. Glutathione peroxidase catalyses the reaction of hydroperoxides with reduced glutathione (GSH) to form glutathione disulphide (GSSG) and the reduction product of the hydroperoxide, hence destroys toxic peroxides [41]. Catalase catalyses the decomposition of H_2O_2 to water and oxygen and thus protects the cell from oxidative damage by H_2O_2 and ˙OH [42].

When some compounds react with others reactive substances to produce a lesser harmful species, they are called antioxidants whereas when they completely neutralize they are called scavengers which are considered as secondary defense mechanism [29]. The negative effects of oxidative stress may be mitigated by antioxidants [43,44]. The commonly used synthetic antioxidants are butylated hydroxyanisole (BHA), butylated hydroxytoluene (BHT), ethoxyquin and propyl gallate. However, since the synthetic antioxidants like butylated hydroxyanisole and butylated hydroxytoluene were suspected for being responsible for liver damage and carcinogenesis [45,46], natural antioxidants that includes phenolic and nitrogen containing compounds and carotenoids, have definitely have an edge over them. These natural antioxidants, which have the capacity to protect the human body from radicals and retard the progress of many chronic diseases [47], are obtained mainly from terrestrial sources. However, in search of new antioxidants, exploration of aquatic habitats has led to the discovery that marine plants and invertebrates also contain antioxidants. Seaweed is considered to be a rich source of antioxidants [48,49]. These natural antioxidants are considered to be safe because of no chemical contamination and are readily accepted by the consumers [50].

Marine algae are exposed to a combination of light and oxygen that leads to the formation of free radicals and other strong oxidizing agents, but the absence of oxidative damage in their structural components and their stability to oxidation suggest they have good antioxidant defense system [51]. The compounds responsible of the antioxidant activity in macroalgae include carotenoids, vitamin E or α-tocopherol and chlorophylls and its derivatives within the lipidic fraction and polyphenols, vitamin C or ascorbic acid and phycobiliproteins within the aqueous fraction [13,52].

Marine algae are now being considered to be a rich source of antioxidants [53]. Some active antioxidant compounds from brown algae were identified to be polyphenols [48], polysachharides and their derivatives [54] and carotenoids [55,56]. Active compounds related to antioxygenic activities usually are aqueous, ethanolic or lipidsoluble extracts. In seaweeds they were either lipoxygenase inhibitors [51] or phlorotannins [57]. The active principles identified in lipid soluble extracts were either phospholipids [58,59], substituted phenols and o-diphenols [60,61] or chlorophyll and related compounds [48,62,63].

Phenols are particularly effective antioxidants for polyunsaturated fatty acids because they can easily transfer a hydrogen atom to lipid peroxyl radicals (reaction (1)) [64,65]. The aryloxyl radical, formed in reaction (l), is usually too sluggish to act as a chain carrier [66,67] but can easily couple with another radical to give non-radical products (reactions (2) and (3)).

$$LOO˙ + ArOH \longrightarrow LOOH + O˙ \tag{1}$$

$$ArO˙ + LOO˙ \longrightarrow \text{non-radical products} \tag{2}$$

$$ArO^· + ArO^· \longrightarrow \text{non-radical products} \tag{3}$$

The mechanism of singlet oxygen quenching by phenols probably involves, in polar solvents, an electron transfer from phenol to singlet oxygen resulting in formation of a radical ion pair [68], the same mechanism has also been observed with aromatic amines [69] and thioanisoles [70]. This process is reversible and the back electron transfer leads to the starting phenol and triplet oxygen [66].

Flavonoids act either by blocking the generation of hypervalent metal forms [71], by scavenging free radicals [72], or by breaking lipid peroxidation chain reactions [73]. Carotenoids are also capable of reacting with radical species [74]. The antioxidant mechanism of carotenoids is attributed to their ability to quench singlet oxygen and scavenge free radicals [75]. The quenching of singlet oxygen by carotenoids has been attributed mainly to physical mechanism [76], where the excess energy of singlet oxygen is transferred to carotenoid. The carotenoid with added energy is excited to triplet state and upon losing the energy as heat relaxes to singlet state without change in the structure as follows,

$$^1O_2· + {}^1Car \longrightarrow {}^3O_2 + {}^3Car·$$

$$^3Car· \longrightarrow {}^1Car + heat$$

The electron rich status of carotenoids makes them more suitable for reaction with the free radicals thus avoiding the use of cellular components by the free radicals for reactions. Krinsky and Yeum [77] reviewed the carotenoid-radical interactions and suggested following three possible interactions,

1. adduct formation:
$$Car + R· \longrightarrow R\text{-}Car$$
2. electron transfer:
$$Car + R· \longrightarrow Car·^+ + R^-$$
3. allylic hydrogen abstraction:
$$Car + R· \longrightarrow Car· + RH$$

5. ANTIOXIDANT ACTIVITY OF SEAWEED POLYPHENOLS

Polyphenols are a structural class of organic chemicals characterized by the presence of large multiples of phenol units. The physiological effects of polyphenolic compounds as chemopreventive or prophylactic agents to reduce the risk of heart and vascular diseases have been receiving a lot of attention [78]. Many epidemiological and *in-vivo* studies have shown that these antioxidants can reduce lipid peroxidation which is involved in atherogenesis, thrombosis, and erectile dysfunction. The only significant class of polyphenolic compounds from marine plants is phlorotannins (Figure 1) or algal polyphenolic compounds, which are major phenolic compounds in brown algae [79]. They have a wide range of molecular sizes (0.4 to 400 kDa) and can occur in variable concentrations (0.5-20% of the dry weight) in brown algae.

Phlorotannins are commonly believed to have defensive or protective functions, e.g., against herbivores, bacteria, fungi, fouling organisms and UV-B radiation, and to function in wound healing processes or in cell-wall construction. The polyphenols of terrestrial plants are gallo- or condensed tannins, are largely based on polymers of 4-8 linked flavan-3-ols, with some esterified to gallic acid whereas, the polyphenols of algae such as the brown kelps are phlorotannins, which can account for between 1% and 10% of the dry weight [80].

Phaeophyceae such as the Laminariales, are known to synthesize UV-inducible polyphenols, i.e. phlorotannins, which absorb in the UVB range (280–320 nm). Also, kelp tissue levels of phlorotannins have been reported to increase during short-term exposure of *Macrocystis integrifolia* to UVA (320–400 nm) and UVB irradiation [79]. Phlorotannins are polymers and oligomers of 1,3,5-trihydroxybenzene (phloroglucinol) [81,82].

Extracts from edible seaweeds such as Dulse (*Palmaria palmata*) shown to be effective antioxidants and capable of inhibiting cancer cell proliferation, has phlorotannins [83]. The total phenolic content four species of brown algae, *Himanthalia elongata, Laminaria saccharina* and *Undaria pinnatifida* was determined, out of which, *H elongata* exhibited the highest content 10.0 g phloroglucinol/kg wet weight of sample [84].

Fucus vesiculosus is a brown seaweed species which is rich in phlorotannins, and which also contains fucoxanthin, a major marine carotenoid. *In-vitro* and *in-vivo* studies on toxicity and antioxidant activity indicated that they are non-toxic and exhibits important antioxidant activity against chemically generated oxidants *in-vitro* and in activated RAW 264.7 macrophages [41].

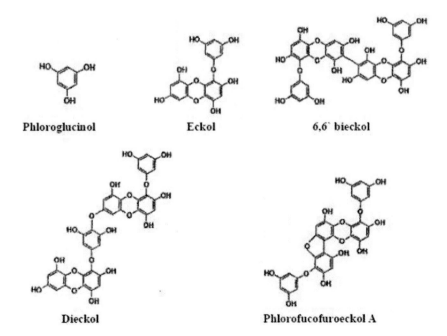

Figure 1. Chemical structure of phlorotannins from seaweeds.

Comparison of polyphenolic content and *in-vitro* antioxidant activities of different seaweeds were made by many workers. Comparison of polyphenolic compounds of brown seaweeds such as *Eisenia bicyclis, Ecklonia Stolonifera, Ecklonia cava, Ecklonia kurome* and *Hizikia fusiformis* with synthetic as well as known antioxidants by 1,1- diphenyl-2-picryhydarzyl (DPPH) radical scavenging activity was done by Kang et al [85]. It was found that these seaweeds were, in fact better than that of well-known antioxidants such as catechin, tocopherol, BHA or BHT. They also showed significant protection against LDL oxidation. *In-vivo* measurement of their reducing power showed that their absorption into the blood stream and clearance kinetics were for polyphenolic compounds. The affinity of polyphenolic compounds toward the exposed connective tissue of damaged blood vessels is thought to further potentiate their effectiveness since these damaged areas are the most susceptible sites for oxidative stress due to inflammatory action [85]. In these extracts, one fraction was a complex of polyphenol, phlorotannin with a molecular weight of 10-400 kDa, while another fractions had 6,6-bieckol, 8,8-bieckol, dieckol and phlorofurofukoeckol [85]. Antioxidant activities of the aqueous extracts of four marine algae, *Caulerpa racemosa var. macrophysa, Gracilaria tenuistipitata var. tenuistipitata, Sargassum sp.*, and *Ulva lactuca*, from the coastal areas in Southern Thailand were determined by Yangthong et al [86] in which *Sargassum* sp showed significantly higher total phenolic content and antioxidant activities than the others.

The antioxidative activity of different solvent fractions of *Polysiphonia urceolata* were measured by DPPH radical scavenging assay and the β-carotene linoleate assay systems, and compared with that of BHT, gallic acid and ascorbic acid. Crude extract and the ethyl acetate-soluble fraction exhibited higher antioxidative activity than BHT. The total phenolic content and also the reducing power showed positive correlations with antioxidant activity and DPPH radical-scavenging activity [87]. Sachindra et al [88] compared the radical scavenging and singlet oxygen quenching activity of different solvent extracts from Indian seaweeds. They observed that the methanol extract and its fractions from brown seaweeds exhibited higher ABTS (2,2'-azinobis 3-ethylbenzothizoline-6-sulfonic acid) radical scavenging activity with more than 90% scavenging in butanol and ethyl acetate fractions and correlated with polyphenol content. It was also observed that ethyl acetate fraction from crude extract showed higher inhibitory activity against hemoglobin induced linoleic acid oxidation and singlet oxygen quenching activity of the crude extract from brown seaweed was lower (<13%) compared to red seaweeds (16.4–20.5%). *Sargassum polycystum* and *Laurencia obtusa* of Seribu Islands waters, Indonesia were tested for antioxidant activity using the thiocyanate method, which showed that only extracts of fresh material showed activity and *L. obtusa* extracts had higher antioxidant activity than those of *S. polycystum* [89].

The dulse and kelp extracts inhibited HeLa cell proliferation in a dose-dependent manner and showed a positive relationship between algal extract polyphenol content and inhibition of HeLa cell proliferation suggesting a link related to extract content of kelp phlorotannins and dulse polyphenols including mycosporine-like amino acids and phenolic acids [17]. Thus algal anticarcinogenicity may involve effects on cell proliferation and antioxidant activity.

Catechins (Figure 2) belong to a small group of polyphenolic compounds, which include catechin (3-hydroxyflavan) and its epimer, and also occur as gallates with gallic acid. Epigallocatechin gallate, epigallocatechin, epicatechin gallate and epicatechin were most important components in terms of antioxidant ability among all catechins. High concentration of epigallocatechin was detected in seaweeds, but epigallocatechin gallate, epicatechin gallate

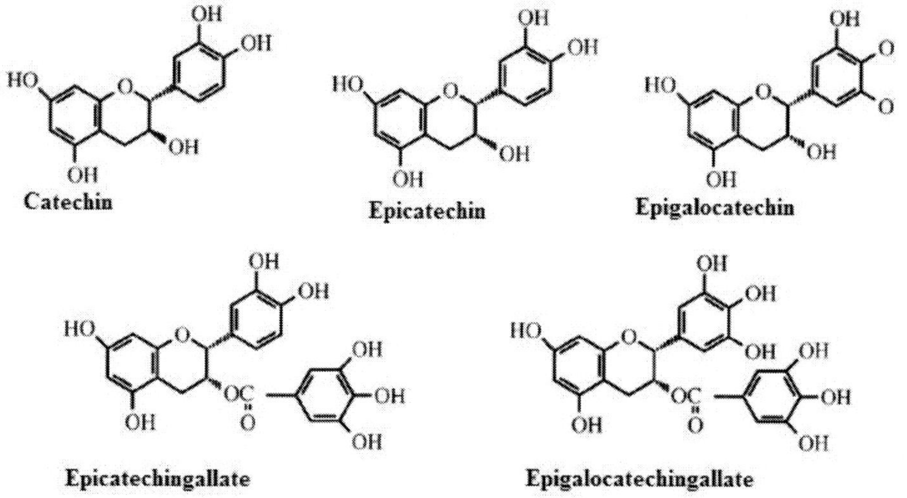

Figure 2. Chemical structure of catechins.

and epicatechin were minor compounds [90]. Highest concentration of epicatechin was found in *Eisenia bicyclis*, whereas, *Padina arborescens* contain epigallocatechin. Out of 27 seaweeds examined in this study only *Geledium elegans* contained catechin. *Eisenia bicyclis* contained catechin and its isomers, whereas *Caulerpa racemosa, Kappaphycus alvarezii, Monostroma nitidum, Undaria pinnatifida* and *Laminaria religiosa* did not contain catechin and its isomers. Catechol was found in all Japanese seaweed samples, except *E. bicyclis* [90]. Extracts of *Hizikia fusiformis* had the best antioxidant power in both absence and presence of Fe^{2+} in fish oil emulsion. The highest of scavenging effects was found in morin, followed by rutin and extract of seaweeds (*H. fusiformis* and *U. pinnatifida*). Catechin and gallic acid had the highest effects of Fe^{2+} chelating, whearas Fe^{2+} chelating effect of seaweed extracts was lower [91]. Dried *Scytosiphon lomentaria* were tested for assays of antioxidant activities, including suppression of hemoglobin-induced linoleic acid peroxidation, reducing power, ferrous ion chelating, DPPH radical-scavenging and scavenging of superoxide anion radical [92]. The water extract of *S. lomentaria* contained total phenols at about 5.5 mg catechin equivalents (CatE)/g dry sample and showed strong antioxidant activities. Bromophenols also play a major key role as an antioxygenic compound in marine red alga [93].

Some of these compounds act in *in-vivo* systems by enhancing the antioxidant enzymes and by other cellular mechanisms. For example, Triphlorethol-A, an open chain trimer of phloroglucinol, is one of phlorotannin components isolated from *E. cava* was found to scavenge intracellular reactive oxygen species (ROS) and thus prevented lipid peroxidation. The radical scavenging activity of triphlorethol-A protected the Chinese hamster lung fibroblast (V79-4) cells exposed to hydrogen peroxide (H_2O_2) against cell death, via the activation of mitogen-activated protein kinases (MAPK) previously called extracellular signal-regulated kinases (ERKs). Further, it reduced the apoptotic cells formation induced by H_2O_2. It also increased the activities of cellular antioxidant enzymes like, superoxide dismutase (SOD), catalase (CAT) and glutathione peroxidase (GPx) suggesting that the scavenging of ROS may be related to the increased antioxidant activity. Hence, it is suggested that triphlorethol-A protects V79-4 cells against H_2O_2 damage by enhancing the cellular

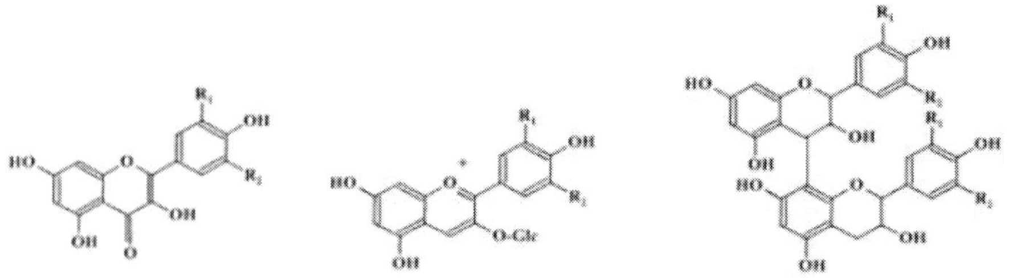

Figure 3. Chemical structure of flavonoids.

antioxidative activity. Therefore, the effects of triphlorethol-A on cell viability might involve dual actions, direct action on oxygen radical scavenging and indirect action through induction of anti-oxidative enzymes [94].

One of the functions of antioxidant is to prevent lipid oxidation. There are various examples suggesting that algal extracts inhibit lipid oxidation. One such is *Grateloupia filicina,* a red alga, the extract of which inhibit oxidation of linoleic acid and fish oil, when used as substrates and also has protecting ability for H_2O_2-induced DNA damage in rat lymphocytes [95]. Lipid oxidation studies have reported that phlorotannin-containing extracts from the kelps *Sargassum kjellmanianum* and *E cava* were effective antioxidants, the high molecular weight fractions containing polymers such as dieckol, phlorofucofuroeckol and 6-60 bieckol conferring greater protection than low molecular weight fractions containing phloroglucinol and eckol [57,96,97]. Moreover, the stable free radical quenching activity of phloroglucinol was considerably less than that of polymers such as phlorofucofuroeckol [96]. Jime´nez-Escrig et al [98] reported that extracts from Laminariales exhibited not only stable free radical scavenging activity, but also ferric ion reducing activity, albeit, the reducing activity was lower than that of the red alga *Porphyra umbilicalis.* This latter evidence may relate to the low levels of free phloroglucinol in kelps [57,80].

Since flavanoids (Figure 3) have several hydroxyl groups in its structure, they are expected to have radical scavenging activity [99]. Flavonoids include flavonols, flavones, proanthocyanidins, anthocyanidins and isoflavonoids [100]. Evidence of flavonoids has been reported from the red algae *Acanthophora spicifera* [101]. Scutellarein 4'- methyl ether (5,6,7-trihydroxy-4'-methoxy flavone) is reported from red alga, *Osmundea pinnatifida* [102]. In the study by Yumiko et al [103] on distribution of flavonoids, they detected hesperidin, catechol, morin, quercitin and myricetin in Japanese seaweeds. Eventhough, the antioxidative activity of flavonoids from seaweeds is not clear, it has been demonstrated that flavonoids in general have free radical scavenging activity [104].

6. ANTIOXIDANT ACTIVITY OF SEAWEED POLYSACCHARIDES AND THEIR DERIVATIVES

The cell walls of marine algae characteristically contain sulphated polysaccharides which are not found in land plants and which may have specific functions in ionic regulation [105]. Red algae produce sulfated galactans, agars and carrageenans, which are the main components of their cell walls. Brown seaweeds comprises of alginate as the major polysaccharide. These sulphated polysaccharides are known to exhibit many biological and physiological activities including anticoagulant, anti-thrombotic, anti-inflammatory, antiviral and antitumour activities [106-108].

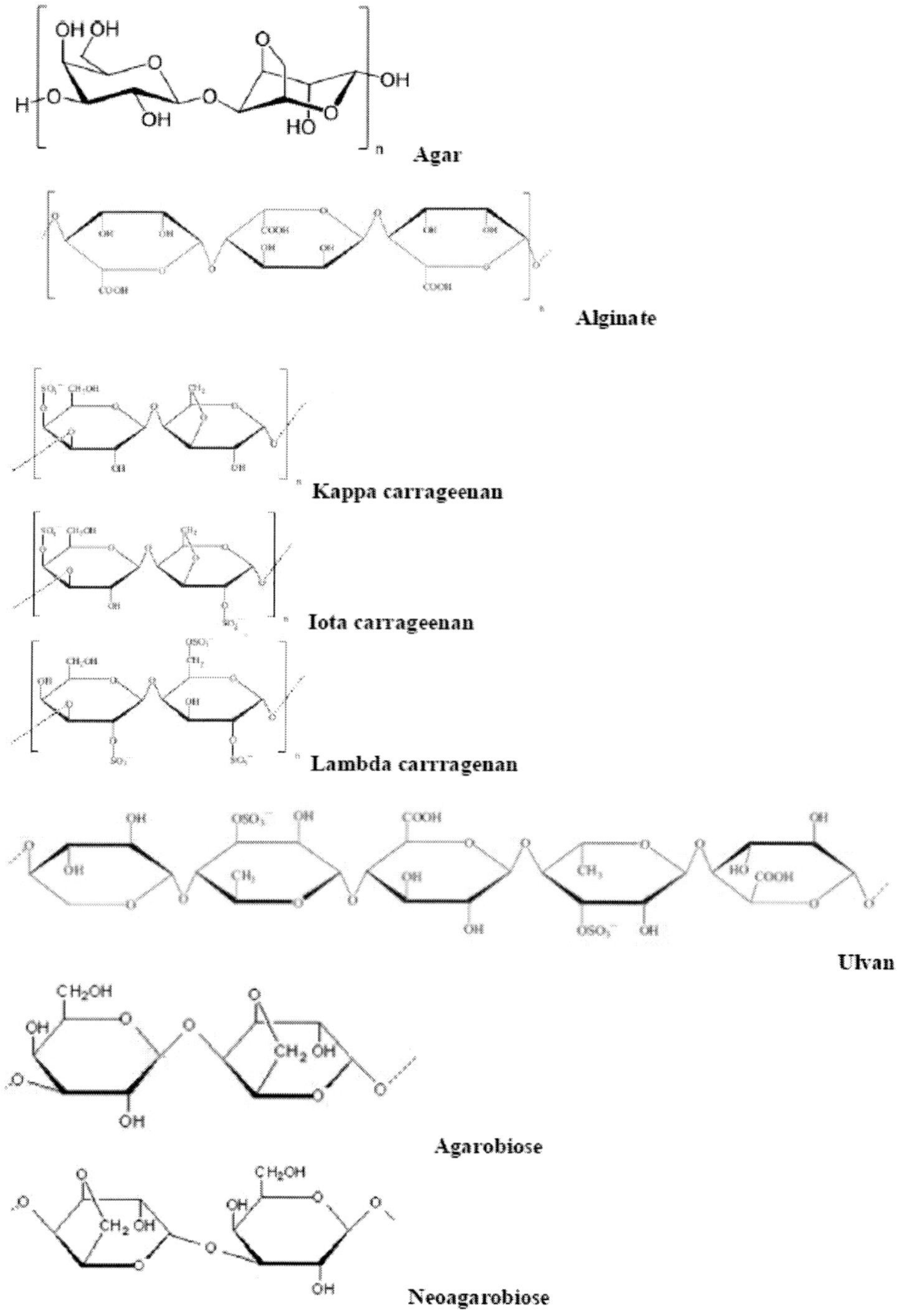

Figure 4. Chemical structure of seaweed polysaccharides and their derivatives.

In recent years, algal polysaccharides (Figure 4) were reported to be useful candidates in the search for an effective, non toxic substance and have been demonstrated to play an important role as free radical scavengers *in-vitro* and antioxidants for the prevention of oxidative damage in living organisms [109-112]. However, the activity of these polysaccharides depends on several structural parameters such as the degree of sulphation, molecular weight, sulphation position, type of sugar and glycosidic branching [113,114]. Some reports reveal that the sulphate and phosphate groups in the polysaccharides lead to differences in their biological activities. It was observed that phosphorylated and sulphated glucans exhibited higher antioxidant activities than that of the glucans and other neutral polysaccharides [115]. Several classes of sulphated polysaccharides from seaweeds including laminaran, alginic acid, fucoidan and other unidentified macromolecules present in the extracts have been shown to possess antioxidant activity [111,116,117].

Oligosaccharides have recently drawn a considerable interest due to medical effects and indications of their very specific interaction in certain carbohydrate systems with larger molecules of similar structure [118]. The heptasaccharide present in the polysaccharides have even more strong biological activity than the polysaccharides themselves [119,120]. Agar and carrageenan oligosaccharide were produced by using enzymes, agarases and carrageenases respectively from bacterial sources and both these enzymes hydrolyse the β-(1-4) linkage of respective substrate [121]. These oligosaccharide exhibit antioxidant, antitumor and cytoprotective effect [122]. Hydrolysis of seaweeds using digestive enzymes such as carbohydrases and proteases, degrade seaweed tissues to release a variety of bioactive compounds and produce some water soluble material which possess bioactivity like antioxidant activity [123].

Oligosaccharides from agar have many functions like hepatoprotective potential [124], anti-oxidative [125] and have potential applications in the food, cosmetic and medical industries [126]. *In-vivo* studies of the antioxidant effects on tissue peroxidative damage induced by carbon tetrachloride in rat model indicated that agaro-oligosaccharides could elevate the activity of superoxide dismutase (SOD), glutathione peroxidase (GSH-Px) and decrease the level of malondialdehyde (MDA), glutamate oxaloacetate transaminase (AST), glutamic pyruvic transaminase (ALT) significantly.

The different derivatives of carrageenan oligosaccharides and unmodified oligosaccharides themselves exhibited different levels of antioxidant activity [127]. Mou et al [128] found that the oral administration of carrageenan oligosaccharides was advantageous to promote the activities of antioxidant enzymes, which might play an important role on antitumor mechanism. Carrageenan oligosaccharides and their derivatives can protect the lymphocytes against UVA injury by increasing their viability, therefore, help in preventing systemic immune suppression caused by UVA. The demonstrated effect of the sulfonated carrageenan on superoxide dismutase and catalase activity suggests that carrageenan oligosaccharide is effective in promoting the antioxidation ability and eliminating danger from free radicals [129].

7. ANTIOXIDANT ACTIVITY OF SEAWEED CAROTENOIDS

One of the important characteristics of carotenoids is their ability to act as antioxidants, thus protecting cells and tissues from damaging effects of free radicals and singlet oxygen. Carotenoids have been found to be important in protecting against diseases and age related phenomena caused by oxidants [130]. Marine carotenoids including those from seaweeds (Figure 5) have been shown exhibit various health benefits such as anticancerous, antiinflamatory, antiobesity and antioxidant activity [56].

Fucoxanthin from the brown seaweed *Hijkia fusiformis* was able to quench organic radicals DPPH and 12-Doxyl-Steraic acid [131]. Fucoxanthin was identified as the active compound responsible for DPPH radical scavenging activity of organic extracts from different edible seaweeds [132]. Fucoxanthin is hydrolyzed to fucoxanthinol, and amarouciaxanthin *in-vivo* [133,134]. Fucoxanthin and its metabolites, fucoxanthinol and halocynthiaxanthin were found to be equivalent to that of α-tocopherol in antioxidant activity [135]. *In-vivo* studies indicated that fucoxanthin restrains oxidative stress induced by retinol deficiency through modulation of Na^+Ka^+-ATPase and antioxidant enzyme activities in rats [136]. The efficiency of fucoxanthin to induce antioxidant enzymes in retinol deficiency rats was found to be higher than that of β-carotene [137].

In order to study the effect of carotenoid rich seaweeds on lipid metabolism, feed containing the different levels of brown seaweed *Undaria pinnatifida* (Wakame) was fed to rats and the activities of enzymes involved in fatty acid synthesis and oxidation were evaluated [138]. Diet containing wakame was found to reduce serum and liver triacylglycerol levels, activities of fatty acid synthesis enzymes like glucose–6-phosphate dehydrogenese and

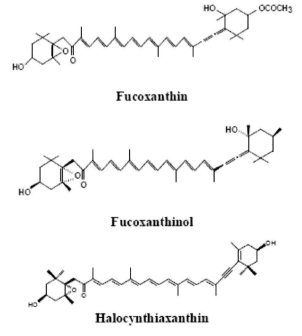

Figure 5. Structure of seaweed carotenoid fucoxanthin and its metabolites.

stimulate the fatty acid oxidation enzymes like hepatic carnitine palmitoylytransferase, acyl-CoA dehydrogenase, acyl-CoA oxidase, enoyl-CoA hydratase and 2,4-dienoyl-CoA reductase. Further, it was reported that dietary fish oil and wakame synergistically reduces the serum and liver triacylglycerol level in rat serum [139].

The antioxidant activity of tocopherols and polyphenols is by donating hydrogen to free radicals producing relatively unreactive antioxidant radicals, thus, acting as chain-breaking antioxidants, while carotenoids can act as antioxidants by quenching singlet oxygen and/or trapping free radicals [135], both resulting in suppression of lipid peroxidation [140]. Airanthi et al [141] demonstrated the synergy in the antioxidant activity of the combination of brown seaweed phenolics and fucoxanthin. The synergistic antioxidant effect of both was observed in soybean PC liposomes also.

8. MISCELLANEOUS ANTIOXIDATIVE COMPONENTS OF SEAWEEDS

Edible seaweeds are known to contain labile antioxidant molecules such as ascorbate and glutathione (GSH) when fresh [142,143], as well as more stable molecules including carotenoids (α- and β-carotene, fucoxanthin), catechins, phlorotannins as mentioned earlier and mycosporine-like amino acids (mycosporine-glycine and –taurine) [144-146] and the tocopherols [142].

Lipid soluble extracts of *Laminaria digitata* and *Himanthalia elongata* synergistically enhance the antioxidant effect of vitamin E when the latter is added to a methyl linoleate model system at a level higher than 0.1 mmol L. Antioxidant activities of lipophilic extracts of the marine algal genus *Cystoseira* in micellar model system were studied by Foti et al [147]. The activity has been ascribed to the presence of tetraprenyltoluquinols (Figure 6) in the extracts, which are tocopherol-like compounds characteristic of these algae. Among those studied, *C amentacea* var. *stricta* is one of the richest in tetraprenyltoluquinols with an intrinsic capability of transferring a hydrogen atom to the peroxyl radical.

The ability of *L digitata*, *H elongata*, *Fucus vesiculosus*, *F serratus* and *Ascophyllum nodosum* to scavenge peroxyl radicals was investigated. A synergistic effect occurred when both algal extracts (1.5 g /L) and vitamin E (0.4 mmol /L) were present, and the effectiveness of the combined antioxidants during the whole induction period was vitamin E effectiveness [148].

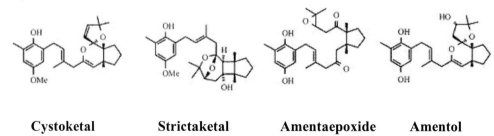

Cystoketal   **Strictaketal**   **Amentaepoxide**   **Amentol**

Figure 6. Structure of tetraprenyltoluquinols from seaweeds.

Lipophilic extracts studied from 16 species of seaweeds showed that antioxidant ability of *R confer* and *S. latiuscula* were comparable with that of BHT and relatively higher than that of propyl gallate and antioxidative property increased with increasing content of unsaturated fatty acid in the extracts [149].

Cystoseira species, having the mean concentration of α-tocopherol around 10 mg/kg of alga (dry weight), have metabolites whose structures reminiscent of those of tocopherols, and meroditerpenoids are present in much higher concentration as in *C stricta* and these metabolites acts as inhibitors of methyl linoleate peroxidation, while their $O_2\cdot$- quenching activity is rather low, similar to α-tocopherol [147]. Dimethylsulphoniopropionate (DMSP) has recently been investigated as an effective antioxidant in marine algal species. Most algal species contain high concentrations of DMSP (100–400 mmol/l) and its concentration inside cells increases rapidly with the intensity of light. It is interesting that *in-vitro* experiments suggest that DMSP can more effectively scavenge hydroxyl radical than glutathione and ascorbate [150]. The lysis of DMSP substantially increases the antioxidant protection in both aqueous and lipid membrane phases within the cells of alga. UV-A and UV-B can stimulate antioxidant capability in many algae species as an adaptive response to the irradiation [151].

Rhodophyta, such as *P palmata* and *Stictosiphonia arbuscula*, have been reported to contain between 0.2 and 0.5 mg ascorbate/g wet wt, whereas Phaeophyceae, such as Laminariales, contain only trace amounts upto 0.17 mg ascorbate/g wet wt [152,153]. Algal tissue levels of GSH are similarly variable between orders, with rhodophyta containing approx. 0.22 mg GSH/g wet wt and Phaeophyceae containing between 19 and 3082 mg GSH/100 g dry wt [143,154].

The brown seaweeds like *Ecklonia cava, Ishige okamurae, Sargassum fullvelum, S horneri, S coreanum, S thunbergii, Scytosipon lomentaria* were enzymatically hydrolyzed to prepare water soluble extracts by using five carbohydrate degrading enzymes and five proteases and their antioxidant effects were evaluated in three different reactive oxygen species assays including superoxide anion, hydroxyl radical, hydrogen peroxide and free radical (DPPH) scavenging [155]. In this investigation, Ultraflo (carbohydrate degrading enzyme) and Alcalase (protease) extract of *S horneri* exhibited strong hydrogen peroxide scavenging activities. Both extracts (Ultraflo and Alcalase extract of *S horneri*) with higher concentrations than 25 µg/ml protected at least more than 50% of DNA damage, caused by an addition of 50 µM H_2O_2 to human lymphocyte cells.

Dulse extracts were not only effective hydroxyl and stable free radical scavengers, but also inhibitors of lipid oxidation and cell proliferation *in-vitro* [156,157]. The antioxidant activity of *P palmata* extracts may be associated with the presence of a unique class of secondary metabolites, the mycosporine-like amino acids (MAAs) which function as UV-absorbing sunscreen molecules in the rhodophyta [157,158]. The MAAs, such as mycosporine-glycine, palythine and palythene, consist of glycine residue and hydroxylated sidechain functions on an aromatic methyl ester core. Algal tissue levels of these metabolites have been reported to have a positive correlation with UV exposure during growth [158].

CONCLUSION

Lipid oxidation involving ROS is one of the major causes for deterioration in food quality. ROS are implicated in many of the diseases related to oxidative stress. Antioxidants have been used as a strategy to prevent deterioration of foods as well as ROS related health implications. Due to the concern on the deleterious effects of synthetic antioxidants, the interest has been towards discovery and application of natural antioxidants. Higher plants and their components are known to be rich source of natural antioxidants. However, the focus has been shifted towards discovery of new antioxidants from marine resources. As the consumption of foods of marine origin are found to be beneficial against oxidative stress related diseases, research efforts have been towards elucidating the protective mechanism of marine foods, which demonstrated that the antioxidative principles in seaweeds is one of the reason for their health benefits.

Polyphenols, carotenoids and polysaccharides are found to be the major antioxidant compounds in seaweeds. Recently, the oligosaccharide derivatives of seaweed polysaccharides have gained much importance as bioactive molecules. Much more research efforts are needed to prepare, purify and characterize these oligosaccharides as potential antioxidative component. Developments in molecular biology have led to preparation of enzymes suitable for preparation of oligosaccharides from seaweed polysaccharides. Many new antioxidative components from seaweeds are being isolated and characterized. It is expected that seaweeds will be become one of the major natural resources for the preparation of low cost and high efficiency antioxidant compounds which can be used for food and biomedical applications.

REFERENCES

[1] McHugh, D. J. (2003). A guide to seaweed industry. *FAO Fish Tech. Pap.*, T441, 118.
[2] Dreher, M. I. (1987). *Handbook of dietary fiber*. New York, Dekker M. and Basel Inc.
[3] Kristchevsky, D. (1998). Dietary Fiber. *An. Rev. Nutri.*, 8, 301-328.
[4] Stephen, A. M., and Cummings, J. H. (1980). Mechanism of action of dietary fibre in the human colon. *Nature*, 284, 283-284.
[5] Goni, I., Valdivieso, L., and Gudiel-Urbano, M. (2002). Capacity of edible seaweeds to modify in vitro starch digestibility of wheat bread. *Nahrung*, 46, 18-20.
[6] Haugan, J. A., and Liaaenjensen, S. (1994). Algal Carotenoids 54. Carotenoids of brown-algae (Phaeophyceae). *Biochem. Syst. Ecol.*, 22, 31-41.
[7] Robic, A., Gaillard, C., Sassi, J. F., Lerat, Y., and Lahaye, M. (2009). Ultrastructure of ulvan: a polysaccharide from green seaweeds. *Biopolymers*, 91, 652-664.
[8] Jensen, A. (1969). Tocopherol content of seaweed and seaweed meal. II. Individual, diurnal and seasonal variations in some Fucaceae. *J. Sci. Food Agric.*, 20, 454-458.
[9] Jensen, A. (1972). The nutritive value of seaweed meal for domestic animals, in *Proc. 7th Int. Symposium Seaweed Res*, Sapporo, Japan, p 7.
[10] Gonzales, A. G., Martin, J. D., Norte, M., Perez, R., Rivera, P., and Ruano, J. Z. (1983). X-ray structure determination of new brominated metabolites isolated from the red seaweeds Laurencia obtuse. *Tetrahedron. Lett.*, 24, 4143-4146.

[11] Mishra, V. K., Temelli, F., Ooraikul Shacklock, P. F., and Craigie,.J. S. (1993). Lipids of the red algae Palmaria palmate. *Botanica Marina*, 36, 169-174.

[12] Kamat, S.Y., Wahidulla, S., D'Souza, L., Naik, C. G., Ambiye, V., and Bhakuni, D. S. (1992). Bioactivity of marine organisms,VI. Antiviral evaluation of marine algal extracts from the Indian coast. *Botanica Marina*, 35, 161-164.

[13] Yuan, Y. V. (2007). *Antioxidant measurement and applications: Antioxidants from edible seaweeds*. In F. Shahidi and C.-T. Ho. (Eds.), ACS symposium series 956, 268-301. Washington, ACS Oxford University press.

[14] Tutor, B. L., Benslimane, F., Gouleau, M. P., Gouygou, J. P., Saadan, B., and Quemeneur. F. (1998). Antioxidant and pro-oxidant activity of brown algae, Laminaria digiatta, Hemanthalia elongata, Fucus vesiculosus and Ascophyllum nodosum. *J. Appl. Phycotechnol.*, 10, 121-129.

[15] Matsukawa, R., Dubinsky, Z., Kishimoto, E., Masaki, K., Matsuda, Y., Takeuchi, T., Chihara, M., Yamamoto, Y., Niki, E., and Karube, I. (1997). A comparison of screening methods for antioxidant activity in seaweeds. *J. Appl. Phycol.*, 9, 29-35.

[16] Gonzalez del Val, A., Platas, G., Basilio, A., Cabello, A., Gorrochategui, J., Suay, I., Vicente, F., Portillo, E., Jimenez del Rio, M.; Reina, G.G., and Pelaez, F. (2001). Screening of antimicrobial activities in red, green and brown macroalgae from Gran Canaria (Canary Islands, Spain). *Int. Microbiol.*, 4, 35-40.

[17] Yuan, Y.V., and Walsh, N. A. (2006). Antioxidant and antiproliferative activities of extracts from a variety of edible seaweeds. *Food Chem. Toxicol.*, 44, 1144-1150.

[18] Kang, J. Y., Khan, M. N. A., Park, N. H., Cho, J.Y., Lee, M. C., Fujii, H., and Hong, Y. K. (2008). Antipyretic, analgesic and anti-inflammatory activities of the seaweed Sargassum fulvellum and Sargassum thunbergii in mice. *J. Ethnopharmacol.*, 116, 187-190.

[19] Pushpamali, W. A., Nikapitiya, C., De Zoysa, M., Whang, I., Kim, S. J., and Lee, J. (2008). Isolation and purification of an anticoagulant from fermented red seaweed Lomentaria catenata. *Carb. Polym.*, 73, 274-279.

[20] Sinha, S., Astani, A., Ghosh, T., Schnitzler, P., and Ray, B. (2009). Polysaccharides from Sargassum tenerrimum: structural features, chemical modification and anti-viral activity. *Phytochem.*, 71, 235-240.

[21] Kwon, M. J., and Nam, T. J. (2006). Porphyran induces apoptosis related signal pathway in AGS gastric cancer cell lines. Life Sci, 79, 1956-1962.

[22] Maruyama, H., Watanabe, K., and Yamamoto, I. (1991). Effect of dietary kelp on lipid peroxidation and glutathione peroxidase activity in livers of rats given breast carcinogen DMBA. *Nutr. Cancer*, 15, 221-228.

[23] Lee, E. J., and Sung, M. K. (2003). Chemoprevention of azoxymethane induced rat colon carcinogenesis by seatangle, a fiber-rich seaweed. *Plant Foods Hum. Nutri.*, 58, 1-8.

[24] Higashi-Okai, K., Otani, S., and Okai, Y. (1999). Potent suppressive effect of a Japanese edible seaweed, Enteromorpha prolifera (Sujiao-nori) on initiation and promotion phases of chemically induced mouse skin tumorigenesis. *Cancer Lett.*, 140, 21-25.

[25] Ames, B. N. (1983). Dietary carcinogens and anticarcinogens: oxygen radicals and degenerative diseases. *Science,* 221, 1256–1264

[26] Ruberto, G., Baratta, M. T., Biondi, D. M., and Amico V. (2001). Antioxidant activity of extracts of the marine algal genus Cystoseira in a micellar model system. *J. Appl. Phycol.*, 13, 403-407.

[27] Kovatcheva, E. G., Koleva, I. I., Ilieva, M., Pavlov, A., Mincheva, M., and Konushlieva, M. (2001). Antioxidant activity of extracts from Lavandula vera MM cell culture. *Food Chem.*, 7, 1069-1077.

[28] Halliwell, B., and Aruoma, O. I. (1991). DNA damage by oxygen-derived species its mechanism and measurement in mammalian systems. *FEBS Lett.*, 281, 9-19.

[29] Bandyopadhyay, U., Das, D., and Banerjee, R. K. (1999). Reactive oxygen species: Oxidative damage and pathogenesis. *Curr. Sci.*, 77, 5-10.

[30] Girotti, A. W. (1990). Photodynamic peroxidation in biological systems. *Photochem. Photobiol.*, 51, 497-509.

[31] Barclay, L. R. C. (1993). Model biomembranes: quantitative studies of peroxidation, antioxidant action, partitioning, and oxidative stress. *Can. J. Chem.*, 71, l-16.

[32] Stadtman, E. R. (1992). Protein oxidation and aging. *Science*, 257, 1220-1224.

[33] Ramanathan, L., and Das, N. P. (1992). Studies on the control of lipid oxidation in ground fish by some polyphenolic products. *J. Agric. Food Chem.*, 40, 17-21.

[34] Gutteridge, J. M. C. (1995). Lipid peroxidation and antioxidants as biomarker of tissue damage. *Clin. Chem.*, 41, 1819-1828.

[35] Kassie, F., Pqrzefall, W., and Knasmuller, S. (2000). Single cell gel electrophoresis assay: a new technique for human biomonitoring studies. *Mut. Res.*, 463, 13-31.

[36] Halliwell, B. (1994). Free radicals, antioxidants, and human disease: curiosity, cause, or consequence? *The Lancet*, 344, 721-724.

[37] Steinberg, D. (1997). Low density lipoprotein oxidation and its pathobiological significance. *J. Biol. Chem.*, 272, 20963-20966.

[38] Ross, R. (1999). Atherosclerosis-an inflammatory disease. *N. Eng. J. Med.*, 340, 115-126.

[39] Zaragoza, M. C., Lopez, D., Saiz, M. P., Poquet, M., Pe´rez, J., Puig-parellada, P., Ma`rmol, F., Simonetti, P., Gardana, C., Lerat, Y., Burtin, P., Inisan, C., Rousseau, I., Besnard, M., and Mitjavila, M. T. (2008). Toxicity and antioxidant activity in vitro and in vivo of two Fucus vesiculosus extracts. *J. Agric. Food Chem.*, 56, 7773-7780.

[40] Fridovich, I. (1983). Superoxide radical: an endogenous toxicant. *Ann. Rev. Pharmacol. Toxicol.*, 23, 239-257.

[41] Chance, B., Sies, H., and Boveris, A. (1979). Hydroperoxide metabolism in mammalian organs. *Physiol. Rev.*, 59, 527-605.

[42] Deisseroth, A., and Dounce, A. L. (1970). Catalase: Physical and chemical properties, mechanism of catalysis, and physiological role. *Physiol. Rev.*, 50, 319-375.

[43] Larson, R. A. (1995). Plant defenses against oxidative stress. *Arch. Insect. Biochem. Physiol.*, 29, 175-186.

[44] Halliwell, B., Aeschbach, R., Löliger, J., and Aruoma, O. I. (1995). The characterization of antioxidants. *Food Chem. Toxicol.*, 33, 601-617.

[45] Witschi, H. P. (1986). Enhanced tumour development by butylated hydroxytoluene (BHT) in the liver, lung and gastro-intestinal tract. *Food Chem. Toxicol.*, 24, 1127-1130.

[46] Grice, H. C. (1988). Safety evaluation of butylated hydroxyanisole from the perspective of effects on forestomach and oesophageal squamous epithelium. *Food Chem. Toxicol.*, 26, 717-723.

[47] Velioglu, Y. S., Mazza, G., Gao, L., and Oomah, B. D. (1998). Antioxidant activity and total phenolics in selected fruits, vegetables, and grain products. *J. Agric. Food Chem.*, 46, 4113-4117.

[48] Cahyana, A. H., Shuto, Y., and Kinoshita, Y. (1992). Pyropheophytin a as an antioxidative substance from the marine alga arame (Eisenia bicyclis). *Biosci. Biotech. Biochem.*, 56, 1533-1535.

[49] Lim, S. N., Cheung, P. C. K., Ooi, V. E. C., and Ang, P. O. (2002). Evaluation of antioxidative activity of extracts from a brown seaweed, Sargassum siliquastrum. *J. Agric. Food Chem.*, 50, 3862-3866.

[50] Pokorny, J. (1991). Natural antioxidants for food use. *Trends Food Sci. Technol.*, 2, 223-227.

[51] Matsukawa, R., Dubinsky, Z., Kishimoto, E., Masaki, K., Matsuda, Y., Takeuchi, T., Chihara, M, Yamamoto, Y., Niki, E., and Karube, I. (1997). A comparison of screening methods for antioxidant activity in seaweeds. *J. Appl. Phycol.*, 9, 29-35.

[52] Plaza, M, Cifuentes, A, and Ibanez, E. (2008). In the search of new functional food ingredients from algae. *Trends Food Sci. Technol.*, 19, 31-39.

[53] Nagai, T., and Yukimoto, T. (2003). Preparation and functional properties of beverages made from sea algae. *Food Chem.*, 81, 327-332.

[54] Shahidi, F., and Zhong, Y. (2007). Antioxidants from marine by-products *In: Maximising the value of marine by-products*. Shahidi F (Ed), pp. 397-412, Cambridge England, Woodhead Publ Ltd.

[55] Yan, X., Chuda, Y., Suzuki, M., and Nagata, T. (1999). Fucoxanthin as the major antioxidant in Hijikia fusiformis, a common edible seaweed. *Biosci. Biotechnol. Biochem.*, 63, 605-607.

[56] Sachindra, N. M., Hosokawa, M., and Miyashita, K, (2007). Biofunctions of marine carotenoids. *In: Biocatalysis and biotechnology for functional foods and industrial products* C.T. Hou and J. Shaw (Eds.), pp. 91-110, New York, CRC Press.

[57] Nakamura, T., Nagayama, K., Uchida, K., and Tanaka, R. (1996). Antioxidant activity of phlorotannins isolated from the brown alga Eisenia bicyclis. Fish. Sci., 62, 923-926.

[58] Kaneda, T., and Ando, H. (1971). Component lipids of purple laver and their antioxygenic activity. *Proc. Intl. Seaweed Symp.*, 7, 553–557.

[59] Fujimoto, K., and Kaneda, T. (1980). Screening test for antioxygenic compounds frommarine algae and fractionation from Eisenia bicyclis and Undaria pinnatifida. *Bull. Japan Soc. Sci. Fish,* 46, 1125-1130.

[60] Fujimoto, K., and Kaneda, T. (1984). Separation of antioxygenic (antioxidant) compounds from marine algae. *Hydrobiologia*, 116, 111-113.

[61] Fujimoto, K., Ohmura, H., and Kaneda, T. (1985). Screening for antioxygenic compoundsin marine algae and bromophenols as effective principles in a red alga Polysiphonia ulceolate. *Bull. Japan Soc. Sci. Fish.*, 51, 1139-1143.

[62] Nishibori, S., and Namiki, K. (1985). Antioxidative activity of seaweed lipids and their utilisation in food. *Kaseigaku Zasshi*, 36, 845-850. (in Japanese).

[63] Nishibori, S., and Namiki, K. (1988). Antioxidative substances in the green fractions of the lipids of aonori (Enteromorpha sp.). *Kaseigaku Zasshi*, 39, 1173-1178. (in Japanese).

[64] Burton, G. W., and Ingold, K. U. (1981). Autoxidation of biological molecules. 1. The antioxidant activity of vitamin E and related chain-breaking phenolic antioxidants in vitro. *J. Am. Chem. Soc.*, 103, 6472-6477.

[65] Pryor, W. A., Cornicehi, J. A., Devall, L. J., Tait, B., Trivedi, B. K., Witiak, D. T., and Wu, M. (1993). A rapid screening test to determine the antioxidant potencies of natural and synthetic antioxidants. *J. Org. Chem.*, 58, 3521-3532.

[66] Breccia, A., Rodgers, M. A. J., and Semeraro, G. (1986). *Oxygen and Sulfur Radical in Chemistry and Medicine, Oxygen Radicals in Chemistry and Biology*. Bologna Edizioni Scientifiche Lo Scarabeo.

[67] Bowry, V. W., and Stocker, R. (1993). Tocopherol-mediated peroxidation. The prooxidant effect of vitamin E on the radical-initiated oxidation of human low-density lipoprotein. *J. Am. Chem. Soc.*, 115, 6029-6044.

[68] Mukai, K., Daifuku,.K., Okabe, K., Tanigaki,T., and Inoue, K. (1991). Structure-activity relationship in the quenching reaction of singlet oxygen by tocopherol (Vitamin E) derivatives and related phenols. Finding of linear correlation between the rates of quenching of singlet oxygen and scavenging of peroxyl and phenoxyl radicals in solution. *J. Org. Chem.*, 56, 4188-4192,

[69] Saito, I., Matsuura,.T., and Inoue, K. (1983). Formation of superoxide ion via one-electron transfer from electron donors to singlet oxygen. *J. Am. Chem. Soc.*, 105, 3200-3206.

[70] Inoue, K., Matsuura, T., and Saito, I. (1985). Importance of single electron-transfer in singlet oxygen reaction in aqueous solution. Oxidation of electron-rich thioanisoles. *Tetrahedron.*, 41, 2177-2181.

[71] Kuo, S. M., Leavitt, P. S., and Lin, C. P. (1998). Dietary flavonoids interact with trace metals and affect metallothionein level in human intestinal cells. *Biol Trace Elem Res.*, 62, 135-153.

[72] Robak, J., and Gryglewski, R. J. (1998). Flavonoids are scavengers of superoxide anions. *Biochem. Pharmacol.*, 37, 837-841.

[73] Cook, N. C., and Samman, S. (1996). Flavonoids. Chemistry, metabolism, cardioprotective effects, and dietary source. *J. Nutr. Biochem.*, 7, 66-76.

[74] Martin, H. D., Ruck, V., Schmidt, M., Sell, S., Beutner, S., Mayer, B., and Walsh, R. (1999). Chemistry of carotenoid oxidation and free radical reactions. *Pure Appl. Chem.*, 71, 2253–2262.

[75] Hirayama, O., Nakamura, K., Hamada, S., and Kobayashi, Y. (1994). Singlet oxygen quenching ability of naturally occurring carotenoids. *Lipids*, 29, 149-150.

[76] Farmillo, A., and Wilkinson, F. (1973). On the mechanism of quenching of singlet oxygen in solution. *Photochem. Photobiol.*, 18, 447-450.

[77] Krinsky, N. I., and Yeum, K. J. (2003). Carotenoid-radical interactions. *Biochem. Biophys. Res. Commu.*, 305, 754-760

[78] Diaz, M. N., Frei, B., Vita, J. A., and Keaney, J. J. (1997). Antioxidants and atherosclerotic heart disease. *N. Eng. J. Med.*, 337, 408-416.

[79] Swanson, A. K., and Druehl, L. D. (2002). Induction, exudation and the UV protective role of kelp phlorotannins. *Aqua. Bot.*, 73, 241-253.

[80] Ragan, M. A., and Glombitza, K. W. (1986). Phlorotannins, brown algal polyphenols. *Prog. Phycolog. Res.*, 4, 129-241.

[81] Koivikko, R., Loponen, J., Pihlaja, K., and Jormalainen, V. (2007). High Performance liquid chromatographic analysis of phlorotannins from the brown algae Fucus vesiculosus. *Phytochem. Anal.*, 18, 326-332.

[82] Koivikko, R., Era"nen, J. K., Loponen, J., and Jormalainen, V. (2008). Variation of phlorotannins among three populations of Fucus Vesiculosus as revealed by HPLC and colorimetric quantification. *J. Chem. Ecol.*, 34, 57-64.

[83] Yuan, Y. V., Carrington, M. F., and Walsh, N. A. (2005). Extracts from dulse (Palmaria palmata) are effective antioxidants and inhibitors of cell proliferation in vitro. *Food Chem. Toxicol.*, 43, 1073-1081.

[84] Rodrıguez-Bernaldo de Quiros, A., Frecha-Ferreiro, S., Vidal-Perez, A. M., and Lopez-Hernandez, J. (2010). Antioxidant compounds in edible brown seaweeds. *Eur. Food Res. Technol.*, 231, 495-498.

[85] Kang, K., Park, Y., Hwang, H. J., Kim, S. H, Lee, J. G., and Shin, H. C. (2003). Antioxidative properties of brown algae polyphenolics and their perspectives as chemopreventive agents against vascular risk factors. *Arch. Pharm. Res.*, 26, 286-293.

[86] Yangthong, M., Towatana, N. H., and Phromkunthong, W. (2009). Antioxidant activities of four edible seaweeds from the southern coast of Thailand. *Plant Foods Hum. Nutr.*, 64, 218-223.

[87] Duan, X. J., Zhang, W. W., Li, X. M., and Wang, B. G. (2006). Evaluation of antioxidant property of extract and fractions obtained from red alga, Polysiphonia urceolata. *Food Chem.*, 95, 37-43.

[88] Sachindra, N. M., Airanthi, M. K. W. A., Hosokawa, M., and Miyashita, K. (2010). Radical scavenging and singlet oxygen quenching activity of extracts from Indian seaweeds. *J. Food Sci. Technol.*, 47, 94-99.

[89] Anggadiredja, J., Andyani, R., Hayati., and Muawanah. (1997). Antioxidant activity of Sargassum polycystum (Phaeophyta) and Laurencia obtusa (Rhodophyta) from Seribu Islands. *J. Appl. Phycol.*, 9, 477-479.

[90] Yoshi, Y., Wang, W., Petillo, D., and Suzuki, T. (2000). Distribution of catechins in Japanese seaweeds. *Fish Sci.*, 66, 998-1000.

[91] Santoso, J., Yoshie, Y., and Suzuki, T. (2004). Polyphenolic compounds from seaweeds: distribution and their antioxidative effect. *Develop. Food Sci.*, 42, 169-177.

[92] Kuda, T., Tsunekawa, M., Hishi, T., and Araki, Y. (2005). Antioxidant properties of dried 'kayamo-nori', a brown alga Scytosiphon lomentaria (Scytosiphonales, Phaeophyceae). *Food Chem.*, 89, 617-622.

[93] Fugimoto, K., Ohmura, H., and Kaneda, T. (1985). Screening for antioxygenic compounds in marine algae and bromophenols as effective principles in red alga Polysiphonia ulceolate. *J. Jap. Soc. Sci. Fish.*, 51, 1139-1143.

[94] Kang, K. A., Lee, K. H., Chae, S., Koh, Y.S., Yoo, B. S., Kim, J. H., Ham, Y. M., Baik, J. S., Lee, N. M., and Hyun, J. W. (2005). Triphlorethol-A from Ecklonia cava protects V79-4 lung fibroblast against hydrogen peroxide induced cell damage. *Free Rad. Res.*, 39, 883-892.

[95] Athukorala, Y., Lee, K. W., Park, E. J., Heo, M. S., Yeo, I. K., Lee, Y. D., and Jeon, Y. J. (2005). Reduction of lipid peroxidation and H_2O_2-mediated DNA damage by a red alga (Grateloupia filicina) methanolic extract. *J. Sci. Food Agric.*, 85, 2341–2348.

[96] Kim, J. A., Lee, J. M., Shin, D. B., and Lee, N. H. (2004). The antioxidant activity and tyrosinase inhibitory activity of phlorotannins in Ecklonia cava. *Food Sci. Biotechnol.,* 13, 476–480.

[97] Yan, X., Li, X., Zhou, C., and Fan, X. (1996). Prevention of fish oil rancidity by phlorotannins from Sargassum kjellmanianum. *J. Appl. Phycol.,* 8, 201–203.

[98] Jimenez-Escrig, A., Jimenez-Jimenez, I., Pulido, R., and Saura-Calixto, F. (2001). Antioxidant activity of fresh and processed edible seaweeds. *J. Sci. Food Agri.,* 81, 530–534.

[99] Qian, J.Y., Mayer, D., and Kuhn, M. (1999). Flavanoids in fine buckwheat (Fagopyrum esculentum Monch) flour and their free radical scavenging activities. *Deutsche Lebensmittel-Rundschau,* 95, 343-349.

[100] Ndhlala, A. R., Kasiyamhuru, A., Mupure, C., Chitindingu, K., Benhura, M. A. and Muchuweti, M. (2007). Phenolic composition of Flacourtia indica, Opuntiamegacantha and Sclerocarya birrea. *Food Chem.,* 103, 82-87.

[101] Zeng, L. M., Wang, C. J., Su, J. Y., Li, D., Owen, N. L., Lu, Y., Lu, N., and Zheng., Q. T. (2001). Flavonoids from the red alga Acanthophora spicifera. *Chinese J. Chem.,* 19, 1097-1100.

[102] Sabina, H., and Aliya, R. (2009). Seaweed as a new source of flavone, Scutellarein 4'-methyl ether. *Pak. J. Bot.,* 41, 1927-1930.

[103] Yumiko, Y. S., Pei, H. Y., and Takeshi, S. (2003). Distribution of flavanoids and related compounds from seaweeds in Japan. *J. Tokyo Uni. Fish.,* 89, 1-6.

[104] Souza, J. G., Tomei, R. R., Kanashiro, A., Kabeya, L. M., Azzolini, A. E. C. S, Dias, D. A., Salvador, M. J., and Lucisano-Valim, Y. M. (2007). Ethanolic crude extract and flavanoids isolated from Alternanthera maritime: Neutrophil chemiluminescence Inhibition and free radical scavenging activity. *Z. Natuforsch.,* 62c,339-347.

[105] Kloareg B., and Quatrano, R. S. (1988). Structure of the cell walls of marine algae and ecophysiological functions of the matrix polysaccharides. *Ocean Mar. Biol. Annu. Rev.,* 26, 259-315.

[106] Alban, S., Schaurete, A., and Franz, G. (2002). Anticoagulant sulphated polysaccharides: part I. Synthesis and structure-activity relationship of new pullulan sulfates. *Carb. Polym.,* 47, 267-276.

[107] Arfors, K. E., and Ley, K. J. (1993). Sulfated polysaccharides in inflammation. *Lab. Clin. Med.,* 121, 201-202.

[108] Caceres, P. J., Carlucci, M. J., Damonte, E. B., Matsuhiro, B., and Zuniga, E. A. (2000). Carrageenans from Chilean samples of Stenogramme interrupta (Phyllophoraceae): Structural analysis and biological activity. *Phytochem.,* 53, 81-86.

[109] Zhang, Q. B., Yu, P. Z., Li, Z. E., Zhang, H., Xu, Z. H., and Li, P. C. (2003). Antioxidant activities of sulfated polysaccharide fractions from Porphyra haitanesis. *J. Appl. Phycol.,* 15, 305-310.

[110] Zhang, Q. B, Li, N., Liu, X., Zhao, Z., Li, Z. E., and Xu, Z. H. (2004). The structure of a sulfated galactan from Porphyra haitanensis and its in vivo antioxidant activity. *Carbohydrate Res.,* 339, 105-111.

[111] Ruperez, P., Ahrazem, O., and Leal, J. A. (2002). Potential antioxidant capacity of sulfated polysaccharides from the edible marine brown seaweed Fucus Vesiculosus. *J. Agric. Food Chem.,* 50, 840-845.

[112] Xue, C. H., Fang, Y., Lin, H., Chen, L., Li, Z. J., Deng, D., and Lu, C. X. (2001). Chemical characters and antioxidative properties of sulfated polysaccharides from Laminaria japonica. *J. Appl. Phycol.*, 13, 67-70.

[113] Melo, M. R. S., Feitosa, J. P. A., Freitas, A. L. P., and Paula, R. C. M. (2002). Isolation and characterization of soluble sulfated polysaccharide from the red seaweed Gracilaria cornea. *Carb. Polym.*, 49, 491-498.

[114] Yang, J. H., Du, Y. M., Wen, Y., Li, T.Y., Hu, L. (2003). Sulfation of Chinese lacquer polysaccharides in different solvents. *Carb. Polym.* 52: 397-403.

[115] Tsipali, E., Whaley, S., Kalbfleisch, J., Ensley, HE., Browder, I. W., and Williams, D. L. (2001). Glucans exhibit weak antioxidant activity, but stimulate macrophage free radical activity. *Free Rad. Biol. Med.*, 30, 393-402.

[116] deSouza, M. C. R., Marques, C. T., Dore, C. M. G, daSilva, F. R. F., Rocha, H. A. O and Leite, E. L. (2007). Antioxidant activities of sulfated polysaccharides from brown and red seaweeds *J. Appl. Phycol.*, 19, 153-160.

[117] Wang, J., Zhang, Q., Zhang, Z., and Li, Z. (2008). Antioxidant activity of sulfated polysaccharide fractions extracted from Laminaria japonica. *Intl. J. Biol. Macromol.*, 42, 127–132.

[118] Caram-Lelham, N., Sundelof, L. O., and Anderrson, T. (1995). Preparative separation of olifosaccharides from carrageenan, sodium hyaluronate and dextran by SuperdexTM 30 prep. grade. *Carb. Res.*, 273, 71-76.

[119] Yan, J., Zong, H., Shen, A., Chen, S., Yin, X., Shen, X., Liu, W., Gu, X., and Gu, J. (2003). The β-(1,6)-branched β-(1,3) glucohexose and its analogues containing an α-(1,3) linked bond have similar stimulatory effects on the mouse spleen as Lentinan. *Intl. Immunopharmacol.*, 3, 1861-1871.

[120] Ning, J., Zhang, W., Yi, Y., Yang, G., Wu, Z., Yi, J., and Kong, F. (2003). Synthesis of β (1,6)- branched- β (1,3) glucohexaose and its analogues containing an α (1,3) linked bond with antitumour activity. *Bioorg. Med. Chem.*, 11, 2193-2203.

[121] Michel, G., Collen, P. N., Barbeyron, T., Czjzek, M., and Helbert, W. (2006). Bioconversion of red seaweed galactans: a focus on bacterial agarases and carrageenases. *Appl. Microbiol. Biotechnol.*, 71, 23-33.

[122] Yuan, H., and Song, J. (2005). Preparation and structural characterization and in vitro antitumor activity of kappa-carrageenan oligosaccharide fraction from Kappaphycus striatum. *J. Appl. Phycol.*, 17, 7-13.

[123] Heo, S. J., Lee, K. W., Song, C. B., and Jeon, Y. J. (2003). Antioxidant activity of enzymatic extract from seaweeds. *Algae*, 18, 71-81.

[124] Chen, H., Yan, X., Zhu, P., and Lin, J. (2006). Antioxidant activity and hepatoprotective potential of agaro-oligosaccharides in vitro and in vivo. *Nutri. J.*, 5, 31-42.

[125] Wang, J., Jiang, X., Mou, H., and Guan, H. (2004). Anti-oxidation of agar oligosaccharides produced by agarase from a marine bacterium. *J. Appl. Phycol.*, 16, 333–340.

[126] Kobayashi, R., Takimasa, M., Suzuki, T., Kirimura, K., and Usami, S. (1997). Neoagarobiose as a novel moisturizer with whitening effect. *Biosci. Biotechnol. Biochem.*, 61, 162–163.

[127] Yuan, H., Song, J., Zhang, W., Li, X., Li, N., and Gao, X. (2006). Antioxidant activity and cytoprotective effect of K-carrageenan oligosaccharides and their different derivatives. *Bioorgan. Med. Chem. Lett.*, 16, 1329–1334.

[128] Mou, H. J., Jiang, X. L., Jiang, X., and Guan, H. S. (2002). Isolation and properties of a carrageenan degrading bacterium. *J. Fish Sci. China*, 9, 251-254.

[129] Haijin, M., Xiaolu, J., and Huashi, G. (2003). A κ-carrageenan derived oligosaccharide prepared by enzymatic degradation containing anti-tumour activity. *J. Appl. Phycol.*, 15, 297-303.

[130] Halliwell, B. (1996). Oxidative stress, nutrition and health. Experimental strategies for optimization of nutritional antioxidant intake in humans. *Free Rad. Res.*, 25, 57-74.

[131] Nishino, H. (1998). Cancer prevention by carotenoids. *Mut. Res.*, 213, 3-13.

[132] Yan, X., Chuda, Y., Suzuki, M., and Nagata, T. (1999). Fucoxanthin as the major antioxidant in Hijkia fusiformis, a common edible seaweed. *Biosci. Biotechnol. Biochem.*, 63, 605-607.

[133] Sugawara, T., Baskaran, V., Tsuzuki, W., and Nagao, A. (2002) Brown algae fucoxanthin is hydrolyzed to fucoxanthinol during absorption by caco-2 human intestinal cells and mice. *J. Nutr.*, 132, 946–951.

[134] Asai, A., Sugawara, T., Ono, H., and Nagao, A. (2004) Biotransformation of fucoxanthinol into amarouciaxanthin in mice and HEPG2 cells: formation and cytotoxicity of fucoxanthin metabolites. *Drug Metabol. Disp.*, 32, 205–211.

[135] Sachindra, N. M., Sato, E., Maeda, H., Hosokawa, M., Niwano, Y., Kohno, M., and Miyashita, K. (2007). Radical scavenging and singlet oxygen quenching activity of marine carotenoid fucoxanthin and its metabolites. *J. Agri. Food Chem.*, 55, 8516 – 8522.

[136] Sangeetha, R. K., Bhaskar, N., and Baskaran, V. (2008). Fucoxanthin restrains oxidative stress induced by retinol deficiency through modulation of Na^+Ka^+-ATPase and antioxidant enzyme activities in rats. *Eur. J. Nutr.*, 47, 432–441.

[137] Sangeetha, R. K., Bhaskar, N., and Baskaran, V. (2009). Comparative effects of β-carotene and fucoxanthin on retinol deficiency induced oxidative stress in rats. *Mol. Cell Biochem.*, 331, 59–67.

[138] Murata, M., Ishihara, K., and Saito, H. (1999). Hepatic fatty acid enzyme activities are stimulated in rats fed the brown seaweed, Undaria pinnatifida (Wakame). *J. Nutr.*, 129, 146-150.

[139] Murata, M., Sano, Y., Ishihara, K., and Uchida, M. (2002). Dietary fish oil and Undaria pinnatifida (Wakame) synergistically decrease rat serum and liver triacylglycerol. *J. Nutr.*, 132, 742 – 747.

[140] Hu, C. C., Lin, J. T., Lu, F. J., Chou, F. P., and Yang, D. J. (2008). Determination of carotenoids in Dunaliella salina cultivated in Taiwan and antioxidant capacity of the algal carotenoid extract. *Food Chem.*, 109, 439–446.

[141] Airanthi, M. K. W., Hosokawa, M., and Miyashita, K. (2011). Comparative antioxidant activity of edible Japanese brown seaweeds. *J. Food Sci.*, 76, 104-111.

[142] Morgan, K. C., Wright, J. L. C., and Simpson, F. J. (1980). Review of chemical constituents of the red algae Palmaria palmate (Dulse). *Econ. Bot.*, 34, 27-50.

[143] Burritt, D. J., Larkindale, J., and Hurd, C. L. (2002). Antioxidant metabolism in the intertidal red seaweed Stictosiphonia arbuscula following dessication. *Planta*, 215, 829–838.

[144] Dunlap, W. C., Masaki, K., Yamamoto, Y., Larsen, R. M., and Karube, I. (1998). A novel antioxidant derived from seaweed. In Y. LeGal and H. Halvorson (Eds.), *New developments in marine biotechnology*, pp. 33-35. New York, Plenum Press.

[145] Nakayama, R., Tamura, Y., Kikuzaki, H., and Nakatani, N. (1999). Antioxidant effect of the constituents of susabinori (Porphyra yezoensis). *J. Am. Oil Chem. Soc.*, 76, 649-653.

[146] Sekikawa, I., Kubota, C., Hiraoki, T., and Tsujino, I. (1986). Isolation and structure of a 357 nm UV absorbing substance, usujirene, from the red alga Palmaria palmata (L.) O. Kuntze. *Jpn. J. Phycol.*, 34, 185-188.

[147] Foti, M., Piattelli, M., Amico, V., and Rubertob, G. (1994). Antioxidant activity of phenolic meroditerpenoids from marine algae. *J. Photochem. Photobiol. B*, 26, 159-164.

[148] Tutour, B. L., Benslimane, F., Gouleau, M. P., Gouygou, J. P., Saadan, B., and Quemeneur, F. (1998). Antioxidant and prooxidant activities of the brown algae, Laminaria digitata, Himanthalia elongata, Fucus vesiculosus, Fucus serratus and Ascophyllum nodosum. *J. Appl. Phycol.*, 10, 121-129.

[149] Huang, H. L., and Wang, B. G. (2004). Antioxidant Capacity and Lipophilic Content of Seaweeds Collected from the Qingdao Coastline. *J. Agric. Food Chem.*, 52, 4993-4997.

[150] Sunda, W., Kieber, D. J., Kiene, R. P., and Huntsman, S. (2002). An antioxidant function of DMSP and DMS in marine algae. *Nature*, 418, 317–320.

[151] Malanga, G., and Puntarulo, S. (1995). Oxidative stress and antioxidant content in Chlorella vulgaris after exposure to ultraviolet-B radiation. *Plant Physiol.*, 94, 672–679.

[152] Aguilera, J., Dummermuth, A., Karsten, U., Schriek, R., and Wiencke, C. (2002). Enzymatic defences against photooxidative stress induced by ultraviolet radiation in Arctic marine macroalgae. *Polar Biol.*, 25, 432-441.

[153] Indergaard, M., and Minsaas, J. (1991). Animal and human nutrition. In M. D. Guiry and G. Blunden (Eds.), *Seaweed Resources in Europe: Uses and Potential*, pp. 21-64. Toronto, John Wiley and Sons Ltd.

[154] Kakinuma, M., Park, C. S., and Amano, H. (2001). Distribution of free L-cysteine and glutathione in seaweeds. *Fish Sci.*, 67, 194-196.

[155] Heo, S. J., Park, E. J., Lee, K. W., and Jeon, Y. J. (2005). Antioxidant activities of enzymatic extracts from brown seaweeds. *Biores. Technol.*, 96, 1613-1623.

[156] Yuan, Y. V., Bone, D. E., and Carrington, M. F. (2005). Antioxidant activity of dulse (Palmaria palmata) extract evaluated in vitro. *Food Chem.*, 91, 485-494.

[157] Yuan, Y. V., Carrington, M. F., and Walsh, N. A. (2005). Extracts from dulse (Palmaria palmata) are effective antioxidants and inhibitors of cell proliferation in vitro. *Food Chem. Toxicol.*, 43, 1073-1081.

[158] Karsten, U., and Wiencke, C. (1999). Factors controlling the formation of UV absorbing mycosporine-like amino acids in the marine red alga Palmaria palmata from Spitsbergen (Norway). *J. Plant Physiol.*, 155, 407-415.

In: Seaweed
Editor: Vitor H. Pomin

ISBN 978-1-61470-878-0
© 2012 Nova Science Publishers, Inc.

Chapter 8

HALOGENATED COMPOUNDS FROM SEAWEED, A BIOLOGICAL OVERVIEW

Clara Grosso, Juliana Vinholes, Patrícia Valentão and Paula B. Andrade[*]

REQUIMTE/Laboratório de Farmacognosia,
Departamento de Química, Faculdade de Farmácia,
Universidade do Porto, R. Aníbal Cunha, Porto, Portugal

ABSTRACT

Seaweeds are a renewable marine resource recognized as a rich provider of valuable compounds. Since they live in a competitive environment, marine algae developed different strategies to survive, including the biosynthesis of a variety of compounds from different metabolic pathways. Indeed, more than 15000 metabolites were already determined, belonging to several groups of primary and secondary metabolites. For this fact the interest of pharmacologists, physiologists and chemists in this group of living organisms has risen. The research on natural products of marine origin has shown that seaweeds are of unlimited potential for textile, fuel, plastics, paint, varnish, cosmetics, pharmaceutical and food industries. The last three take advantage from the several biological properties attributed to seaweeds.

The present chapter will focus on seaweeds' halogenated compounds that are naturally produced by Phaeophyta, Chlorophyta and Rhodophyta. Halogenation can occur in several classes of metabolites, like indoles, terpenes, acetogenins, phenols, fatty acids and volatile hydrocarbons. The incorporation of a halogen in such compounds can induce/increase their biological properties, like antimicrobial and antitumor. Consequently, the consumption of seaweeds and their based products constitute a benefit to human health and therefore they are seen as a potential medicinal food of the 21st century.

---

[*] E-mail: pandrade@ff.up.pt

1. INTRODUCTION

Eukaryotic algae constitute a broad group of living organisms, which, perhaps, is only to be expected due to the extraordinary range of habitats in which they can live. They are important components of many ecosystems, from hot springs to snow, and encompass freshwater and marine habitats [1].

The global exploitation of seaweeds, which is mainly based on the culture of edible species or on the production of hydrocolloids (agar, carrageenan and alginate), is a multi-billion dollar industry. The broad range of application of these hydrocolloids in food and cosmetic industries is due to their physical properties, such as gelling, water-retention and ability to emulsify [2]. However, in recent years pharmaceutical industries also started to go through another direction, searching for new natural bioactive compounds from seaweeds [3].

Seaweeds live in a competitive environment where they have to fight against different sources of threat, namely competition for space, attack by pathogens and/or growth of epiphytes and endophytes [4]. For this reason, they developed different structural, morphological and chemical strategies to survive [5]. For the pharmaceutical industry, this last strategy is the most interesting one, since these living organisms synthesize a wide variety of primary and secondary compounds from different metabolic pathways.

Despite the great variety of metabolites produced by seaweeds, this chapter will only concern seaweeds' halogenated compounds (HC). It is estimated that more than 1000 HC have already been described in seaweeds [6]. Curiously, bromine is the most common halogen found in marine natural products, although its concentration in seawater is lower than that of chlorine [7]. The biosynthesis of HC has been recently reviewed [8-10], and for the majority of the compounds biohalogenation involves the oxidation of halide (chlorine, bromine and iodine) by specialized enzymes (haloperoxidases) and hydrogen peroxide (H_2O_2) [8-10]. However, the existence of other enzymes responsible for halogenation has also been reported [11-14].

Distinct HC production patterns are observed among the three macroalgal lineages: these compounds represent 90 % of those reported for red alga (Rhodophyta), 7 % for green alga (Chlorophyta) and only 1 % for brown alga (Phaeophyta) [15]. HC are widespread by several different classes of primary and secondary metabolites, including indoles, terpenes, acetogenins, phenols, fatty acids and volatile halogenated hydrocarbons. They are part of a defense system against microorganism infections, herbivory, allelopathy, detrimental fouling by different kinds of epiphytes or excess of self-generated hypochlorite and hydrogen peroxide [16,17]. The diversity of HCs was originally considered as a useful chemotaxonomic tool, but the geographic and seasonal variation observed in the chemical composition of seaweeds reduced their chemotaxonomic value [7].

The diversity on the production of primary and secondary metabolites by algae, in order to guarantee their survival, may be foreseen as an optimized mechanism in the species evolution, particularly influencing certain biological targets like DNA, enzymes, receptors and membranes. As a consequence, these defensive toxins may also be used in therapeutics to bind to relevant targets, turning them useful in pharmacological terms. Therefore, studies covering the biological potential of extracts from red, green and brown species have been published [18-23], and a broad spectrum of activities was observed. Special attention has been devoted to the biological activity of HCs, since the presence of halogen seems to improve the molecule activity [24]. Thus, an increased search for this kind of compounds was noticed in the last years and a number of new HCs have been isolated from seaweed, indicating these organisms as a potent source of highly bioactive metabolites that might represent useful leads in the development of new pharmaceutical drugs. The present chapter aims to overview the various classes of HC found in seaweed, with special attention to those with biological interest.

2. HALOGENATED COMPOUNDS

2.1. Indole Alkaloids

Alkaloids are rare in marine organisms and algae; however, some halogenated indole alkaloids have been found. The majority of this type of compounds was isolated from red algae and is characterized by the presence of an indole group substituted by bromine, chlorine and, in a few cases, sulphur is present together with bromine (Figure 1) [25]. Bromoindoles and *N*-methylbromoindoles structures, such as 2,3,5,6-tetrabromo-1-*H*-indole (Figure 1B) and 2,3,5,6-tetrabromo-1-methyl-1-*H*-indole, and the sulphur-containing bromoalkaloids 2,4,6-tribromo-3-(methylthio)-1-*H*-indole and 4,6-dibromo-2-(methylsulphinyl)-3-(methylthio)-1-*H*-indole were isolated from *Laurencia* (Rhodomelaceae) genus. Among this genus *Laurencia brongniarti* is an exceptional producer of such compounds, especially of sulphur-containing ones [26]. Other compounds include those isolated from *Rhodophyllis membranaceae*, namely polyhalogenated indoles like 4,7-dibromo-2,3-dichloro-1-*H*-indole (Figure 1C), and from *Chaetomorpha basiretorsa* the bromobisindole 4,4',6,6'-tetrabromo-2-(methylsulphinyl)-2'-(methylthio)-1*H*,1'*H*-3,3'-bisindole (Figure 1A) [26].

Concerning the biological activity of halogenated indole alkaloids, 4,4',6,6'-tetrabromo-2-(methylsulphinyl)-2'-(methylthio)-1*H*,1'*H*-3,3'-bisindole (Figure 1A) was cytotoxic against HT-29 (human colon adenocarcinoma grade II) and P-388 (mouse leukemia) cells [26]. 2,3,5,6-Tetrabromo-1-*H*-indole (Figure 1B) was reported has having antimicrobial activity against *Bacillus subtilis* and *Saccharomyces cerevisiae*. In addition, the presence of polyhalogenated indoles like 4,7-dibromo-2,3-dichloro-1-*H*-indole (Figure 1C) in the crude extract of *Rhodophyllis membranacea* was related to its strong antifungal activity [27].

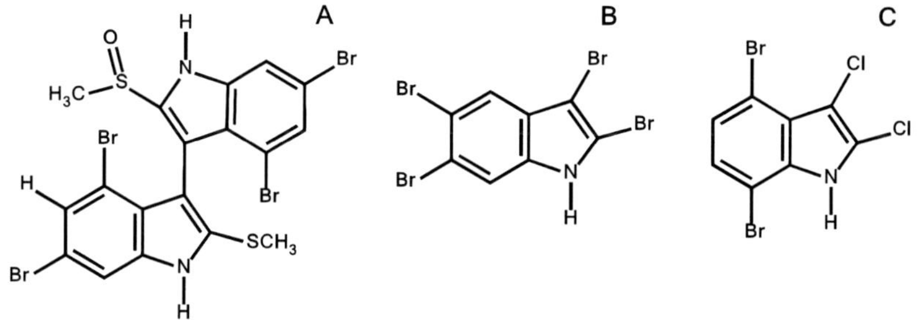

Figure 1. Halogenated indole alkaloids isolated from algae with biological activity. A- 4,4',6,6'-tetrabromo-2-(methylsulphinyl)-2'-(methylthio)-1*H*,1'*H*-3,3'-bisindole; B- 2,3,5,6-tetrabromo-1-*H*-indole; C- 4,7-dibromo-2,3-dichloro-1-*H*-indole.

2.2. Terpenes

Terpenes represent the major class of HC produced by seaweeds, and are found as monoterpenes (C_{10}), sesquiterpenes (C_{15}), diterpenes (C_{20}), and triterpenes (C_{30}) (Figures 2-4). Among seaweeds macrophytic red algae are the richer source of different and unique halogenated terpenes, being *Laurencia* genus the principal producer of such compounds [28, 29]. Halogenated diterpenes, followed by sesquiterpenes are the most abundant in this genus [30], but few mono and triterpenes are also found. The sesquiterpenes structures in *Laurencia* are typically cyclic, often polycyclic like laurinterol (Figure 2K), and not so frequently spiro-ring fusions can also be present, such as in elatol (Figure 4E) [31,32]. Concerning higher terpenes, diterpenes usually are dibrominated polycycles, such as neorogioldiol B, while triterpenes frequently are polyethers like thyresenol A (Figure 2Q) [33,34].

The distinction of the different Rhodophyta families is based on their monoterpenes profile. Whereas Rhizophyllidaceae, Plocamiaceae and Delesseriaceae are the leading producers, few monoterpenes are reported for Rhodomelaceae [35]. Among the monoterpenes described in Rhodophyta, highly halogenated compounds with linear and cyclic structures are described, like (Z)-6-bromo-3-(bromomethylene)-2-chloro-7-methylocta-1,6-diene (Figure 2D), halomon (Figure 2A) and plocamene D [7,35,36].

As previously mentioned, the contribution of HC in green and brown algae is low, although some compounds are reported, like the bicyclic monohalogenated sesquiterpenes of *Neomeris annulata* (Chlorophyta) 1(R)-bromo-ent-maaliol, (2S,4R,4aR,8aS)-2-bromo-6-isopropyl-4,8a-dimethyl-1,2,3,4,4a,7,8,8a-octahydronaphthalen-4a-ol and neomeranol [37], while sesquiterpenes with selinane structures, as 1-bromoselin-4(14),11-diene and 9-bromoselin-4(14),11-diene, have been recently reported in the brown alga *Dictyopteris divaricata* [38].

As previously mentioned, the reactions of halogen addition are catalyzed by haloperoxidases, which oxidize Br^- and Cl^- to "Br^+" and "Cl^+" in the presence of H_2O_2. The enzyme involved in terpenes halogenation is mainly a vanadium bromoperoxidase (V-BrPO), which was detected in different brown and red algae, like *Ascophyllum nodosum* and *Corallina pilulifera*, respectively [39, 40]. This was confirmed by Carter-Franklin and Butler [41] who tested the bromination catalyses of the sesquiterpene (E)-(+)-nerolidol in the presence of V-BrPO, potassium bromide (KBr) and hydrogen peroxide. The products of this reaction were α-, β-, and γ-snyderol, which have already been described as natural products in seaweed species. Therefore, they confirmed that the isolated enzyme catalyses the sesquiterpenes production *in vivo* [41].

2.2.1. Biological Activities

Halogenated terpenes exhibit a great variety of biological activities. Nevertheless this chapter will focus on the potential therapeutic application of such compounds as anticancer, antimicrobial, antitrypanosomal, antileishamanial and antiplasmodial agents.

2.2.1.1. Anticancer

As stated before, seaweeds are able to produce secondary metabolites of extremely interesting structural diversity. As a result, a number of natural compounds with unique mechanisms of action have been identified and some of them revealed to be very potent anticancer agents. In the beginning of the nineties, Fuller and coworkers [36] isolated a polyhalogenated acyclic monoterpene from the red alga *Portieria hornemannii*, named halomon (Figure 2A), which was considered to be a good candidate as antitumor agent. Other compounds structurally similar to halomon were also isolated and their citotoxicity was compared to that of halomon. Similar results were obtained for (3*S*,6*S*)-7-bromo-3-(bromomethyl)-2,3,6-trichloro-7-methyloct-1-ene (Figure 2B), (*S*)-3-(bromomethyl)-2,3,6-trichloro-7-methylocta-1,6-diene, (*S*)-6-bromo-3-(bromomethyl)-2,3-dichloro-7-methylocta-1,6-diene and (3*S*,6*R*)-6-bromo-3-(bromomethyl)-3,7-dichloro-7-methyloct-1-ene. It was noticed that the halogen at position 7 was not essential for their cytotoxicity; on the other hand, halogenation in position 6 is fundamental, as the analogues (*S*)-3-(bromomethyl)-2,3-dichloro-7-methylocta-1,6-diene (Figure 2C) and (*S*)-3-(bromomethyl)-3-chloro-7-methylocta-1,6-diene were less active [36]. Other perspective on cancer treatment is associated to DNA methyltransferase inhibitors that have recently been recognized as promising chemotherapeutic or preventive agents. Halomon (Figure 2A) and (*Z*)-6-bromo-3-(bromomethylene)-2-chloro-7-methylocta-1,6-diene (Figure 2D), from Madagascar *P. hornemannii*, revealed to inhibit this enzyme at concentrations of 1.25 and 1.65 µM, respectively [42].

Plocamium genus is a source of bioactive polyhalogenated monoterpenes. As example, furoplocamioid C (Figure 2E), prefuroplocamioid (Figure 2F), pirene (Figure 2G), and (1*R*,2*R*,4*R*,5*R*)-1,2,4-trichloro-5-((*E*)-2-chlorovinyl)-1,5-dimethylcyclohexane (Figure 2H), isolated from *Plocamium cartilagineum,* showed selective cytotoxicity against CT26 (murine colon adenocarcinoma), SW480 (human colon adenocarcinoma) and HeLa (human cervical adenocarcinoma) cell lines. The action of pirene (Figure 2G) is specific and irreversible to SW480 cells, which overexpress the transmembrane P-glycoprotein frequently related to chemoresistance [43]. Other unusual compounds were described in this genus, like the halogenated monoterpene aldehyde 4,6-dibromo-3,7-dimethylocta-2,7-dienal (Figure 2I), isolated from *Plocamium corallorhzia*, and aplysiaterpenoid A (Figure 2J) from *Plocamium telfiriae*. The former showed good activity toward esophageal cancer cell line WHCO1 (IC$_{50}$=7.5 µM) [44], while the last was reported with mild cytotoxicity against various tumor cell lines, with IC$_{50}$ values of 10 µg/ml for L1210 mouse lymphoma, 15.3 µg/ml for QG-90 human lung carcinoma, and 30.2 µg/ml for MCF-7 human breast cancer [45].

Among red seaweeds *Laurencia* genus gives the major contribution for bioactive compounds. Cytotoxic HC isolated from *Laurencia obtusa* and *Laurencia microcladia* include laurinterol (Figure 2K) and perforenol B (Figure 2L), which were tested against human tumor cell lines K562 (a chronic myelogenous leukemia cell line), MCF7 (derived from mammary adenocarcinoma), PC3 (derived from prostate adenocarcinoma), HeLa and A431 (derived from epidermoid carcinoma). Values of IC$_{50}$ in the range 16-107 µM were found for these compounds [46]. In addition, the study suggested that bromine is essential for the cytotoxic activity, since perforenol B (Figure 2L) was highly active when compared to compounds structurally similar like perforenone A (Figure 2M) and 3-epi-perforenone A, but without the halogen [46]. Moreover, other two cytotoxic compounds were isolated from *L. obtusa*, the diterpenes neorogioldiol B (Figure 2N) and prevezol B. Neorogioldiol B was more potent against HeLa and PC3 cell lines (IC$_{50}$ of 34.4 and 50.8 µM, respectively), while prevezol B had significant cytotoxicity against A431, and K562 cell lines (IC$_{50}$ of 65.2 and 76.4 µM, respectively) [33].

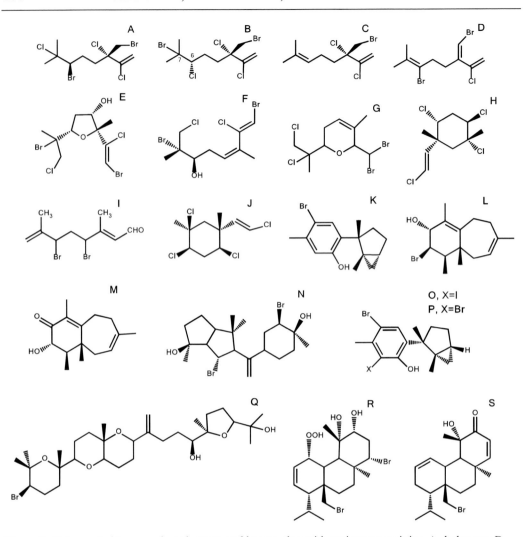

Figure 2. Halogenated terpenes in red, green and brown alga with anticancer activity. A- halomon; B- (3S,6S)-7-bromo-3-(bromomethyl)-2,3,6-trichloro-7-methyloct-1-ene; C- (S)-3-(bromomethyl)-2,3-dichloro-7-methylocta-1,6-diene; D- (Z)-6-bromo-3-(bromomethylene)-2-chloro-7-methylocta-1,6-diene; E- furoplocamioid C; F- prefuroplocamioid; G- pirene; H- (1R,2R,4R,5R)-1,2,4-trichloro-5-((E)-2-chlorovinyl)-1,5-dimethylcyclohexane; I- 4,6-dibromo-3,7-dimethylocta-2,7-dienal; J- aplysiaterpenoid A; K- laurinterol; L- perforenol B; M- perforenone A; N- neorogioldiol B; O- 4-bromo-6-((1S,2R,5R)-1,2-dimethylbicyclo[3.1.0]hexan-2-yl)-2-iodo-3-methylphenol; P- 2,4-dibromo-6-((1S,2R,5R)-1,2-dimethylbicyclo[3.1.0]hexan-2-yl)-3-methylphenol; Q- thyresenol A; R- 1(S)-hydroperoxy-12(R)-hydroxy-bromosphaerol B; S- sphaerococcenol A.

Compounds 4-bromo-6-((1S,2R,5R)-1,2-dimethylbicyclo[3.1.0]hexan-2-yl)-2-iodo-3-methylphenol (Figure 2O) and 2,4-dibromo-6-((1S,2R,5R)-1,2-dimethylbicyclo[3.1.0]hexan-2-yl)-3-methylphenol (Figure 2P), isolated from *L. microcladia* were reported as having relatively high cytotoxicity against HT29, MCF7, PC3, HeLa and A431 cell lines. Comparing with results obtained in the past with similar metabolites the authors suggested that, apart from the halogen, the presence of the aromatic hydroxyl group is responsible for cytotoxicity increase [47].

Polyether squalene derivatives like thyresenol A (Figure 2Q), thyresenol B and dehydrothyrsiferol, isolated from *Laurencia viridis*, were potent cytotoxic agents for P388

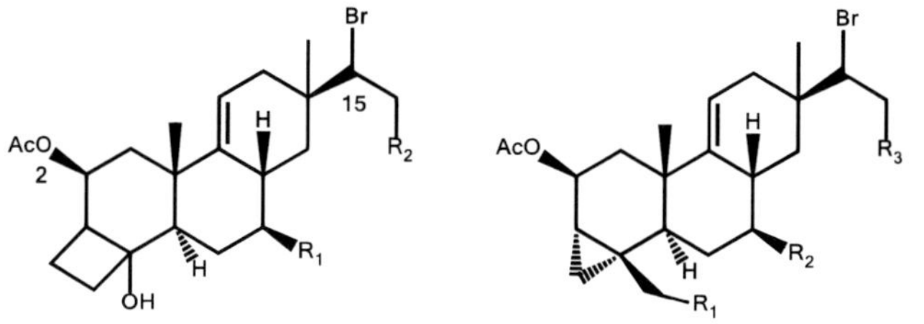

	R1	R2
A	OH	OH
B	OH	OAc
C	OAc	OAc

	R1	R2	R3
D	OH	OH	OAc
E	OH	OAc	OAc
F	H	OH	OH
G	H	OAc	OH

Figure 3. Brominated diterpenes of the isoparguerene and parguerene series isolated from the red alga *Jania rubens*. A- isoparguerol; B- isoparguerol-16-acetate; C- isoparguerol-7,16-diacetate; D- parguerol-16-acetate; E- parguerol-7,16-diacetate; F- deoxyparguerol; G- deoxyparguerol-7-acetate.

cell lines. The last compound, originally found in *Laurencia pinnatifida*, was reported as apoptosis inducer in estrogenic-dependent and -independent breast cancer cells [34,48].

The red algae *Sphaerococcus coronopifolius* provided three brominated diterpenoids metabolites that were tested against two human lung cancer cell lines (NSCLC-N6-L16 and A549). Compound 1(S)-hydroperoxy-12(R)-hydroxy-bromosphaerol-B (Figure 2R), and the structurally similar 1(S)-hydroperoxy-12(S)-hydroxy-bromosphaerol-B and 14(R)-hydroxy-13,14-dihydro-sphaerococcenol A were the most potent compounds, with IC$_{50}$ values lower than 12 µg/ml [49]. More recently, the same author have isolated and identified new compounds from the same matrix. Sphaerococcenol A (Figure 2S) and 12(S)-hydroxy-bromosphaerol along with the previously identified 14(R)-hydroxy-13,14-dihydro-sphaerococcenol A revealed to be very potent against one human apoptosis-resistant cell line, U373 (glioblastoma), and the human apoptosis-sensitive PC-3 (prostate) and LoVo (colon cancer) cell lines [50].

Brominated diterpenes from isoparguerene and parguerene series isolated from the red alga *Jania rubens* were reported as antitumor agents [51]. Isoparguerol (Figure 3A), isoparguerol-16-acetate (Figure 3B), isoparguerol-7,16-diacetate (Figure 3C), parguerol-16-acetate (Figure 3D), parguerol-7,16-diacetate (Figure 3E), deoxyparguerol (Figure 3F) and deoxyparguerol-7-acetate (Figure 3G) were tested against Ehrlich ascites carcinoma cells at 50 and 100 µg/ml. The inhibition observed varied from 50 to 100 %. It was previously reported by Takeda et al. [52] that the cytotoxic effect of these series is related to the acetoxy group in position 2 and bromine in position 15.

2.2.1.2. Antibacterial

Common gram-positive clinical pathogens are showing an increasing trend for resistance to conventional antimicrobial agents. New drugs with potent antibacterial activities are urgently needed to remediate this problem.

Several terpenes were tested as antibacterial agents, being *Laurencia* genus the major contributor of such compounds. For instance, the sesquiterpene chamigrane epoxide (Figure 4A), isolated from five different species of *Laurencia*, showed strong activity against *Staphylococcus aureus*, *Staphylococcus* sp. and *Salmonella* sp. (MICs of 125 µg/l), while a higher MIC was observed against *Vibrio cholerae* (200 µg/l) [53]. Additionally, laurinterol

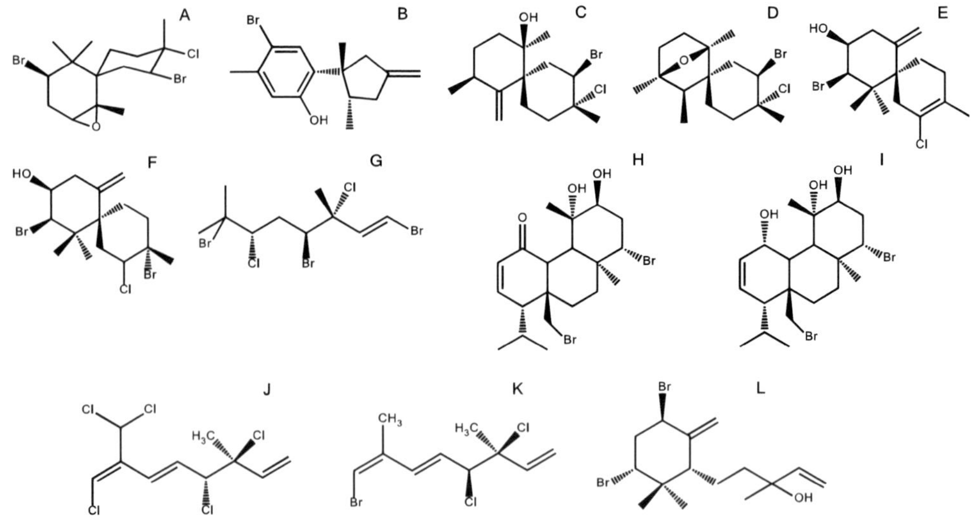

Figure 4. Terpenes with antimicrobial activity. A- chamigrane epoxide; B- allolaurinterol; C- pannosanol; D- pannosane; E- elatol; F- iso-obtusol; G- anverene; H- bromosphaerone; I- 12(S)-hydroxybromosphaerodiol; J- (1E,3E,5R,6S)-1,5,6-tricholoro-2-(dichloromethyl)-6-methylocta-1,3,7-triene; K- (1Z,3E,5S,6S)-1-bromo-5,6-dichloro-2,6-dimethylocta-1,3,7-triene; L- (8R)-8-bromo-10-epi-β-snyderol.

and allolaurinterol (Figure 2K and 4B, respectively), at 3.13 µg/ml, have been described with potent bactericidal activity against three strains of methicillin-resistant *S. aureus*. They were also effective against three strains of vancomycin-susceptible *Enterococcus*, at 3.13 µg/ml and 6.25 µg/ml, respectively [31].

Pannosanol (Figure 4C), the major compound isolated from the Malaysian *Laurencia pannosa*, was active against *Proteus mirabilis*, *V. cholerae* and *Chromobacterium violaceum*, while pannosane (Figure 4D) was active only against the latter [54].

Elatol (Figure 4E) and iso-obtusol (Figure 4F) isolated from *Laurencia majuscula* were evaluated against a bacteria panel. Elatol was very potent against *Staphylococcus epidermidis*, *Klebsiella pneumoniae* and *Salmonella* sp., while iso-obtusol exhibited significant activity against the last two bacteria. These compounds revealed to be more potent than some of the commercial antibiotics used as reference [32].

Anverene (Figure 4G), obtained from *P. cartilagineum* red alga, showed modest but selective activity against vancomycin-resistant *Enterococci faecium* [55].

Two bromoditerpenes isolated from *S. coronopifolius* (red algae), bromosphaerone and 12(S)-hydroxybromosphaerodiol (Figures 4H and 4I, respectively), showed antibacterial activity against *S. aureus*, with MICs of 0.104 and 0.146 µM, respectively [56].

2.2.1.3. Antiplasmodial

Malaria is an infectious disease caused by a protozoan of the genus *Plasmodium*, which is transmitted to humans by an infected mosquito. In the search of compounds able to treat this disease, halogenated monoterpenes isolated from *Plocamium cornutum*, (1E,3E,5R,6S)-1,5,6-tricholoro-2-(dichloromethyl)-6-methylocta-1,3,7-triene (Figure 4J) and (5R,6S,E)-5,6-dichloro-2-(dichloromethyl)-6-methylocta-1,3,7-triene, were tested, revealing to be active.

The presence of a dichloromethyl moiety in position 2 seems to be fundamental for the antiplasmodial capacity, since (1Z,3E,5S,6S)-1-bromo-5,6-dichloro-2,6-dimethylocta-1,3,7-triene (Figure 4K), lacking this group, was the less active compound [57].

Sphaerococcenol A (Figure 2S) was considered the compound responsible for the antimalarial activity of *S. coronopifolius* extract against chloroquine resistant *Plasmodium falciparum* FCB1 strains, showing an IC$_{50}$ of 1 μM [56]. In addition, (8R)-8-bromo-10-epi-β-snyderol (Figure 4L) from *L. obtusa*, revealed moderate activity against D6 and chloroquine-resistant clone W2 of *P. faciparum* [58].

2.2.1.4. Antitrypanosomal

Trypanosomiasis, or Chagas' disease, is a debilitating illness caused by *Trypanosoma cruzi*, with high prevalence in Latin America, affecting, in average, around 14 million people. Elatol (Figure 4E) was tested against *T. cruzi* at different life-cycle stages (epimastigote in the insect gut, trypamastigote infection stage, and amastigote in the host). Remarkable results were reported for elatol as inhibitor of trypamastigote and amastigotes stages, with IC$_{50}$ values of 1.38 and 1.01 μM, respectively. In both cases this compound was more efficient than the reference drugs [59].

2.2.1.5. Antileishamanial

Leishmania amazonensis is a flagellated protozoan parasite that causes human cutaneous leishmaniasis, an infectious disease with debilitating and disfiguring actions. Elatol (Figure 4E) has been recently reported with a dose-dependent antileishmanial activity against the promastigote and amastigotes forms of *L. amazonensis*. In the first case promastigotes were inhibited with and IC$_{50}$ of 4.0 μM, while the decrease of amastigotes reached an IC$_{50}$ value (0.45 μM) close to that of the reference drug amphotericin B (0.31 μM) [60].

2.3. Acetogenins

Acetogenins constitute a C$_{15}$ nonterpenoid class of compounds unique to seaweeds of the genus *Laurencia*. Compounds described in these red seaweeds show remarkable structural variations, including unusual cyclic ethers with different ring sizes, with five- to nine-membered central acetogenic oxygen, an enyne or allene side chain, and at least one bromine atom [61].

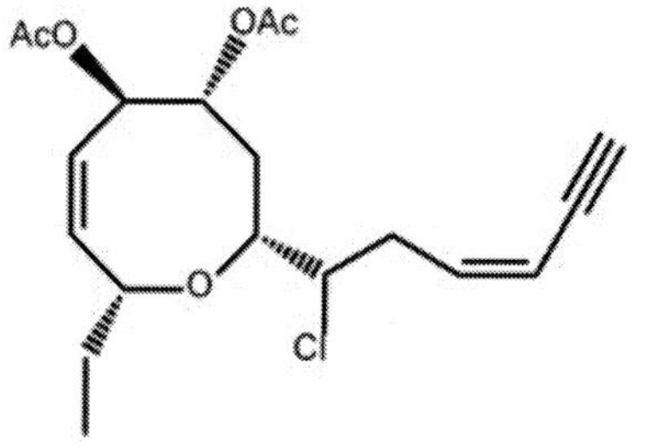

Figure 5. (3Z,7R,9R,10R,13R)-9,10-Diacetoxy-6-chlorolauthisa-3,11-dien-1-yne, a C$_{15}$ acetogenin with antistaphylococcal activity.

Regarding biological activities, C_{15} acetogenins isolated from *Laurencia glandulifera* were reported with significant antistaphylococcal activity. Apart from the halogen group, found in all compounds, the presence of two acetyl groups in (3Z,7R,9R,10R,13R)-9,10-diacetoxy-6-chlorolauthisa-3,11-dien-1-yne (Figure 5) seems to increase its lipophilic character, improving its cellular bioavailability and, thus, its activity [62].

2.4. Phenolic Compounds

Several classes of halogenated phenolic compounds have been described in extracts of seaweeds belonging to Rhodophyta, Phaeophyta and Chlorophyta phyla (Figure 6). They are predominantly brominated, although some iodinated and chlorinated compounds were also found.

2.4.1. Bromophenols

Several simple and complex bromophenols are reported in bibliography. Concerning the first category, monomeric bromophenols, such as 2,4-dibromophenol (Figure 6F), 2,4,6-tribromophenol (Figure 6E) and dibromo-iodophenol were identified in the brown seaweeds *Eisenia bicyclis* and *Ecklonia kurome* [63]. Whitfield [64] determined 2- and 4-bromophenol, 2,4- and 2,6-dibromophenol, and 2,4,6-tribromophenol contents in 49 species of seaweeds from eastern Australia. It was possible to conclude that among the analysed samples, the red and green ones showed higher amount of bromophenols.

Red seaweeds are also rich in more complex bromophenols. For instance, 3,5-dibromo-4-methoxy-phenylacetic acid, 2-methoxy-3-(3',5'-dibromo-4'-methoxyphenyl)acrylic acid [65] and 2,6-dibromo-3,5-dihydroxyphenylacetic acid [66] were detected in extracts of *Halopytis incurvus*. Katsui et al. [67] and Weinstein et al. [68] reported the presence of 5,6-dibromoprotocatechualdehyde, 2,3-dibromo-4,5-dihydroxybenzyl methyl ether, dipotassium 2,3-dibromo-5-hydroxy benzyl-1',4-disulphate and lanosol (Figure 6I) in *Rhodomela larix*. Other more complex compounds were identified in the red seaweed *Rytiphlea tinctoria*, namely, 2,4-dibromo-1,3,5-trimethoxy-benzene, 5,6,3',5'-tetrabromo-3,4,2',4',6'-pentamethoxydiphenylmethane, 5,6-dibromo-3,4-dimethoxy-benzyl alcohol and its ethyl ether [69].

The ethanolic extract of the brown alga *Leathesia nana* revealed the presence of (E)-3-(2,3-dibromo-4,5-dihydroxyphenyl)-2-methyl-propenal [70].

2.4.2. Phloroglucinol and Derivatives

Only a few halogenated phloroglucinols have been isolated from macroalgae. For instance, iodophloroglucinol and bromophloroglucinol were identified in the brown seaweed *Eisenia arborea* [71], while bromo- and chloro-derivatives were detected in *Rhabdonia verticillata* (Rhodophyta): 2,4-dibromo-1,3,5-trihydroxybenzene (=dibromophloroglucinol), 2-chloro-1,3,5-trihydroxybenzene, 2-bromo-1,3,5-trihydroxybenzene, 2-bromo-4-chloro-1,3,5-trihydroxybenzene, 2,4-dibromo-6-chloro-1,3,5-trihydroxybenzene and 2,4,6-tribromo-1,3,5-trihydroxybenzene [72].

Phlorotannins are polyphenols formed by the polymerization of phloroglucinol units and are only known to occur in brown algae [8, 73]. They are grouped into four subclasses:

phlorotannins with an ether linkage (fuhalols and phlorethols); with a phenyl linkage (fucols); with both ether and phenyl linkage (fucophlorethols); and with a dibenzodioxin linkage (eckols and carmalols). Examples of the first class are bromotriphlorethol A1 and bromotriphlorethol A2 found in *Cystophora congesta* [74]. *Carpophyllum angustifolium*, on the other hand, possesses phlorethols, fuhalols and fucols [75]. Monosubstituted eckols like 4'-bromoeckol and 4'-iodoeckol were described in *E. arborea* [71].

2.4.3. Biological Activities

Due to their high structural variability halogenated phenolic compounds possess several biological activities, such as antioxidant, antimicrobial, antidiabetic, anticoagulation, and cytotoxicity. For instance, the methanolic extract of the red seaweed *Symphyocladia latiuscula*, containing 2,3,6-tribromo-4,5-dihydroxybenzyl methyl ether (Figure 6A), (2*R*)-2-(2,3,6-tribromo-4,5-dihydroxybenzyl)-cyclohexanone (=symphyoketone) (Figure 6B) and 2,3,6-tribromo-4,5-dihydroxybenzyl alcohol (Figure 6C), showed antiradical activity, with the last two being 2-fold more potent than L-ascorbic acid [76]. However, Lee et al. [77] observed that the debrominated forms of the bromophenols obtained from an extract of the red seaweed *Tichocarpus crinitus* were more active than their precursors.

Oh et al. [78] studied the *in vitro* antimicrobial capacity of six bromophenols from the red seaweed *Odonthalia corymbifera* and concluded that di-phenolic compounds were active against Gram-positive and Gram-negative organisms, excepting *Escherichia coli*, whereas the mono-phenolic metabolites were inactive. Xu et al. [79] also observed the antibacterial effect of five bromophenols isolated from the red seaweed *Rhodomela confervoides* and bis (2,3-dibromo-4,5-dihydroxybenzyl)ether (Figure 6D) revealed to be the most active.

The antidiabetic potential of seaweeds was revealed in α-glucosidase inhibitory *in vitro* assays [80-82]. In a study performed by Kim et al. [81], 2,4,6-tribromophenol (Figure 6E) showed to be more active than 2,4-dibromophenol (Figure 6F) against *S. cerevisiae* and *B. stearothermophilus* α-glucosidases. The inhibitory potential of the bromophenol increased with increasing degree of bromo-substitution per benzene ring and with decreasing degree of methyl-substitution. This was also observed by Kurihara et al. [80], working with 6-bromo-3-(2,5-dibromo-6-hydroxy-3-(hydroxymethyl)phenoxy)-4-(hydroxymethyl)benzene-1,2-diol (Figure 6G) and bis(2,3-dibromo-4,5-dihydroxybenzyl) ether (Figure 6D), isolated from the red seaweed *O. corymbifera*. This last was also found in the red seaweed *Polyopes lancifolia* and showed to be more active than acarbose and voglibose (positive controls) against yeast (*S. cerevisiae*) and bacterial (*B. stearothermophilus*) α-glucosidases, but less effective regarding human α-glucosidase [82].

Thrombin, the ultimate proteinase of the coagulation cascade, is an attractive target for the treatment of a variety of cardiovascular diseases. With this respect, (+)-3-(2,3-dibromo-4,5-dihydroxy-phenyl)-4-bromo-5,6-dihydroxy-1,3-dihydroiso-benzofuran (Figure 6H), isolated from the brown seaweed *L. nana*, exhibited almost seven-fold higher inhibitory activity compared with argatruban, a positive control [83].

The cytotoxicity of the methanolic extract of the red seaweed *Polysiphonia lanosa*, as well as of its chloroformic and hexane fractions, were assessed against the human cancer cell line DLD-1 (adenocarcinoma). The chloroformic fraction was more active than the whole methanolic extract (IC_{50} = 4.59 µg/ml *vs* IC_{50} = 39.7 µg/ml).

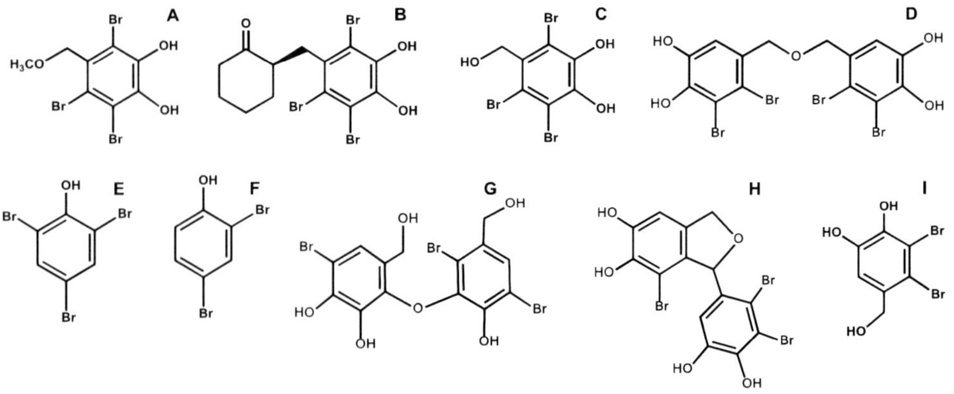

Figure 6. Some bioactive halogenated phenolics. A- 2,3,6-tribromo-4,5-dihydroxybenzyl methyl ether; B- (2R)-2-(2,3,6-tribromo-4,5-dihydroxybenzyl)-cyclohexanone (=symphyoketone); C- 2,3,6-tribromo-4,5-dihydroxybenzyl alcohol; D- bis (2,3-dibromo-4,5-dihydroxybenzyl)ether; E- 2,4,6-tribromophenol; F- 2,4-dibromophenol; G- 6-bromo-3-(2,5-dibromo-6-hydroxy-3-(hydroxymethyl)phenoxy)-4-(hydroxymethyl)benzene-1,2-diol; H- (+)-3-(2,3-dibromo-4,5-dihydroxy-phenyl)-4-bromo-5,6-dihydroxy-1,3-dihydroiso-benzofuran; I- lanosol.

The authors also isolated some brominated compounds from this active fraction and reported the strong citotoxicity of lanosol (Figure 6I), with a IC$_{50}$ of 18.3 µg/ml, methyl and n-propyl ethers of lanosol (IC$_{50}$ = 14.6 and 12.4 µg/ml, respectively), and of lanosol aldehyde (IC$_{50}$ = 30.9 µg/ml). Again, it was concluded that compounds with two phenolic groups were more active than those with just one [84].

2.5. Fatty Acids and Derivatives

Brown and red seaweeds present a quite similar overall composition of saturated and unsaturated fatty acids, with a high proportion of long chain (C$_{18}$ and C$_{20}$) polyunsaturated forms (PUFA) [1, 8]. The presence of 20% or more of fatty acids, such as arachidonic, eicosapentaenoic (EPA) and docosahexaenoic (DHA), which is quite common in seaweeds, is of great interest because of the recognized need for very long chain PUFA in healthy diets [1]. The major saturated fatty acid in seaweeds is palmitic acid [1].

As stated by Dembitsky and Srebnik [85], the presence of a halogen atom (Cl, Br or I) in the fatty alkyl chain causes considerable changes in the physico-chemical characteristics of the fatty acid and, consequently, increases its reactivity and changes the conformation of biological membranes.

Up to now, several brominated fatty acids, as well as different classes of derivatives, have been identified in red seaweeds. For instance, *Asparagopsis taxiformis* and *Asparagopsis armata* produce a number of brominated acetic and acrylic acids [85,86]. *Bonnemaisonia nootkana*, *Bonnemaisonia hamifera*, and *Trailliella intricate* synthesize brominated forms of 2-heptenoic, 2-nonenoic, 2-n-butylacrylic and 2-n-hexylacrylic acids [87].

Concerning chlorinated metabolites, a C$_{13}$ chlorinated fatty acid was identified in the red alga *P. cartilagineum* by Řezanka and Dembitsky [88], while different chlorosulpholipids were found in seaweeds belonging to Chlorophyta, Phaeophyta and Rhodophyta [85, 89]. Several types of chlorinated oxylipins were isolated from brown seaweeds, namely from

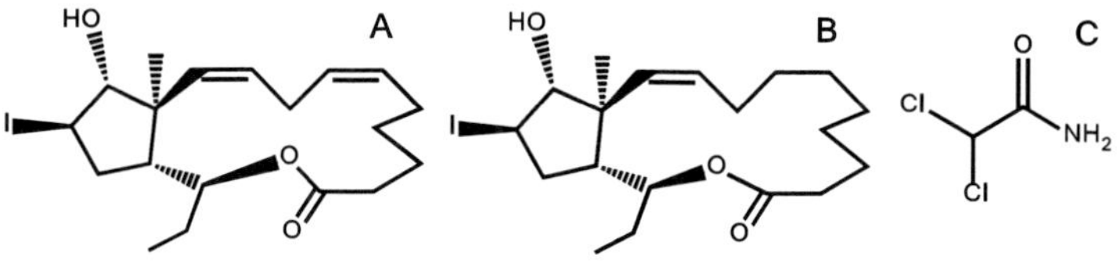

Figure 7. Some bioactive halogenated fatty acids and fatty acid derivatives. A- eiseniaiodide A; B- eiseniaiodide B; C- dichloroacetamide.

Egregia menziesii [90]. Additionally, Kousaka et al. [91] isolated ecklonialactone derivatives from *E. bicyclis*, namely eiseniaiodide A and eiseniaiodide B (Figures 7A and 7B, respectively) among a whole series of C_{18} oxylipins. Other chlorinated compounds include carboxylic acids derivatives (from acetic and acrylic acids) described in *A. taxiformis* [92], and dichloroacetamide (Figure 7C) found in *Marginisporum aberrans* [93], both red seaweeds.

On the other hand, natural iodinated fatty acids and their derivatives are much rarer than the chlorinated and/or brominated ones. For instance, simple iodinated acetic acid (mono- and diiodo-acetic acids) and acrylic acid (mono- and diiodo-acrylic acids), and their ethyl esters have been found in the red seaweeds *A. taxiformis* [86,92] and *A. armata* [86].

2.5.1. Biological Activities

Concerning these metabolites, Kousaka et al. [91] tested halogenated oxylipins from *E. bicyclis* against *B. subtilis* and *S. aureus*, but no improvement of the activities of any of the halogenated forms, namely eiseniaiodide A and eiseniaiodide B (Figures 7A and 7B, respectively), over the non-halogenated ones was noticed.

On the other hand, Khaskin et al. [94] reported a moderate activity of dichloroacetamide (Figure 7C) against *Botorytis cinerea* and *Alternaria radicina*.

2.6. Low Molecular Weight Volatile Halogenated Organic Compounds (VHOC)

Marine ecosystems are an important source of low molecular weight volatile halogenated organic compounds (VHOC) that includes several chlorinated, brominated, iodinated and fluorinated halocarbons. Their origin is either anthropogenic, such as bromomethane (CH_3Br), chlorofluorocarbons (CFC), chloroform ($CHCl_3$), dichloromethane (CH_2Cl_2), tetrachloromethane (CCl_4), tetrachloroethylene ($Cl_2C=CCl_2$) and trichloroethylene ($ClCH=CCl_2$) that result from human activities like the use of pesticides and antifreezing agents, and/or biogenic, resulting from living organisms, like brown and red seaweeds (Table 1) [6,7,40,95-99]. However, some compounds, such as $ClCH=CCl_2$, $Cl_2C=CCl_2$ and $CHCl_3$, can have both origins [5,97].

Seaweeds are reported to produce hundreds of ng to µg/g (dry basis/day) of brominated halomethanes and iodomethane in the Atlantic [100] and the Pacific [95] oceans. For instance, Newman and Gschwend [101] found that $CHBr_3$ and CH_2Br_2, produced by *A. nodosum*, were released in the range of 28-520 ng/g dry wt/day, and up to 98 µg/g dry wt/day, respectively. Additionally, Gribble [102] estimated that the annual global emission of CH_3Cl from the oceans to the atmosphere can reach up to 5×10^{12} g/year. Usually, the rates of

VHOC's released are higher for tropical algae than for temperate ones [97, 99], and even lower for polar ecosystems (1000-fold less) [103]. Moreover, red algae are the minor sources of VHOC, as found by Nightingale et al. [5].

The intensive research on VHOC's has been motivated mostly by concerns over their role in atmosphere. It is known that brominated and iodinated hydrocarbons can take part in the global greenhouse effect, since their instable radicals can destroy both tropospheric and stratospheric ozone [16,104,105].

Halogen uptake and production of halocompounds by seaweeds are thought to involve vanadium-dependent haloperoxidases and methyltransferase enzymes. Polyhalomethanes (di and tri-halomethanes) are a byproduct of the halogenation of certain organic molecules by haloperoxidases, whereas methyl halides (monohalomethanes) are products of methyltransferase activity [105].

Coralline algae, such as *C. pilulifera* and *Lithophyllum yessoens* (both red seaweed), suppressed the growth of microalgae and other algae on their surface by releasing $CHBr_3$. By this mechanism the algae could obtain a sufficient supply of sunlight and nutrition [40]. McConnell and Fenical [86] reported that the extracts of *A. taxiformis* (a red alga) exhibit considerable antibacterial activity.

Table 1. Low molecular weight volatile halogenated organic compounds (VHOC) with biogenic origin

Halogen	
Bromine	1,2-Dibromoethane ($BrCH_2CH_2Br$)
	Bromoethane (CH_3CH_2Br)
	Bromoform ($CHBr_3$)
	Bromomethane (CH_3Br)
	Bromopropane ($CH_3CH_2CH_2Br$)
	Dibromomethane (CH_2Br_2)
Chlorine	Chloroform ($CHCl_3$)
	Chloromethane (CH_3Cl)
	Dichloromethane (CH_2Cl_2)
	Tetrachloroethylene ($Cl_2C=CCl_2$)
	Trichloroethylene ($ClCH=CCl_2$)
Iodine	Diiodomethane (CH_2I_2)
	Iodoethane (CH_3CH_2I)
	Iodomethane (CH_3I)
	Iodobutane ($CH_3CHICH_2CH_3$)
	Iodopropane (CH_3CHICH_3)
Mixed	Bromochloroiodomethane (CHBrClI)
	Bromochloromethane (CH_2BrCl)
	Bromodichloromethane ($CHBrCl_2$)
	Chloroiodomethane (CH_2ClI)
	Chlorodibromomethane ($CHClBr_2$)
	Dibromochloromethane ($CHBrCl_2$)

CONCLUSION

Seaweeds present important roles in aquatic ecosytems. First, their high nutritional values place them at the bottom of the food chain in all aquatic ecosystems [24]. Secondly, they are responsible for about half of the O_2 production to the atmosphere [24] and finally, since they live in an extreme competitive environment, they produce several classes of defence metabolites that have been poorly or have not been yet exploited at all by scientific community to find their role as new human drugs. In fact, researchers have been devoted more attention to land plant ecosystems than to the aquatic ones. However, from those seaweed compounds that have already been described and tested in some *in vitro* or *in vivo* bioassays, it can be seen that the incorporation of halogen atoms may increase their bioactivity. Some promising examples of halogenated compounds were presented in this overview, which also demonstrated that seaweeds can constitute a benefit to human health and can be regarded as a potential medicinal food.

REFERENCES

[1] Harwood, J. L. and Guschina, I. A. (2009). The versatility of algae and their lipid metabolism. *Biochimie,* 91, 679-684.

[2] Renn, D. (1997). Biotechnology and the red seaweed polysaccharide industry: status, needs and prospects. *Trends in Biotechnology*, 15, 9-14.

[3] Smit, A. J. (2004). Medicinal and pharmaceutical uses of seaweed natural products: A review. Journal of Applied Phycology, 16, 245-262.

[4] Fletcher, R. L. (1995). Epiphytism and fouling in *Gracilaria* cultivation: an overview. *Journal of Applied Phycology*, 7, 325-333.

[5] Nightingale, P. D., Malin, G. and Liss, P. S. (1995). Production of chloroform and other low-molecular-weight halocarbons by some species of macroalgae. *Limnology and Oceanography*, 40, 680-689.

[6] Gribble, G. W. (1998). Naturally occurring organohalogen compounds. *Accounts of Chemical Research,* 31, 141-152.

[7] Kladi, M., Vagias, C. and Roussis, V. (2004). Volatile halogenated metabolites from marine red algae. *Phytochemical Reviews*, 3, 337-366.

[8] La Barre, S., Potin, P., Leblanc, C. and Delage, L. (2010). The halogenated metabolism of brown algae (Phaeophyta), its biological importance and its environmental significance. *Marine Drugs*, 8, 988-1010.

[9] Bernroitner, M., Zamocky, M., Furtmüller, P. G., Peschek, G. A. and Obinger, C. (2009). Occurrence, phylogeny, structure, and function of catalases and peroxidases in cyanobacteria. *Journal of Experimental Botany*, 60, 423-440.

[10] Butler, A. and Sandy, M. (2009). Mechanistic considerations of halogenating enzymes. *Nature,* 460, 848-854.

[11] Wuosmaa, A. M. and Hager, L. P. (1990). Methyl chloride transferase: a carbocation route for biosynthesis of halometabolites. *Science*, 249, 160-162.

[12] Chen, X. and van Pée, K. H. (2008). Catalytic mechanisms, basic roles, and biotechnological and environmental significance of halogenating enzymes. *Acta Biochimica et Biophysica Sinica*, 40, 183-193.

[13] Deng, H. and O'Hagan, D. (2008). The fluorinase, the chlorinase and the duf-62 enzymes. *Current Opinion in Chemical Biology*. 12, 582-592.

[14] Blasiak, L. C. and Drennan, C. L. (2009). Structural perspective on enzymatic halogenation. *Accounts of Chemical Research*, 42, 147-155.

[15] Harper, M. K., Bugni, T. S., Copp, B. R., James, R. D., Lindsay, B. S., Richardson, A. D., Schnabel, P. C., Tasdemir, D., VanWagoner, R. M., Verbitzki, S. M. and Ireland, C. M. (2001). Introduction to the chemical ecology of marine natural products. *In Marine Chemical Ecology*. (McClintock, J.B., Baker, B.J., Eds.). CRC: Boca Raton, FL, USA, 3-71.

[16] Goodwin, K. D., North, W. J. and Lidstrom, M. E. (1997). Production of bromoform and dibromomethane by Giant Kelp: Factors affecting release and comparison to anthropogenic bromine sources. *Limnology and Oceanography*, 42, 1725-1734.

[17] Dworjanyn, S. A., De Nys, R. and Steinberg, P. D. (1999). Localisation and surface quantification of secondary metabolites in the red alga *Delisea pulchra*. *Marine Biology*, 133, 727-736.

[18] Taskin, E., Caki, Z., Ozturk, M. and Taskin, E. (2010). Assessment of in vitro antitumoral and antimicrobial activities of marine algae harvested from the eastern Mediterranean sea. *African Journal of Biotechnology*, 9, 4272-4277.

[19] Genovese, G., Tedone, L. Hamann, M. T. and Morabito, M. (2009). The Mediterranean red alga *Asparagopsis*: a source of compounds against *Leishmania*. *Marine Drugs*, 7, 361-366.

[20] Zubia, M., Robledo, D. and Freile-Pelegrin, Y. (2007). Antioxidant activities in tropical marine macroalgae from the Yucatan Peninsula, Mexico. *Journal of Applied Phycology*, 19, 449-458.

[21] Salvador, N., Garreta, A. G., Lavelli, L. and Ribera, M. A. (2007). Antimicrobial activity of Iberian macroalgae. *Scientia Marina*, 71, 101-113.

[22] Stirk, W. A., Reinecke, D. L. and van Staden, J. (2007). Seasonal variation in antifungal, antibacterial and acetylcholinesterase activity in seven South African seaweeds. *Journal of Applied Phycology*, 19, 271-276.

[23] del Val, A. G., Platas, G., Basilio, A., Cabello, A., Gorrochategui, J., Suay, I., Vicente, F., Portillo, E., del Rio, M. J., Reina, G. G. and Peláez, F. (2001). Screening of antimicrobial activities in red, green and brown macroalgae from Gran Canaria (Canary Islands, Spain). *International Microbiology*, 4, 35-40.

[24] Cardozo, K. H. M., Guaratini, T., Barros, M. P., Falcão, V. R., Tonon, A. P., Lopes, N. P., Campos, S., Torres, M. A., Souza, A. O., Colepicolo, P. and Pinto, E. (2007). Metabolites from algae with economical impact. *Comparative Biochemistry and Physiology Part C: Toxicology and Pharmacology*, 146, 60-78.

[25] Güven, K. C., Percot, A. and Sezik, E. (2010). Alkaloids in marine algae. *Marine Drugs*, 8, 269-284.

[26] El-Gamal, A. A., Wang, W.-L. and Duh, C.-Y. (2005). Sulfur-containing polybromoindoles from the Formosan red alga *Laurencia brongniartii*. *Journal of Natural Products*, 68, 815-817.

[27] Brennan, M. R. and Erickson, K. L. (1978). Polyhalogenated indoles from the marine alga *Rhodophyllis membranacea* Harvey. *Tetrahedron Letters*, 19, 1637-1640.

[28] Fenical, W. (1975). Halogenation in the Rhodophyta: a review. *Journal of Phycology* 11, 245-259.

[29] Gribble, G. W. (1999). The diversity of naturally occurring organobromine compounds. *Chemical Society Reviews*, 28, 335-346.

[30] Davyt, D., Fernandez, R., Suescun, L., Mombrú, A. W., Saldaña, J., Domínguez, L., Fujii, M. T. and Manta, E. (2006). Bisabolanes from the red alga *Laurencia scoparia*. *Journal of Natural Products*, 69, 1113-1116.

[31] Vairappan, C. S., Kawamoto, T., Miwa, H. and Suzuki M. (2004). Potent antibacterial activity of halogenated compounds against antibiotic-resistant bacteria. *Planta Medica*, 70, 1087-1090.

[32] Vairappan, C. S. (2003). Potent antibacterial activity of halogenated metabolites from Malaysian red algae, *Laurencia majuscula* (Rhodomelaceae, Ceramiales). *Biomolecular Engineering*, 20, 255-259.

[33] Iliopoulou, D., Mihopoulos, N., Vigias, C., Papazafiri, P. and Roussis, V. (2003). Novel cytotoxic brominated diterpenes from the red alga *Laurencia obtusa*. *The Journal of Organic Chemistry*, 68, 7667-7674.

[34] Norte, M., Fernandez, J. J., Souto, M. L., Gavin, J. A. and García-Grávalos, M. D. (1997). Thyrsenols A and B two unusual polyether squalene derivatives. *Tetrahedron*, 53, 3173-3178.

[35] Fuller, R. W., Cardellina, J. H., Kato, Y., Brinen, L. S., Clardy, J., Snader, K. M. and Boyd, M. R. (1992). A pentahalogenated monoterpene from the red alga *Portieria hornemannii* produces a novel cytotoxicity profile against a diverse panel of human tumor cell lines. *Journal of Medicinal Chemistry*, 35, 3007-3011.

[36] Fuller, R. W., Cardellina, J. H., Jurek, J., Scheuer, P. J., Alvarado-Lindner, B., McGuire, M., Gray, G. N., Steiner, J. R. and Clardy J. (1994). Isolation and structure/activity features of halomon-related antitumor monoterpenes from the red alga *Portieria hornemannii*. *Journal of Medicinal Chemistry*, 37, 4407-4411.

[37] Barnekow, D. E., Cardellina, J. H., Zektzer, A. S. and Martin, G. E. (1989). Novel cytotoxic and phytotoxic halogenated sesquiterpenes from the green alga *Neomeris annulata*. *Journal of the American Chemical Society*, 111, 3511-3517.

[38] Ji, N.-Y., Wen, W., Li, X.-M., Xue, Q.-Z., Xiao, H.-L. and Wang, B.-G. (2009). Brominated selinane sesquiterpenes from the marine brown alga *Dictyopteris divaricata*. *Marine Drugs*, 7, 355-360.

[39] Vilter, H. (1984). Peroxidases from Phaeophyceae: a vanadium (V)-dependent peroxidase from Ascophyllum nodosum. *Phytochemistry*, 23, 1387-1390.

[40] Ohsawa, N., Ogata, Y., Okada, N and Itoh, N. (2001). Physiological function of bromoperoxidase in the marine red alga, *Corallina pilulifera*: production of bromoform as an allelochemical and the simultaneous elimination of hydrogen peroxide. *Phytochemistry*, 58, 683-692.

[41] Carter-Franklin, J. N. and Butler, A. (2004). Vanadium bromoperoxidase-catalyzed biosynthesis of halogenated marine natural products. *Journal of the American Chemical Society*, 126, 15060-15066.

[42] Andrianasolo, E. H., France, D., Cornell-Kennon, S. and Gerwick, W. H. (2006). DNA methyl transferase inhibiting halogenated monoterpenes from the Madagascar red marine alga *Portieria hornemannii*. *Journal of Natural Products*, 69, 576-579.

[43] de Inés, C., Argandoña, V. H., Rovirosa, J., San-Martín, A., Díaz-Marrero, A. R., Cueto, M. and González-Coloma, A. (2004). Cytotoxic activity of halogenated monoterpenes from *Plocamium cartilagineum*. Zeitschrift für Naturforschung. C, *Journal of Biosciences*, 59, 339-344.

[44] Mann, M. G. A., Mkwananzi, H. B., Antunes, E. M., Whibley, C. E., Hendricks, D. T., Bolton, J. J. and Beukes, D. R. (2007). Halogenated monoterpene aldehydes from the South African marine alga *Plocamium corallorhiza*. *Journal of Natural Products*, 70, 596-599.

[45] Miyamoto T. (2006). Selected bioactive compounds from Japanese Anaspideans and Nudibranchs. In *Progress in Molecular and Subcellukar Biology, Subseries Marine Molecular Biotechnology*. (Cimino, G. and M. Gavagnin Eds.). Springer-Verlag Berlin Heidelberg, 199-214.

[46] Kladi, M., Xenaki, H., Vagias, C., Papazafiri, P. and Roussis, V. (2006). New cytotoxic sesquiterpenes from the red algae *Laurencia obtusa* and *Laurencia microcladia*. *Tetrahedron*, 62, 182-189.

[47] Kladi, M., Vagias, C., Papazafiri, P., Furnari, G., Serio, D. and Roussis, V. (2007). New sesquiterpenes from the red alga *Laurencia microcladia*. *Tetrahedron*, 63, 7606-7611.

[48] Pec, M. K., Aguirre, A., Moser-Their, K., Fernandez, J. J., Souto, M. L., Dorta, J., Diáz-González, F. and Villar, J. (2003). Induction of apoptosis in estrogen dependent and independent breast cancer cells by the marine terpenoid dehydrothyrsiferol. *Biochemical Pharmacology*, 65, 1451-1461.

[49] Smyrniotopoulos, V., Quesada, A., Vagias, C., Moreau, D., Roussakis, C. and Roussis, V. (2008). Cytotoxic bromoditerpenes from the red alga *Sphaerococcus coronopifolius*. *Tetrahedron*, 64, 5184-5190.

[50] Smyrniotopoulos, V., Vagias, C., Bruyère, C., Lamoral-Theys, D., Kiss, R. and Roussis, V. (2010). Structure and in vitro antitumor activity evaluation of brominated diterpenes from the red alga *Sphaerococcus coronopifolius*. *Bioorganic and Medicinal Chemistry*, 18, 1321-1330.

[51] Awad, N. E. (2004). Bioactive brominated diterpenes from the marine red alga *Jania Rubens* (L.) Lamx. *Phytotherapy Research*, 18, 275-279.

[52] Takeda, S., Kurosawa, E., Komiyama, K. and Suzuki, T. (1990). The structures of cytotoxic diterpenes cointaining bromine from the marine red alga *Laurencia obtusa* (Hudson) Lamouroux. *Bulletin of the Chemical Society of Japan*, 63, 3066-3072.

[53] Vairappan, C. S., Ishii, T., Lee, T. K., Suzuki, M. and Zhaoqi, Z. (2010). Antibacterial activities of a new brominated diterpene from Borneon *Laurencia* spp. *Marine Drugs*, 8, 1743-1749.

[54] Suzuki, M., Daitoh, M., Vairappan, C. S., Abe, T. and Masuda, M. (2001). Novel Halogenated metabolites from the Malaysian *Laurencia pan*nosa. *Journal of Natural Products*, 64, 597-602.

[55] Ankisetty, S., Nandiraju, S., Win, H., Park, Y. C., Amsler, C. D., McClintock, J. B., Baker, J. A., Diyabalanage, T. K., Pasaribu, A., Singh, M. P., Maiese, W. M., Walsh, R. D., Zaworotko, M. J. and Baker, B. J. (2004). Chemical investigation of predator-

deterred macroalgae from the Antarctic Peninsula. *Journal of Natural Products*, 67, 1295-1302.

[56] Etahiri, S., Bultel-Poncé, V., Caux, C. and Guyot, M. (2001). New bromoditerpenes from the red alga *Sphaerococcus coronopifolius*. *Journal of Natural Products* 64, 1024-1027.

[57] Afolayan, A. F., Mann, M. G. A., Lategan, C. A., Smith, P. J., Bolton, J. J. and Beukes, D. R. (2009). Antiplasmodial halogenated monoterpenes from the marine red alga *Plocamium cornutum*. *Phytochemistry*, 70, 597-600.

[58] Topcu, G., Aydogmus, Z., Imre, S., Gören, A. C., Pezzuto, J. M., Clement, J. A. and Kingston, D. G. I. (2003). Brominated sesquiterpenes from the red alga *Laurencia obtusa*. *Journal of Natural Products*, 66, 1505-1508.

[59] Veiga-Santos, P., Pellizzaro-Rocha, K. J., Santos, A. O., Ueda-Nakamura, T., Dias Filho, B. P., Silva, S. O., Sudatti, D. B., Bianco, E. M., Pereira, R. C. and Nakamura, C. V. (2010). In vitro anti-trypanosomal activity of elatol isolated from red seaweed *Laurencia dendroidea*. *Parasitology*, 137, 1661-1670.

[60] Santos, A. O., Veiga-Santos, P., Ueda-Nakamura, Dias Filho, B. P., Sudatti, D. B., Bianco, E. M., Pereira, R. C. and Nakamura, C. V. (2010). Effect of elatol, isolated from red seaweed *Laurencia dendroidea*, on *Leishmania amazonensis*. *Drugs*, 8, 2733-2743.

[61] Dembitsky, V. M., Tolstikov, A. G. and Tolstikov, G. A. (2003). Natural halogenated non-terpenic C15-acetogenins of sea organisms. *Chemistry for Sustainable Development*, 11, 329-339.

[62] Kladi, M., Vagias, C., Stavri, M., Rahman, M. M., Gibbons, S. and Roussis, V. (2008). C15 acetogenins with antistaphylococcal activity from the red alga *Laurencia glandulifera*. *Phytochemistry Letters*, 1, 31-36.

[63] Shibata, T., Hama, Y., Miyasaki, T., Ito, M. and Nakamura, T. (2006). Extracellular secretion of phenolic substances from living brown algae. *Journal of Applied Phycology*, 18, 787-794.

[64] Whitfield, F. B., Helidoniotis, F., Shaw, K. J. and Svoronos, D. (1999). Distribution of bromophenols in species of marine algae from Eastern Australia. *Journal of Agricultural and Food Chemistry*, 47, 2367-2373.

[65] Chantraine, J.-M., Combaut, G. and Teste, J. (1973). Phenols bromes d'une algue rouge, *Halopytis incurvus*: acides carboxyliques. *Phytochemistry*, 12, 1793-1796.

[66] de Nanteuil, G. and Mastagli, P. (1981). A bromophenol in the red alga *Halopitys incurvus*. *Phytochemistry*, 20, 1750-1751.

[67] Katsui, N., Suzuki, Y., Kitamura, S. and Irie, T. (1967). 5,6-Dibromoprotocatechualdehyde and 2,3-dibromo-4,5-dihydroxybenzyl methyl ether. New dibromophenols from *Rhodomela larix*. *Tetrahedron*, 23, 1185-1188.

[68] Weinstein, B., Rold, T. L., Harrell Jr., C. E., Burns III, M. W. and Waaland, J. R. (1975). Reexamination of the bromophenols in the red alga *Rhodomela larix*. *Phytochemistry*, 14, 2667-2670.

[69] Chevolot-Mangueur, A.-M., Cave, A., Potier, P., Teste, J., Chiaroni, A. and Riche, C. (1976). Composés bromes de *Rytiphlea tinctoria* (Rhodophyceae). *Phytochemistry*, 15, 767-771.

[70] Xu, X. L., Fan, X., Song, F. H., Zhao, J. L., Han, L. J. and Shi, J. G. (2004). A new bromophenol from the brown alga *Leathesia nana*. *Chinese Chemical Letters*, 15, 661-663.

[71] Glombitza, K.-W. and Gerstberger, G. (1985). Phlorotannins with dibenzodioxin structural elements from the brown alga *Eisenia arborea*. Phytochemistry, 24, 543-551.

[72] Blackman, A. J. and Matthews, D. J. (1982). Halogenated phloroglucinols from *Rhabdonia verticillata*. *Phytochemistry*, 21, 2141-2142.

[73] Amsler, C. D. and Fairhead, V. A. (2005). Defensive and sensory chemical ecology of brown algae. *Advances in Botanical Research*, 43:1-91.

[74] Koch, M. and Gregson, R. P. (1984). Brominated phlorethols and nonhalogenated phlorotannins from the brown alga *Cystophora congesta*. *Phytochemistry*, 23, 2633-2637.

[75] Glombitza, K.-W. and Schmidt, A. (1999). Nonhalogenated and halogenated phlorotannins from the brown alga *Carpophyllum angustifolium*. *Journal of Natural Products*, 62, 1238-1240.

[76] Choi, J. S., Park, H. J., Jung, H. A., Chung, H. Y., Jung, J. H. and Choi, W. C. (2000). A cyclohexanonyl bromophenol from the red alga *Symphyocladia latiuscula*. *Journal of Natural Products*, 63, 1705-1706.

[77] Lee, J. H., Lee, T.-K., Kang, R.-S., Shin, H. J. and Lee, H.-S. (2007). The in vitro antioxidant activities of the bromophenols from the red alga *Tichocarpus crinitus* and phenolic derivatives. *Journal of the Korean Magnetic Resonance Society*, 11, 56-63.

[78] Oh, K.-B., Lee, J. H., Chung, S.-C., Shin, J., Shin, H. J., Kim, H.-K. and Lee, H.-S. (2008). Antimicrobial activities of the bromophenols from the red alga *Odonthalia corymbifera* and some synthetic derivatives. *Bioorganic and Medicinal Chemistry Letters*, 18, 104-108.

[79] Xu, N., Fan, X., Yan, X., Li, X., Niu, R. and Tseng, C. K. (2003). Antibacterial bromophenols from the marine red alga *Rhodomela confervoides*. *Phytochemistry*, 62, 1221-1224.

[80] Kurihara, H., Mitani, T., Kawabata, J. and Takahashi, K. (1999). Two new bromophenols from the red alga *Odonthalia corymbifera*. *Journal of Natural Products*, 62, 882-884.

[81] Kim, K. Y., Nam, K. A., Kurihara, H. and Kim, S. M. (2008). Potent α-glucosidase inhibitors purified from the red alga *Grateloupia elliptica*. *Phytochemistry*, 69, 2820-2825.

[82] Kim, K. Y., Nguyen, T. H., Kurihara, H. and Kim, S. M. (2010). α-Glucosidase inhibitory activity of bromophenol purified from the red alga *Polyopes lancifolia*. *Journal of Food Science*, 75, H145-H150.

[83] Shi, D., Li, X., Li, J., Guo, S., Su, H., and Fan, X. Antithrombotic effects of bromophenol, an alga-derived thrombin inhibitor. *Chinese Journal of Oceanology and Limnology*, 28, 96-98.

[84] Shoeib, N. A., Bibby, M. C., Blunden, G., Linley, P. A., Swaine, D. J., Wheelhouse, R. T. and Wright, C. W. (2004). In vitro cytotoxic activities of the major bromophenols of the red alga *Polysiphonia lanosa*, and some novel synthetic isomers. *Journal of Natural Products*, 67, 1445-1449.

[85] Dembitsky, V. M. and Srebnik, M. (2002). Natural halogenated fatty acids: their analogues and derivatives. *Progress in Lipid Research*, 41, 315-367.

[86] McConnell, O. and Fenical, W. (1977). Halogen chemistry of the red alga *Asparagopsis*. *Phytochemistry*, 16, 367-374.
[87] McConnell, O. and Fenical, W. (1980). Halogen chemistry of the red alga *Bonnemaisonia*. *Phytochemistry*, 19, 233-47.
[88] Řezanka T. and Dembitsky, V. M. (2001). Polyhalogenated homosesquiterpenic fatty acids from *Plocamium cartilagineum*. *Phytochemistry*, 57, 607-611.
[89] Mercer, E. J. and Davies, C. L. (1979). Distribution of chlorosulpholipids in algae. *Phytochemistry*, 18, 457-462.
[90] Todd, J. S., Proteau, P. J. and Gerwick, W. H. (1993). Egregiachlorides A-C: New chlorinated oxylipins from the marine brown alga *Egregia menziesii*. *Tetrahedron Letters*, 34, 7689-7692.
[91] Kousaka, K., Ogi, N., Akazawa, Y., Fujieda, M., Yamamoto, Y., Takada, Y. and Kimura, J. (2003). Novel oxylipin metabolites from the brown alga *Eisenia bicyclis*. *Journal of Natural Products*, 66, 1318-1323.
[92] Woolard, F. X., Moore, R. E. and Roller, P. P. (1979). Halogenated acetic and acrylic acids from the red alga *Asparagopsis taxiformis*. *Phytochemistry*, 18, 617-620.
[93] Ohta, K. and Takagi, M. (1977). Antimicrobial compounds of the marine red alga *Marginisporum aberrans*. *Phytochemistry*, 16, 1085-1086.
[94] Khaskin, I. G., Shomova, E. A. and Stapler, A. L. (1967). Fungicidic activity of some aromatic dichloro acetamide derivatives. *Mikrobiologiya*, 36, 1019-1023.
[95] Manley, S. L., Goodwin, K. and North, W. J. (1992). Laboratory production of bromoform, methylene bromide and methyl iodide by macroalgae and distribution in nearshore southern California waters. *Limnology and Oceanography*, 37, 1652-1659.
[96] Collén, J., Ekdahl, A., Abrahamsson, K. and Pedersén, M. (1994). The involvement of hydrogen peroxide in the production of volatile halogenated compounds by *Meristiella gelidium*. *Phytochemistry*, 36, 1197-1202.
[97] Abrahamsson, K., Ekdahl, A., Collén J. and Pedersén, M. (1995). Marine algae-a source of trichloroethylene and perchloroethylene. *Limnology and Oceanography*, 40, 1321-1326.
[98] Laturnus, F., Wiencke, C. and Klöser, H. (1996). Antarctic macroalgae- sources of volatile halogenated organic compounds. *Marine Environmental Research*, 41, 169-181.
[99] Ekdahl, A., Pedersen, M. and Abrahamsson, K. (1998). A study of the diurnal variation of biogenic volatile halocarbons. *Marine Chemistry*, 63, 1-8.
[100] Gschwend, P. M., Macfarlane, J. K. and Newman, K. A. (1985). Volatile halogenated organic compounds released to seawater from temperate marine macroalgae. *Science*, 227, 1033-1035.
[101] Newman, K. A. and Gschwend, P. M. (1987). A method for quantitative determination of volatile organic compounds in marine macroalgae. *Limnology and Oceanography*, 32, 702-708.
[102] Gribble, G. W. (1992). Naturally occurring organohalogen compounds- a survey. *Journal of Natural Products*, 55, 1353-1395.
[103] Schall, C., Laturnus, F. and Heumann, K. G. (1994). Biogenic volatile organoiodine and organobromine compounds released from polar macroalgae. *Chemosphere*, 28, 1315-1324.

[104] Laturnus, F. (1995). Release of volatile halogenated organic compounds by unialgal cultures of polar macroalgae. *Chemosphere,* 31, 3387-3395.
[105] Manley, S. L. (2002). Phytogenesis of halomethanes: a product of selection or a metabolic accident? *Biogeochemistry,* 60, 163-180.

In: Seaweed
Editor: Vitor H. Pomin

ISBN 978-1-61470-878-0
© 2012 Nova Science Publishers, Inc.

Chapter 9

SARGASSUM WIGHTII – A NATURE'S GIFT FROM THE OCEAN

Anthony Josephine[1,] and Sekar Ashok Kumar[2]*

[1]Department of Marine Biotechnology,
National Institute of Ocean Technology, NIOT Campus,
Velachery, Pallikaranai, Chennai, India
[2]Centre for Biotechnology, AC Tech campus, Anna University,
Chennai, India

ABSTRACT

Seaweeds or marine algae have long been made up a key part of the Asian diet and are also consumed in other parts of the world. The relative longevity and health of Okinawan Japanese population have been attributed to the consumption of marine algae in their diet. The antique tradition and daily routine of consuming seaweeds in Asian countries has made possible a huge number of epidemiological researches to screen the health benefits coupled to seaweed consumption and for centuries, brown seaweeds has been hailed as a natural answer to a lengthier and a healthier life. Brown algae belong to a very large group called the heterokonts, most of which are colored flagellates. A notable example is Sargassum, which creates unique habitats in the Sargasso Sea (hence the name Sargassum). *Sargassum wightii* is one such species with diverse biological and pharmacological properties. Recently, sulphated polysaccharides from marine brown algae are receiving continuous attention, and especially as an antioxidant, sulphated polysaccharides have piqued the interest of many scientists and researchers as one of the ocean's greatest treasures. In line with this scenario, this review highlights the habitat, economical value and medicinal importance of *Sargassum wightii*, with special emphasis on the significance of sulphated polysaccharides from *Sargassum wightii*.

---

[*]E-mail: ajose_joy@yahoo.co.in.

1. INTRODUCTION

The application of modern genomic methodologies to novel model organisms is opening new avenues of research in marine biology. Japanese and Korean populations are the biggest consumers of seaweed products in the world. The relative long life, wellbeing and health of Okinawan Japanese populations have largely been credited to the consumption of marine algae in their diets [1]. Report from the Bureau of Fisheries has pointed out that marine plants are extensively utilized in France, Ireland, Scotland and other European countries, in the East Indies and China. Seaweeds enter exclusively and extensively into the dietary pattern of the Japanese [2]. It is recognized that the overall content of seaweeds in certain traditional Asian diets contribute to the low incidence of cancer, particularly breast cancer [3].

Algae are not really plants. They belong to a separate category called protists. The algal plants or seaweeds are commonly classified into four principal groups like the green algae or chlorophyceae, the blue-green algae or cyanophyceae, the brown algae or phaeophyceae and the red algae or rhodophyceae. Nearly, about 1500 species of brown algae are almost exclusively found in marine habitats, and especially the browns include the largest of the seaweeds. This was also in agreement with Matsukawa [4], who reported that the antioxidant activity of brown algae was superior to that of red or green groups. The brown algae have chlorophyll a and c, as well as carotenes and xanthophylls. Brown algal cells have a single nuclei and thylakoids in the chloroplast. Cell walls are composed of cellulose layered with polysaccharides. They contain the pigment fucoxanthin, which is responsible for the distinctive greenish-brown color that characterizes their name. They play an important role in marine environments. For instance Macrocystis, a member of the Laminariales or kelps, may reach 60 metres in length, and forms prominent underwater forests. Another notable example is Sargassum, which creates unique habitats in the Sargasso Sea (hence the name Sargassum).

2. FEATURES OF SARGASSUM WIGHTII (S. WIGHTII)

Sargassum species are free floating and they are not attached to any substratum. Off the coast of the United States, just south of Bermuda, a sea is named as "Sargasso Sea", wherein it is estimated that about 7 million tons of sargassum live in the sea. Sargassum reproduces by asexual means; a piece breaks off and becomes a separate plant. Sargassum is commonly found in the beach drift near sargassum beds where they are also known as Gulfweed, and colloquially as the *weed of deceit*, a term also used to include all seaweed species washed up on shore. Sargassum species are found throughout the tropical areas of the world and they grow subtidally attached to coral, rocks or shells in moderately exposed or sheltered rocky or pebble areas. In some cases (e.g. the Sargasso Sea) there are floating populations of sargassum and are differentiated into holdfast, stipes, fronds and fruiting bodies. Some species have berry like gas-filled bladders to keep the plants afloat, thus promoting photosynthesis. Sargassum species is particularly tenacious with fast growth and high reproductive rates and an ability to spread vegetatively.

S. wightii is one such species found extensively and unnoticeably, especially along the coast of Rameswaram, India. Reports suggest that Sargassum species are found to be rich in sulphated polysaccharides. Studies by Abdel-Fattah *et al.* [5] proposed that the polysaccharide part of the sargassum molecule appears to be constructed of a backbone composed of (1→4)-linked β-D-glucuronic acid and β-D-mannose residues. To this backbone, heteropolymeric, partially sulphated branches are attached, comprising various proportions of (1→4)-linked β-D-galactose, β-D-galactose 6-sulphate, and β-D-galactose 3,6-disulphate residues, (1→2)-linked α-L-fucose 4-sulphate residues, and (1→3)-linked β-D-xylose residues. The fucose and xylose residues appear to be nearer to the periphery of the branches, whereas the galactose residues are nearer to the backbone of the polysaccharide. Figure 1 portrays the general picture of *S. wightii*.

3. ECONOMICAL VALUE OF SULPHATED POLYSACCHARIDES

Sulphated polysaccharides are like a closely guarded secret kept by those coastal societies who have enjoyed the wonderful health benefits of the polysaccharide for centuries. Scientific research suggests that many health conditions could be impacted in a positive way by the regular consumption of sulphated polysaccharides. Sulphated polysaccharides are a class of compounds containing hemi-ester sulphate groups in their sugar residues. Sulphated polysaccharides are known to possess a wide range of pharmacological and biomedicinal properties such as anti-coagulant, anti-tumor, hypoglycaemic, antioxidant, anti-lipemic, anti-ulcerogenic and anti-inflammatory actions [6, 7]. Sulphated polysaccharides from marine algae are naturally found glycosaminoglycans (GAGs) which exhibit similar properties to that of heparin [8, 9], which is a well known anti-coagulant drug. It has only been recently that researches have identified sulphated polysaccharides as the true source of health and wellness benefits of oceanic wonder.

Figure 1. Sargassum wightii (a drawn picture).

3.1. Seaweed Collection

S. wightii is ubiquitous in nature. However, it is extensively found in Mandapam, Gulf of Mannar region, Rameswaram, India. After collection, the seaweed sample was washed in seawater and fresh water thoroughly to remove the contamination. The sample was then air dried in shade, coarsely powdered and used for further analysis.

3.2. Extraction of Sulphated Polysaccharides

Sulphated polysaccharides can be extracted from *S. wightii* according to the method of Vieira *et al.* [10] with slight modifications [11]. 1 g of the powdered seaweed was weighed and soaked fully in acetone and kept for 24 h at 4 °C. The seaweed residue obtained after filtration was allowed to air dry and suspended in 30 ml of 0.1 M sodium acetate buffer (pH 6.0) containing 100 mg papain, 5 mM EDTA and 5 mM cysteine. It was then incubated at 60 °C for 24 h. The incubation mixture was centrifuged at 2000 × g for 15 min at 10 °C and the supernatant obtained was precipitated with 1.6 ml of 10% cetyl pyridinium chloride (CPC) and incubated overnight at room temperature. The mixture was again centrifuged and the anionic sulphated polysaccharides in the pellet were dissolved in 2 M NaCl:ethanol (100:15 v/v) solution and was precipitated with 95% ethanol. After overnight incubation at 4 °C, the final precipitate was dried at 60 °C for 2 h, dissolved in 30 ml of distilled water and lyophilized. The sulphated polysaccharides thus extracted was dissolved in physiological saline and passed through a 0.2 µm sterile filter and then used for the study.

3.3. Confirmation of Sulphated Polysaccharides in the Extract of *S. Wightii*

Sulphated polysaccharide extract contained the following components: total sugar (69.26%), protein (3.01%), uronic acid (15.62%) and sulphate (12.09%). The above analysis indicates that the sulphated polysaccharides extracted from *S. wightii* are rich in polysaccharides with lesser contamination and has a substantial amount of sulphate content in it [12].

1,9-dimethyl methylene blue assay is an assay that can be used to confirm the presence of sulphated GAGs. Fucoidan (from Sigma Chemical company), a fucose-rich sulphated polysaccharide from marine brown algae (*Fucus vesiculosus*) which has also been found to exhibit renoprotective effect [13] can be compared with the extract obtained. Moreover, previous studies have shown that sulphated polysaccharide extracts from Sargassum species and fucoidan exhibit certain similarities with that of heparin [14], which is a standard and potent nephroprotective agent among the sulphated GAGs known [15]. Hence considering heparin as a standard, fucoidan was found to contain 78.53% of sulphated GAGs and *S. wightii* was found to contain 74.21% of sulphated GAGs. Hence, these polysaccharides have been designated as sulphated polysaccharides [12].

3.4. Fourier Transform-Infrared Spectroscopy

The Fourier transform-infrared spectrum (FT-IR) was recorded with an IR spectrophotometer, between 400 and 4000 cm^{-1}. The samples (10 mg) were analyzed as a potassium bromide (KBr) pellet. The FT-IR spectrum of sulphated polysaccharides from *S. wightii* was compared with that of sulphated polysaccharides from *Fucus vesiculosus* (Fucoidan, Sigma chemicals).

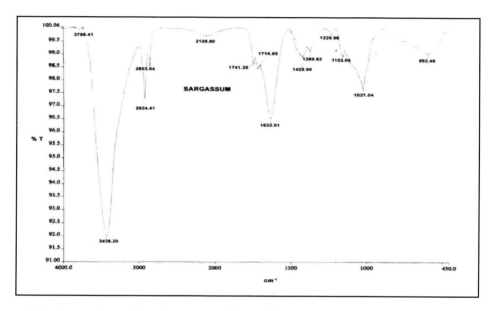

Figure 2. Fourier transform-infrared spectrum of *S. wightii*.

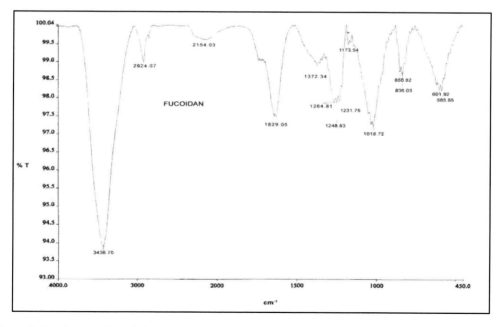

Figure 3. Fourier transform-infrared spectrum of *Fucus vesiculosus*.

Moreover, the FT-IR data analysis also showed that the peak obtained with sulphated polysaccharides from *S. wightii* (Figure 2) closely resembles to that of fucoidan (Figure 3).

In FT-IR spectra, intense broad absorption band at 3436 cm^{-1} and relatively narrow band at 2924 cm^{-1} in both *S. wightii* and *Fucus vesiculosus* indicate the presence of 'OH' group and aliphatic 'CH' bond, respectively. Also, absorption at 1632 cm^{-1} (Sulphated polysaccharides from *S. wightii*) and 1629 cm^{-1} (Sulphated polysaccharides from *Fucus vesiculosus*) is attributed to the aromatic ring, C = C groups. However, absorption bands formed at 1264-1231 cm^{-1} and 585 cm^{-1} for *Fucus vesiculosus* may indicate the presence of sulphate groups. Similarly, the sulphated polysaccharides from *S. wightii* have absorption bands at 1229 cm^{-1} and 592 cm^{-1}, which may correspond to the sulphate groups. Moreover, it has been reported in previous studies that absorptions between 1225-1255 cm^{-1} specifies the presence of sulphate groups [16]. Secondary alcohol, C-OH group is represented at 1021 cm^{-1} for sulphated polysaccharides from *S. wightii* and 1018 cm^{-1} for sulphated polysaccharides from *Fucus vesiculosus*.

FT-IR studies revealed characteristic absorption bands of sulphated polysaccharides, parallel to that described by Chevolot *et al.* [17]. Lloyd and Dodson [18] have demonstrated that absorption band of strong intensity at 1240 cm^{-1}, may be attributed to the S = O stretching vibration and is associated with the spatial distribution of sulphated groups. Besides, Matsuhiro and Miller [19] have shown that the normal FT-IR spectra of the polysaccharides show bands at 1260, 836, 615 and 580 cm^{-1}, which are assigned to sulphate groups. Moreover, the FT-IR spectra of the heteropolysaccharides of the brown marine algae *Padina gymnospora* showed an intense absorption band at 1264 cm^{-1} (S = O) common to all the sulphate groups [20]. Hence, the FT-IR data indicates that the sulphated polysaccharides extracted from *S. wightii* bear close similarity with that of fucoidan, and the absorption bands correspond to polysaccharides with sulphate groups.

3.5. Potential Antioxidant Activity of Sulphated Polysaccharides by the Ferric Reducing/Antioxidant Power (FRAP) Assay

The FRAP assay offers a putative index of antioxidant potential of the sample. This method allows the determination of the ferric reducing ability [μmol Fe (III) converted into Fe (II)] of the samples in aqueous solutions, which serves as a measure of their antioxidant power. It has been well established that polysaccharides possess strong antioxidant activity [21, 22]. Additionally, sulphation of the polysaccharides was found to further increase its antioxidant capacity against various antioxidant systems *in vitro* [23]. Hence, the antioxidant activity of sulphated polysaccharides from *S. wightii* was assessed under *in vitro* condition [12]. Sulphated polysaccharides were able to reduce the 2, 4, 6-Tri (2-pyridyl-5-triazine) (TPTZ) resulting in the formation of blue coloured complex. The antioxidant activity was found to be 66.27 and 107.25 μmol Fe(II)/g at 4 min and 8 min, respectively. The above observation allows determination of the ferric reducing ability of sulphated polysaccharides, as more the reducing ability, the more the antioxidant power. Previous report shows that Sargassum species were found to exhibit the highest antioxidant activity [24]. This was also in agreement with Matsukawa [4], who reported that the antioxidant activity of brown algae was superior to that of red or green groups. Hence, this assay proves that sulphated polysaccharides from *S. wightii* exhibits potent antioxidant effect.

4. MEDICINAL IMPORTANCE OF SARGASSUM SPECIES

Sargassum was found to possess various biological properties, which includes the following,

- Exhibits the highest anti-vaccinia virus activity, reducing about 65% of the plaques formed by the vaccinia virus [25].
- Removes chromium from the tannery effluent, which would be of greater advantage in the tanning industry for the reuse of the effluent to prepare the tanning agent. It has been estimated that this species exhibited a maximum uptake of about 35 mg of chromium per gram of seaweed [26].
- Suggested as a partial substitute for fish meal in formulated diets of *Labeo rohita*. It has been recorded to possess good food conversion ratio, food assimilation efficiency, protein efficiency ratio and better nutrient digestibility [27].
- Shows a marked response on the growth and biochemical constituents of the plant *Vigna sinensis*. About 100% germination was recorded with the seeds soaked in the aqueous extracts of *S. wightii*. It promoted the seedling growth, shoot length, root length, chlorophyll and carotenoids content, etc [28].
- Improves the biosorption of nickel (II) and copper (II) ions [29, 30].
- In China, Sargassum is cultivated and cleaned for use as an herbal remedy. It is being consumed as a Seaweed Sargassum Tea and is said to remove excess phlegm.
- It is a well recognized sources of iodine
- Widely used for the treatment of cancer
- In Korea, diet rich in seaweeds are given to the new mothers for the first month after birth, which is believed to provide many health benefits to the mother and the child [31].
- Used as a detoxifying agent
- Used as an anti-viral agent
- Used as a cholesterol lowering therapy in rats [32], which might probably, be due to the stimulation of the liver enzymes, or by blocking the macrophage scavenger receptor that is involved in LDL uptake.
- About 3 g of decosahexaenoic acid, 5 g of seaweed (wakame) powder and 50 mg of isoflavonoids from soyabean were given daily for 10 weeks to immigrants in Brazil, who were at high risk for developing diseases. It has been reported that the above combinations reduced the blood pressure and cholesterol levels, suppressed the urinary markers of bone resorption and attenuated the tendency toward diabetes [1].
- Ingesting 3.6 g per day of Undaria (wakame) for 4 weeks resulted in a 14 mm Hg drop in systolic blood pressure in Asian patients, who had hypertension [33].

4.1. Therapeutic Applications of Sulphated Polysaccharides from *S. Wightii*

Brown algae are important commercially because of their polysaccharides content. These phycopolysaccharides have broad applications in food, pharmaceuticals and cosmetics, and as nutritional supplements. Epidemiological studies conducted in those societies where brown seaweed is consumed on a regular basis have identified lower rates of many diseases plaguing our society, and the protective effect is mainly attributed to the sulphated polysaccharides present in them. Some of the valuable medicinal properties of sulphated polysaccharides from *S. wightii* are given below.

4.2. Sulphated Polysaccharides as an Antioxidant

Co-administration of sulphated polysaccharides from *S. wightii* to Cyclosporine A (CsA – a nephrotoxic drug) -induced rats significantly minimized the oxidants level. It has been demonstrated that CsA increases hydroxyl species generation along with an increase in thiobarbituric acid reactive substances and loss of protein thiols, which may also play a pivotal role in nephrotoxicity in experimental animals [34]. Importantly, sulphated polysaccharides maintained the oxidative balance by boosting up the antioxidant status and reducing the oxidant production. The ability of enzymatic extracts from Sargassum species to scavenge free radicals might be attributed to the protective effect of inhibiting the reactive oxygen species (ROS) production [11]. Brown algae were found to be rich in potent natural antioxidants like vitamins C and E, which have the ability to quench free radicals. Furthermore, it has also been documented that brown algae containing α- and γ-tocopherols increases the production of vasodilators, which plays an important role in the prevention of cardiovascular disease [35]. Also, Mori *et al.* [36] detailed that *S. microcanthum* inhibits free radical induced lipid peroxidation (LPO) under *in vitro* condition. The stability of seaweeds to oxidation during storage also suggests that their cells have protective antioxidative defense systems [37, 4]. Algal polysaccharides have been demonstrated to play an important role as free-radical scavengers *in vitro* and as antioxidants for the prevention of oxidative damage in living organisms [38]. Ruperez *et al.* [16] have documented that sulphated polysaccharides exhibit a potent antioxidant capacity.

ROS and reactive nitrogen species (RNS) are known to play a central role in the maintenance of vascular homeostasis and injury. Studies conducted demonstrated that CsA-challenged nephrotoxic rats showed enhanced nitrosative stress, while sulphated polysaccharides treated CsA-induced animals showed significant recovery from increased nitrosative stress [39]. Moreover, it directly inhibited the peroxynitrate⁻ formation by scavenging superoxide radical and thereby reducing the availability of superoxides to combine with nitric oxide to form peroxynitrate. There is a long list of evidences supporting the free radical (especially, superoxide radical) scavenging effect of algal polysaccharides [40] and especially from Sargassum species [41]. Earlier reports documents that low molecular weight heparin was found to prevent the nitrosative stress aggravated during adriamycin-induced nephrotoxicity [42]. Thus, sulphated polysaccharides from *S. wightii* are also proved to be an effective agent in protecting the renal tissue from nitrosative stress.

4.3. Sulphated Polysaccharides as a Protective Agent against Cellular Macromolecular Damages

Lipids, proteins and DNA are the important targets of ROS/RNS and their oxidative products are known to exert diverse biological effects. Thus, it has been suggested that excessive free radicals are potential toxic hazards to various biological molecules through LPO [43], DNA damage [44] and inhibition of protein synthesis [45]. This is more evident from enhanced levels of malondialdehyde (MDA), 8-hydroxy deoxyguanosine (8-OHdG) and protein carbonyls coupled with depleted protein thiols; indicative of LPO, DNA damage and protein oxidation in the CsA treated groups. Sulphated polysaccharides established its antioxidant effect by significantly preventing the oxidation of lipids, DNA, and protein [39]. However, sulphated polysaccharides administration considerably decreased protein carbonyl formation and restored the status of protein thiols, which might probably be due to its effect of replenishing glutathione (GSH) level. It has been documented that low density lipoprotein (LDL) oxidation was considerably prevented in the presence of sulphated polysaccharides, by minimizing the production of thiobarbituric acid reactive substances, an indicative of LPO [46]. Further, the absence of oxidative damage in the structural components (Polyunsaturated fatty acids) of seaweeds and their resistance to oxidation during storage suggests that their cells have protective antioxidative defense system [37]. Thus a significant decline in the levels of lipid peroxides in sulphated polysaccharides given rats suggest that it may have the ability to protect the renal tissue from free radical injury induced by CsA. This is in consonance with a report showing that the ethanolic and aqueous extracts of *S. polycystum*, significantly prevented acetaminophen-induced LPO, and was further concluded that the protective effect might due to their free radical scavenging property [47]. Further, the ability of sulphated polysaccharides to reduce the oxidation of protein thiols and formation of protein carbonyls were reported by Veena *et al.* [48], and the protective outcome has been attributed to its antioxidant effect and its potential to replenish the GSH level.

4.4. Sulphated Polysaccharides as Hepatoprotective

Sulphated polysaccharides were also found to exhibit hepatoprotective activity. Moreover, the aqueous extracts of *S. henslowianum* and *S. siliquastrum* proved to be a promising hepatoprotective agent against carbon tetrachloride-induced liver injury [49]. It is probable that the curative action on hepatic injury by the crude seaweed may have been due to their antioxidant activities, which act as scavengers of free radicals such as superoxide and alkoxy radicals [50] and preserve the structural integrity of the plasma cellular membrane of the hepatocytes, thereby protecting them from breakage by reactive metabolites. Recently, the hot water extract of *S. polycystum* has been proved to exhibit profound antioxidant [51] and anti-hyperlipidemic [52] effects, and thereby found to protect the rats from acetaminophen-induced hepatotoxicity. Hepatotoxicity induced by CsA is revealed by severe abnormal histological findings. Interestingly, sulphated polysaccharides from *S. wightii* had significantly protected the liver tissue from inflammation around portal triad (Triaditis) with severe patchy microvesicular fatty degeneration. Furthermore, sulphated polysaccharides also augmented the liver antioxidant status to a significant extent [53], thus proving to be hepatoprotective.

4.5. Sulphated Polysaccharides as Nephroprotective

Seaweed polysaccharides are generally designated as sulphated polysaccharides due to their high sulphate content. They are found to accumulate in the kidneys, which further substantiate their renoprotective effects [54], and it has also been documented that the administration of sulphated polysaccharides through subcutaneous route would especially be important for the long term therapeutic use. Furthermore, sulphated polysaccharides treatment was found to show a beneficial effect against diabetic nephropathy by significantly minimizing the tubular lesions and other complications associated with the above conditions. Earlier reports further confirm that sulphated polysaccharides like sodium pentosan polysulphate and low molecular weight heparin exhibited a profound beneficial role against urolithiasis and adriamycin-induced nephrotoxicity [55, 15, 56]. Electron microscopic analysis of CsA-induced rat kidney revealed abnormal glomerular basement thickening indicating severe damage in the glomerulus. However, sulphated polysaccharides, being naturally occurring GAGs replenished the glomerular basement membrane and thereby protected the glomerulus [57].

4.6. Sulphated Polysaccharides as Anti-Hyperlipidemic Agent

Sulphated polysaccharides were also shown to possess hypolipidemic action. Sulphated polysaccharides co-administration to CsA-provoked rats enhanced HDL level by 1.31-fold and lowered the LDL and VLDL concentrations, thereby bringing out its anti-atherosclerotic effect. Sulphated polysaccharides supplementation showed typical inhibition of LDL oxidation, which might be probably through their antioxidative effect and thereby appreciably lowered the risk of cardiovascular diseases [58]. Jimenez-Escrig *et al.* [59] have observed that the organic extracts from brown algae exhibited good efficiency in the *in vitro* inhibition of LDL oxidation. Similarly, fucoidan was also found to play a better role in inhibiting human LDL oxidation *in vitro* [46]. These reports strongly emphasize the antioxidant potential of sulphated polysaccharides. Further, sulphated polysaccharides from *S. wightii* also showed substantial modulation on the activities of Lecithin:cholesterol acyl transferase (LCAT) and Lipoprotein lipase (LPL), thereby bringing out its positive role as hypotriglyceridemic agent [58]. Enormous report proves the correlation that the hypolipidemic effect of *S. polycystum* extract is attributed to the presence of sulphated polysaccharides [52]. The influence of sulphated polysaccharides on hepatic cholesterol 17 α-hydroxylase activity suggested the possible mechanism of cholesterol conversion to bile acid, and thereby effectively lowered serum cholesterol levels. Similar observation was reported by Pengzhan *et al.* [60] wherein ulvan, a sulphated polysaccharide and its fractions U1 and U2, showed a significant reduction of LDL cholesterol coupled with elevated serum HDL cholesterol and decreased triglycerides. Since, there exists a well-built relationship between hyperlipidemia and glomerulosclerosis, sulphated polysaccharides from *S. wightii* by acting as an anti-lipemic agent, potentially reduced the jeopardy of hyperlipidemia-associated glomerular dysfunction, as well.

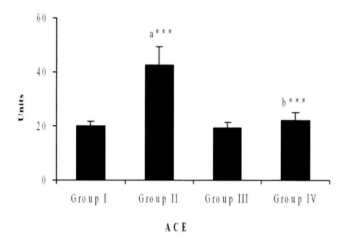

Figure 4. Effect of CsA and sulphated polysaccharides on the activity of ACE in the renal tissue. Values are expressed as mean ± S.D. for six rats in each group. Units: ACE: Units/mg protein, one unit is equal to the amount of enzyme catalyzing the formation of 1 μM of hippuric acid from hippuryl-L-histidyl-L-leucine (HHL) in 1 min at 37 °C. Group I-Control; Group II-CsA; Group III-Sulphated polysaccharides; Group IV-CsA+sulphated polysaccharides. Comparisons are made between: a-Group I and Groups II, III, IV; b-Group II and Group IV. The symbol (***) represents statistical significance at $P < 0.001$.

4.7. Sulphated Polysaccharides as a Protective Agent from Mitochondrial Damage

Recently, sulphated polymannuroguluronate has been reported to combat the oxidative damage to mitochondria, effectively. In fact, mitochondria suffer oxidative damage more easily owing to their continual exposure to ROS, which contributes to the apoptosis of cells. Indeed, sulphated polysaccharides are used as a remedy for diverse diseases allied with impaired energy metabolism and that their dietary supplementation could bestow potent antioxidative defense and ameliorate oxidative stress [61]. It has also been established that sulphated polysaccharides stimulate both oxidative and non-oxidative glucose metabolism, thereby increasing the ATP synthase activity [62], which in combination with increased glucose utilization would likely to enhance the overall cellular metabolism. All these notions together with the fact that sulphated polysaccharides simultaneously enhanced the mitochondrial membrane potential and mitochondrial respiratory chain enzymes activities [63], led to deduce that sulphated polysaccharides might exert antioxidative actions. Interestingly, sulphated polysaccharides from *S. wightii* prevented CsA-induced morphological changes like mitochondrial swelling, autophagy by lysosomes, clumping of nuclear chromatin and dispersion of ribosomes to a remarkable extent [12].

4.8. Sulphated Polysaccharides as Anti-Hypertensive

Zhu *et al.* [64] have shown that D-polymannuronic sulphate exert antihypertensive effect and further demonstrated that the effect might be involved both in increasing the generation of vasodilators and in decreasing the production of angiotensin II and endothelin-1 (ET-1) *in vivo*. Reports also document that fucoidan, a sulphated polysaccharide exhibited profound renoprotective effect by increasing the renal blood flow [65] and inhibiting proteinuria in active heymann nephritis [66]. It has been previously reported that *Undaria,* a brown algae inhibited the activity of angiotensin converting enzyme (ACE) both *in vivo* and *in vitro* [67]. It has also been suggested that ingesting potassium-loaded seaweed fibers countered hypertension successfully [68]. A report also documents that low molecular weight hyaluronate when given during transplantation, is found to prolong the transplanted kidney survival rate [69]. Sulphated polysaccharides administration to CsA-given rats was found to substantially decrease the activity of ACE (Figure 4) and thereby angiotensin II.

Also, sulphated polysaccharides considerably reduced the level of ET-1 (a potent vasoconstrictor) mRNA expression, analyzed by Reverse transcriptase polymerase chain reaction (RT-PCR) (Figure 5). The PCR products were electrophoretically separated and its images were photographed and the expression of each target gene was standardized with internal control gene (RPS 16) expression and represented as a ratio.

ET-1 is known to induce increased renal vascular resistance, reduced renal blood flow and various effects on glomerular filtration rate (GFR) [70]. The anti-hypertensive activity of sulphated polysaccharides was reported by Zhu *et al.* [64], based on their effect of decreasing the production of angiotensin II and ET-1 under *in vivo* conditions. It has been established that the antihypertensive peptides were identified from brown algae, *Undaria pinnatifida* (Wakame) and they have been found to battle strongly against hypertension [67], similar to that of a potent antihypertensive drug, captopril. Recently, the antihypertensive effect of wakame was demonstrated in hypertensive patients also [33].

Moreover, a study by Brinkkoetter *et al.* [71] have concluded that through inhibiting the ACE activity and angiotensin II production, the glomerular basement membrane permselectivity can be restored, which could potentially bring down the proteinuric effects in patients with diabetic nephropathy. Thus, the sulphated polysaccharides from *S. wightii*, by bringing down the activity of ACE (thereby angiotensin II levels to normal), was able to prevent glomerular basement membrane damage, and reduce proteinuria [57]. In addition, sulphated polysaccharides were found to inhibit ROS production effectively [72], which might also be attributed to its anti-hypertensive effect. The effect of sulphated polysaccharides on the inhibition of ACE activity and ET-1 expression, through its free radical scavenging effect, offers an explanation for the beneficial effect of inhibiting NADPH oxidase activity and thereby reduction of the risk of enhanced oxidative stress.

On the other hand, sulphated polysaccharides treatment also significantly restored the level of transforming growth factor-β (TGF-β_1), which is a key fibrogenetic cytokine involved in the fibrosis of a number of chronic diseases of the kidney and other organs, to near normal (Figure 6). Moreover, it has been found that both angiotensin II and TGF-β_1 acts coherently in inducing mesangial matrix protein secretion [73], and in fact it seems that angiotensin II may be acting through stimulation of TGF-β_1 production. Conversely, studies have demonstrated that TGF-β_1 also increases the production of ET-1. The histological hallmark of chronic kidney rejection is progressive fibrosis, wherein extracellular matrix turnover plays an important role. Renal matrix degradation is mainly regulated by two proteolytic systems: the matrix metalloproteinase (MMP) system and the plasminogen activating system.

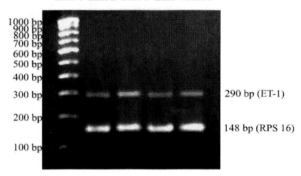

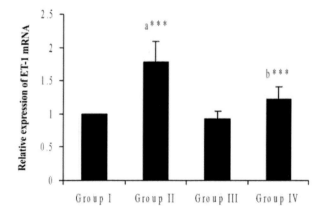

Figure 5. RT-PCR analysis of ET-1 mRNA expression in the kidney tissue of control and experimental groups. The 290 and 148 bp fragments represent ET-1 transcript and RPS 16 as internal standard, respectively. The graph represents the relative densitometric intensity levels of ET-1 mRNA expression compared with RPS 16 mRNA. Values are expressed as mean ± S.D. for six rats in each group. Comparisons are made between: a-Group I and Groups II, III, IV; b-Group II and Group IV. The symbol (***) represents statistical significance at P < 0.001.

Many diseases have been associated with an imbalance of extracellular matrix synthesis and degradation, which may result in the accumulation of extracellular matrix molecules [74]. In the kidney, these processes can lead to interstitial fibrosis. Sulphated polysaccharides treated CsA-induced group shows marked recovery in the expression of MMP-2, assessed by immunohistochemical analysis (Figure 7) and Plasminogen activator inhibitor-1 (PAI-1), assessed by RT-PCR (Figure 8). MMP-2 degrade collagen IV and V and denatured forms of collagen I, contributing to basement membrane turnover and interstitial matrix turnover.

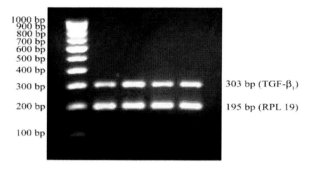

Lane 1: 100 bp DNA ladder
Lane 2: Control
Lane 3: CsA
Lane 4: Sulphated polysaccharides
Lane 5: CsA+Sulphated polysaccharides

Ratio of TGF-β_1 mRNA expression to RPL 19 mRNA

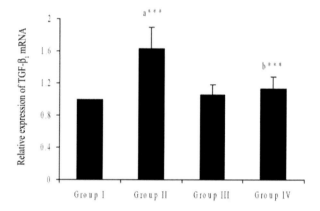

Figure 6. RT-PCR analysis of TGF-β1 mRNA expression in the kidney tissue of control and experimental groups. The 303 and 195 bp fragments represent TGF-β1 transcript and RPL 19 as internal standard, respectively. The graph represents the relative densitometric intensity levels of TGF-β1 mRNA expression compared with RPL 19 mRNA. Values are expressed as mean ± S.D. for six rats in each group. Comparisons are made between: a-Group I and Groups II, III, IV; b-Group II and Group IV. The symbol (***) represents statistical significance at P < 0.001.

The ultimate prognostic marker to indicate fibrosis is the accumulation of extracellular matrix proteins, especially collagen. Sulphated polysaccharides also prevented abnormal accumulation of collagen remarkably (Figure 9). Hence, by inhibiting the primary check, the ROS generation and by curtailing the abnormalities in the expression of angiotensin II, NADPH oxidase, ET-1, TGF-β_1 and matrix metalloproteinase system, sulphated polysaccharides shows its potential in preventing the onset of renal fibrosis, thereby proving to be a promising nephroprotective and anti-hypertensive agent. Ceol *et al.* [75] have reported that GAGs and modified heparin prevents proteinuria, mesangial matrix expansion and increased glomerular and tubular expression of TGF-β_1 mRNA in long term diabetic rats.

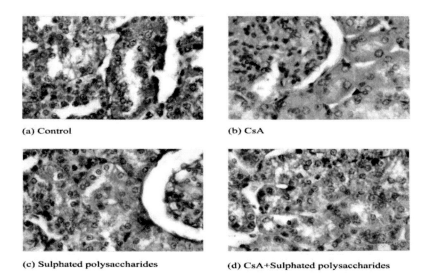

Figure 7. Immunohistochemical analysis of MMP-2 expression in the control and experimental groups. Control and drug control rat shows normal cytoplasmic expression of MMP-2 (7a and 7c); CsA treated rats show less staining, indicating decreased expression of MMP-2 (7b); Sulphated polysaccharides treated CsA induced rats (7d) restored the expression of MMP-2, near to that of control.

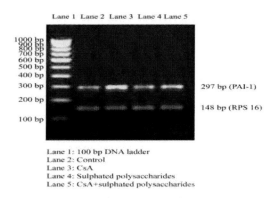

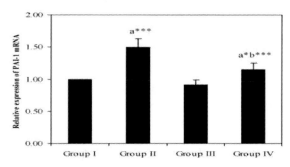

Figure 8. RT-PCR analysis of PAI-1 level in the control and experimental rats. Values are expressed as mean ± S.D. for six rats in each group. I-Control; Group II-CsA; Group III-Sulphated polysaccharides; Group IV-CsA+sulphated polysaccharides. Comparisons are made between: a-Group I and Groups II, III, IV; b-Group II and Group IV. The symbol (***) represents statistical significance at $P < 0.001$. and $P < 0.05$, respectively.

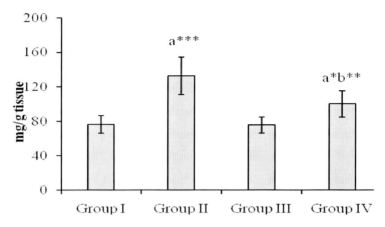

Figure 9. Effect of CsA and sulphated polysaccharides on the level of collagen in the renal tissue. Values are expressed as mean ±S.D. for six rats in each group. Group I-Control; Group II-CsA; Group III-Sulphated polysaccharides; Group IV-CsA+sulphated polysaccharides. Comparisons are made between: a – Group I and Groups II, III, IV; b – Group II and Group IV. The symbols (***), (**) and (*) represent statistical significance at P< 0.001, P< 0.01 and P< 0.05, respectively.

4.9. Sulphated Polysaccharides as Anti-Apoptotic

Reports show the efficacy of sulphated polysaacharides as an anti-apoptotic agent against CsA-induced apoptosis in LLC-PK1 cell line (Porcine kidney epithelial cells). Confocal microscopic analysis for mitochondrial membrane potential of LLC-PK1 cells using JC-1 dye is displayed in Figure 1.10. JC-1 is a mitochondrial transmembrane potential sensitive fluorescent carbocyanine dye with a net positive surface charge [76]. This property facilitates the dye to selectively accumulate in the mitochondria under conditions of intact transmembrane potential. Under normal conditions of intact mitochondrial membrane potential, the dye gets aggregated within mitochondrial matrix and there occurs a shift in spectral emission from green (534 nm) to red (596 nm). Red fluorescence characteristic of JC-1 dimer and intact mitochondria was observed in the control cells (Figure 10a and c). Whereas, cells treated with CsA showed a loss in red fluorescence indicating dissipation of mitochondrial membrane potential (Figure 10b).

CsA-induced loss of mitochondrial membrane potential was significantly prevented by concomitant administration of sulphated polysaccharides, as reflected by the restoration of the red fluorescence, which is distinctive to intact mitochondria (Figure 10d). Sulphated polysaccharides by significantly abolishing the ROS production and maintaining the level of intracellular calcium concentration, effectively re-establishes the mitochondrial membrane potential. Further, a report by Miao et al. [63] suggests that sulphated polysaccharide dramatically increased mitochondrial membrane potential. Since, it is also conceivable that compounds increasing ATP supply will exert anti-apoptotic functions [77], the effect of sulphated polysaccharides in increasing the energetic status, by improving the activities of electron transport chain enzymes might also contribute to its anti-apoptotic effect.

Furthermore, Justo et al., [78] have also suggested that the initial mitochondrial injury leads to cytochrome c release, which then activates caspases. Hence, cytochrome c release has been recognized to be the central player of apoptosis. In Figure 11, sulphated polysaccharides co-treatment to CsA exposed LLC-PK1 cells significantly ($P < 0.01$) inhibited the cytochrome c release by maintaining the mitochondrial membrane potential. Since, cytochrome c release is the major leading pathway of apoptosis, drugs targeting and modulating the cytochrome c level would be of utmost significance. In this way, an antioxidant like sulphated polysaccharides that could prevent the cytochrome c level, also raises the possibility that it could as well prevent apoptosis. Moreover, Miao *et al*. [63] have documented that sulphated polysaccharides exhibited anti-apoptotic effect via targeting mitochondria by scavenging free radicals and decreasing ROS accumulation, and thus prevented release of cytochrome c from mitochondria.

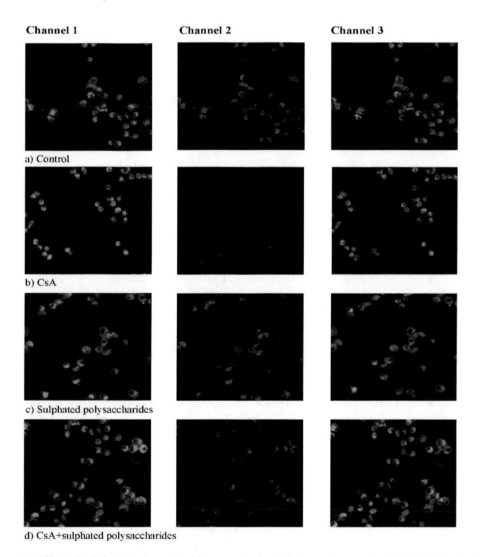

Figure 10. Effect of sulphated polysaccharides on mitochondrial membrane potential in CsA-induced LLC-PK1 cells (100x). The channels 1-3 represent the green, red and merged fields, respectively. CsA-induced LLC-PK1 cells show decrease in red fluorescence, characteristic of loss of membrane potential (b1-3); Sulphated polysaccharides treated CsA exposed LLC-PK1 cells restored the mitochondrial membrane potential (d1-3); Control and drug control treated LLC-Pk1 cells show normal fluorescene (a and c).

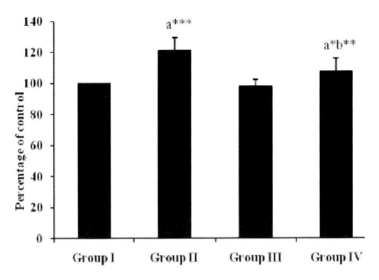

Figure 11. Effect of sulphated polysaccharides on cytochrome c levels in CsA-induced LLC-PK1 cells. All the experiments were done in triplicates. Average of three values are represented in the graph. Comparisons are made between: a – Group I and Groups II, III, IV; b – Group II and Group IV. The symbols (***), (**) and (*) represent statistical significance at P< 0.001, P< 0.01 and P< 0.05, respectively.

Nuclear condensation and DNA fragmentation are generally considered to be the hallmark of apoptosis. In that way, sulphated polysaccharides significantly prevented apoptotic nuclei, by inhibiting nuclear condensation and DNA fragmentation (data not shown). This observation can be positively correlated with the study of Ishikawa and Kitamura [79], which showed the anti-apoptotic effect of heparin. Heparin was found to effectively inhibit nuclear condensation and fragmentation by minimizing the staining of cells with Hoechst stain, reducing the number of TUNEL positive cells and preventing DNA fragmentation by inhibiting the ladder formation in agarose gel electrophoresis. Thus, they have documented that heparin could inhibit glomerular cell apoptosis induced by various chemicals and oxidant-induced apoptosis in Madin-Darby Canine Kidney epithelial cells. Also, Widlak and Garrard [80] have found that heparin inhibits the apoptotic endonuclease DFF40/CAD, which binds specifically to double stranded DNA and causes fragmentation. The above reports strongly support the anti-apoptotic effect of sulphated polysaccharides.

CONCLUSION

Sulphated polysaccharides from marine brown algae, *Sargassum wightii*, were found to show a promising therapeutic role against various ailments; especially, it exhibited considerable protective effect against nephrotoxicity. However, extensive studies might help us to underscore further hidden mechanism and to explore the biological significance and therapeutical potential of sulphated polysaccharides from *Sargassum wightii*. In conclusion, this review offers a potential new therapeutic approach for a healthier and lengthier life.

REFERENCES

[1] Yamori, Y; Miura, A; Taira, K. Implications from and for food cultures for cardiovascular diseases: Japanese food particularly Okinawan diets. *Asia Pac. J. Clin. Nutr.,* 2001, 10, 144-5.

[2] Smith, HM. The seaweed industries of Japan. *Bulletin of the US Bureau of Fisheries,* 1905, 24, 135-65.

[3] Kanke, Y; Iitoi, Y; Iwasaki, M; Iwase, Y; Iwama, M; Kimira, M; Takahashi, T; Tsugane, S; Watanabe, S; Akabane, M. Effects of human diets of two different Japanese populations on cancer incidence in rat hepatic drug-metabolizing and antioxidant enzyme systems. *Nutr. Cancer,* 1996, 26, 63-71.

[4] Matsukawa, R. A comparison of screening methods for antioxidant activity in seaweeds. *J. Appl. Phycol.,* 1997, 9, 29-35.

[5] Abdel-Fattah, AF; Magdel-Din, MH; Salem, HM. Constitution of Sargassan, a sulphated heteropolysaccharide from Sargassum linifolium. *Carb. res.,* 1974, 33, 209-15.

[6] Chizhov, AO; Dell, A; Morris, HR; Haslam, SM; McDowell, RA; Shashkov, AS; Nifant'ev, NE; Khatuntseva, EA; Usov, AI. A study of fucoidan from the brown seaweed Chorda filum. *Carbohydr. Res.,* 1999, 320, 108-19.

[7] Berteau, O; Mulloy, B. Sulfated fucans, fresh perspectives: structures, functions, and biological properties of sulfated fucans and an overview of enzymes active toward this class of polysaccharide. *Glycobiology,* 2003, 13, 29R-40R.

[8] Witvrouw, M; De Clercq, E. Sulfated polysaccharides extracted from sea algae as potential antiviral drugs. *Gen. Pharmacol.,* 1997, 29, 497-511.

[9] Schaeffer, DJ Krylov, VS. Anti-HIV activity of extracts and compounds from algae and cyanobacteria. *Ecotoxicol. Environ. Saf.,* 2000, 45, 208-27.

[10] Vieira, RP; Mulloy, B; Mourao, PA. Structure of a fucose-branched chondroitin sulfate from sea cucumber. Evidence for the presence of 3-O-sulfo-beta-D-glucuronosyl residues. *J. Biol. Chem.,* 1991, 266, 13530-6.

[11] Josephine, A; Veena, C.K; Amudha, G; Preetha, S.P; Varalakshmi, P. Evaluating the effect of sulphated polysaccharides on cyclosporine a induced oxidative renal injury. *Mol. Cell Biochem.,* 2006, 287, 101-8.

[12] Josephine, A; Amudha, G; Veena, C.K; Preetha, S.P; Rajeswari, A; Varalakshmi, P. Beneficial effects of sulphated polysaccharides from Sargassum wightii against mitochondrial alterations induced by Cyclosporine A in rat kidney. *Mol. Nutr. Food Res.,* 2007a, 51, 1413-22.

[13] Zhang, Q; Li, Z; Xu, Z; Niu, X; Zhang, H. Effects of fucoidan on chronic renal failure in rats. *Planta Med.,* 2003a, 69, 537-41.

[14] Shanmugam, M; Mody, KH. Heparinoid-active sulphated polysaccharides from marine algae as potential blood anticoagulant agents. *Current Science,* 2000, 79, 1672-83.

[15] Deepa, PR; Varalakshmi, P. The cytoprotective role of a low-molecular-weight heparin fragment studied in an experimental model of glomerulotoxicity. *Eur. J. Pharmacol.,* 2003, 478, 199-205.

[16] Ruperez, P; Ahrazem, O; Leal, JA. Potential antioxidant capacity of sulfated polysaccharides from the edible marine brown seaweed Fucus vesiculosus. *J. Agric. Food Chem.*, 2002, 50, 840-5.

[17] Chevolot, L; Foucault, A; Chaubet, F; Kervarec, N; Sinquin, C; Fisher, AM; Boisson-Vidal, C. Further data on the structure of brown seaweed fucans: relationships with anticoagulant activity. *Carbohydr. Res.*, 1999, 319, 154-65.

[18] Lloyd, AG; Dodgson, KS. Infra-red spectra of carbohydrate sulphate esters. *Nature*, 1959, 184, 548-9.

[19] Matsuhiro, B; Miller, LG. Soluble polysaccharides from rhodymenia: characterization by FT-IR spectroscopy. *Bol. Soc. Chil. Quim.*, 2002, 47, 265-71.

[20] Silva, TM; Alves, LG; de Queiroz, KC; Santos, MG; Marques, CT; Chavante, SF; Rocha, HA; Leite, EL. Partial characterization and anticoagulant activity of a heterofucan from the brown seaweed Padina gymnospora. *Braz. J. Med. Biol. Res.*, 2005, 38, 523-33.

[21] Volpi, N; Tarugi, P. The protective effect on Cu^{2+}- and AAPH-mediated oxidation of human low-density lipoproteins depends on glycosaminoglycan structure. *Biochimie*, 1999, 81, 955-63.

[22] Campo, GM; Avenoso, A; Campo, S; Ferlazzo, AM; Micali, C; Zanghi, L; Calatroni, A. Hyaluronic acid and chondroitin-4-sulphate treatment reduces damage in carbon tetrachloride-induced acute rat liver injury. *Life Sci.*, 2004, 74, 1289-305.

[23] Yang, XB; Gao, XD; Han, F; Tan, RX. Sulfation of a polysaccharide produced by a marine filamentous fungus Phoma herbarum YS4108 alters its antioxidant properties in vitro. *Biochim. Biophys. Acta*, 2005, 1725, 120-7.

[24] Yan, X; Nagata, T; Fan, X. Antioxidative activities in some common seaweeds. *Plant Foods Hum. Nutr.*, 1998, 52, 253-62.

[25] Premanathan, M; Kathiresan, K; Chandra, K; Bajpai, SK. In vitro anti-vaccinia virus activity of some marine plants. *Indian J. Med. Res.*, 1994, 99, 236-8.

[26] Aravindhan, R; Madhan, B; Rao, JR; Nair, BU; Ramasami, T. Bioaccumulation of chromium from tannery wastewater: an approach for chrome recovery and reuse. *Environ. Sci. Technol.*, 2004, 38, 300-6.

[27] Bindu, MS; Sobha, V. Conversion efficiency and nutrient digestibility of certain seaweed diets by laboratory reared Labeo rohita (Hamilton). *Indian J. Exp. Biol.*, 2004, 42, 1239-44.

[28] Sivasankari, S; Venkatesalu, V; Anantharaj, M; Chandrasekaran, M. Effect of seaweed extracts on the growth and biochemical constituents of Vigna sinensis. *Bioresour. Technol.*, 2006, 97, 1745-51.

[29] Vijayaraghavan, K; Padmesh, TV; Palanivelu, K; Velan, M. Biosorption of nickel(II) ions onto Sargassum wightii: application of two-parameter and three-parameter isotherm models. *J. Hazard. Mater*, 2006, 133, 304-8.

[30] Vijayaraghavan, K; Prabu, D. Potential of Sargassum wightii biomass for copper(II) removal from aqueous solutions: application of different mathematical models to batch and continuous biosorption data. *J. Hazard. Mater*, 2006, 137, 558-64.

[31] Moon, S; Kim, J. Iodine content of human milk and dietary iodine intake of Korean lactating mothers. *Int. J. Food Sci. Nutr.*, 1999, 50, 165-71.

[32] Iritani, N; Nogi, J. Effect of spinach and wakame on cholesterol turnover in the rat. *Atherosclerosis*, 1972, 15, 87-92.

[33] Nakano, T; Hidaka, H; Uchida, J; Nakajima, K; Hata, Y. Hypotensive effects of wakame. *J. Jpn. Soc. Clin. Nutr.*, 1998, 20, 92.

[34] Zhong, Z; Arteel, GE; Connor, HD; Yin, M; Frankenberg, MV; Stachlewitz, RF; Raleigh, JA; Mason, RP; Thurman, RG. Cyclosporin A increases hypoxia and free radical production in rat kidneys: prevention by dietary glycine. *Am. J. Physiol.*, 1998, 275, F595-604.

[35] Solimabi, VJ; Kamat, SY. Distribution of Tocopherol (Vitamin E) in marine algae from Goa, West Coast of India. *Indian J. Mar. Sci.*, 1985, 14, 228-29.

[36] Mori, J; Matsunaga, T; Takahashi, S; Hasegawa, C; Saito H. Inhibitory activity on lipid peroxidation of extracts from marine brown alga. *Phytother. Res.*, 2003, 17, 549-51.

[37] Ramarathnam, N; Ochi, H; Osawa, T; Kawakishi, S. The contribution of plant food antioxidants to human health. *Trends Food Sci. Technol.*, 1995, 6, 75-82.

[38] Zhang, Q; Li, N; Liu, X; Zhao, Z; Li, Z; Xu, Z. The structure of a sulfated galactan from Porphyra haitanensis and its in vivo antioxidant activity. *Carbohydr. Res*, 2004, 339, 105-11.

[39] Josephine, A; Amudha, G; Veena, C.K; Preetha, S.P; Varalakshmi, P. Oxidative and nitrosative stress mediated renal cellular damage induced by Cyclosporine A: Role of sulphated polysaccharides. *Biol. Pharm. Bull.*, 2007b, 30, 1254-9.

[40] Kuda, T; Tsunekawa, M; Hishi, T; Arakki, Y. Antioxidant properties of dried 'kayamonori', a brown alga Scytosiphon lomentaria (Scytosiphonales, Phaeophyceae). *Food Chem*, 2005, 89, 617-22.

[41] Zhang, EX; Yu, LJ; Xiao, X. Studies on oxygen free radical-scavenging effect of polysaccharide from Sargassum thunbergii. *Chinese J. Mar. Drugs*, 1995, 53, 1–4.

[42] Deepa, PR; Varalakshmi, P. Influence of a low-molecular-weight heparin derivative on the nitric oxide levels and apoptotic DNA damage in adriamycin-induced cardiac and renal toxicity. *Toxicology*, 2006, 217, 176-83.

[43] Halliwell, B; Chirico, S. Lipid peroxidation: its mechanism, measurement, and significance. *Am. J. Clin. Nutr.*, 1993, 57, 715S-724S.

[44] Halliwell, B; Aruoma, OI. DNA damage by oxygen-derived species. Its mechanism and measurement in mammalian systems. *FEBS Lett.*, 1991, 281, 9-19.

[45] Martin, H; Dean, M. Identification of a thioredoxin-related protein associated with plasma membranes. *Biochem. Biophys. Res. Commun.* 1991, 175, 123-8.

[46] Xue, C; Fang, Y; Lin, H; Chen, L; Li, Z; Deng, D; Lu, C. Chemical characters and antioxidative properties of sulfated polysaccharides from Laminaria japonica. *J. Appl. Phycol.*, 2001, 13, 67-70.

[47] Raghavendran, HR; Sathivel, A; Devaki, T. Protective effect of Sargassum polycystum (brown alga) against acetaminophen-induced lipid peroxidation in rats. *Phytother. Res.*, 2005b, 19, 113-5.

[48] Veena, CK; Josephine, A; Preetha, SP; Varalakshmi, P; Sundarapandiyan, R. Renal peroxidative changes mediated by oxalate: the protective role of fucoidan. *Life Sci.*, 2006, 79, 1789-95.

[49] Wong, CK; Ooi, VE; Ang, PO. Protective effects of seaweeds against liver injury caused by carbon tetrachloride in rats. *Chemosphere*, 2000, 41, 173-6.

[50] Ooi, VE. Hepatoprotective effect of some edible mushrooms. *Phytother. Res.*, 1996, 10, 536-8.

[51] Raghavendran, HR; Sathivel, A; Devagi, T. Hepatoprotective nature of seaweed alcoholic extract on acetaminophen induced hepatic oxidative stress. *J. Health Science,* 2004, 50, 42-6.

[52] Raghavendran, HR; Sathivel, A; Devaki, T. Effect of Sargassum polycystum (Phaeophyceae)-sulphated polysaccharide extract against acetaminophen-induced hyperlipidemia during toxic hepatitis in experimental rats. *Mol. Cell Biochem.*, 2005a, 276, 89-96.

[53] Josephine, A; Nithya, K; Amudha, G; Veena, C.K; Preetha, S.P; Varalakshmi, P. Role of sulphated polysaccharides from Sargassum Wightii in Cyclosporine A-induced oxidative liver injury in rats. *BMC Pharmacol.*, 2008, 8: 4.

[54] Guimaraes, MA; Mourao, PA. Urinary excretion of sulfated polysaccharides administered to Wistar rats suggests a renal permselectivity to these polymers based on molecular size. *Biochim. Biophys. Acta*, 1997, 1335, 161-72.

[55] Subha, K; Baskar, R; Varalakshmi, P. Biochemical changes in kidneys of normal and stone forming rats with sodium pentosan polysulphate. *Biochem. Int.,* 1992, 26, 357-65.

[56] Rajeswari, A; Varalakshmi, P. Low molecular weight heparin protection against oxalate-induced oxidative renal insult. *Clin. Chim. Acta,* 2006, 370, 108-14.

[57] Josephine, A; Veena, C.K; Amudha, G; Preetha, S.P; Sundarapandian R; Varalakshmi, P. Sulphated polysaccharides: new insight in the prevention of Cyclosporine A-induced glomerular injury. *Basic Clin. Pharmacol. Toxicol.* 2007c; 101: 9-15.

[58] Josephine, A; Veena, C.K; Amudha, G; Preetha, S.P; Varalakshmi, P. Protective role of sulphated polysaccharides in abating the hyperlipidemic nephropathy provoked by cyclosporine A. *Arch. Toxicol.*, 2007d, 81, 371-9.

[59] Jimenez-Escrig, A; Jimenez-Jimenez, I; Pulido, R; Saura-Calixto, F. Antioxidant activity of fresh and processed edible seaweeds. *J. Sci. Food Agric.*, 2001, 81, 530-534.

[60] Pengzhan, Y; Ning, L; Xiguang, L; Gefei, Z; Quanbin, Z; Pengcheng, L. Antihyperlipidemic effects of different molecular weight sulfated polysaccharides from Ulva pertusa (Chlorophyta). *Pharmacol. Res.*, 2003, 48, 543-9.

[61] Zhang, Q; Li, N; Zhou, G; Lu, X; Xu, Z; Li, Z. In vivo antioxidant activity of polysaccharide fraction from Porphyra haitanesis (Rhodephyta) in aging mice. *Pharmacol. Res,* 2003b, 48, 151-5.

[62] Xue, C; Yu, G; Hirata, T; Terao, J; Lin, H. Antioxidative activities of several marine polysaccharides evaluated in a phosphatidylcholine-liposomal suspension and organic solvents. *Biosci. Biotechnol. Biochem.*, 1998, 62, 206-9.

[63] Miao, B; Li, J; Fu, X; Gan, L; Xin, X; Geng, M. Sulfated polymannuroguluronate, a novel anti-AIDS drug candidate, inhibits T cell apoptosis by combating oxidative damage of mitochondria. *Mol. Pharmacol.*, 2005, 68, 1716-27.

[64] Zhu, HB; Geng, MY; Guan, HS; Zhang, JT. Antihypertensive effects of D-polymannuronic sulfate and its related mechanisms in renovascular hypertensive rats. *Acta Pharmacol. Sin*, 2000, 21, 727-32.

[65] Bojakowski, K; Abramczyk, P; Bojakowska, M; Zwolinska, A; Przybylski, J; Gaciong, Z. Fucoidan improves the renal blood flow in the early stage of renal ischemia/reperfusion injury in the rat. *J. Physiol. Pharmacol.*, 2001, 52, 137-43.

[66] Zhang, Q; Li, N; Zhao, T; Qi, H; Xu, Z; Li, Z. Fucoidan inhibits the development of proteinuria in active Heymann nephritis. *Phytother. Res*, 2005, 19, 50-3.

[67] Suetsuna, K; Nakano, T. Identification of an antihypertensive peptide from peptic digest of wakame Undaria pinnatifida. *J. Nutr. Biochem.*, 2000, 11, 450-4.

[68] Krotkiewski, M; Aurell, M; Holm, G; Grimby, G; Szczepanik, J. Effects of a sodium-potassium ion-exchanging seaweed preparation in mild hypertension. *Am. J. Hypertens.*, 1991, 4, 483-8.

[69] Knoflach, A; Azuma, H; Magee, C; Denton, M; Murphy, B; Iyengar, A; Buelow, R; Sayegh, MH. Immunomodulatory functions of low-molecular weight hyaluronate in an acute rat renal allograft rejection model. *J. Am. Soc. Nephrol.*, 1999, 10, 1059-66.

[70] Pei, Y; Chan, C; Cattran, D; Cardella, C; Zaltzman, J; Lopez, M; Tong, J; Schachter, R; Maurer, J. Sustained vasoconstriction associated with daily cyclosporine dose in heart and lung transplant recipients: potential pathophysiologic role of endothelin. *J. Lab. Clin. Med.*, 1995, 125, 113-9.

[71] Brinkkoetter, PT; Holtgrefe, S; van der Woude, FJ; Yard, BA. Angiotensin II type 1-receptor mediated changes in heparan sulfate proteoglycans in human SV40 transformed podocytes. *J. Am. Soc. Nephrol.*, 2004, 15, 33-40.

[72] Park, PJ; Heo, SJ; Park, EJ; Kim, SK; Byun, HG; Jeon, BT; Jeon, YJ. Reactive oxygen scavenging effect of enzymatic extracts from Sargassum thunbergii. *J Agric. Food Chem.*, 2005, 53, 6666-72.

[73] Weiss, RH; Ramirez, A. TGF-beta- and angiotensin-II-induced mesangial matrix protein secretion is mediated by protein kinase C. *Nephrol. Dial. Transplant.*, 1998, 13, 2804-13.

[74] Arthur, MJ. Fibrosis and altered matrix degradation. *Digestion*, 1998, 59, 376-80.

[75] Ceol, M; Gambaro, G; Sauer, U; Baggio, B; Anglani, F; Forino, M; Facchin, S; Bordin, L; Weigert, C; Nerlich, A; Schleicher, ED. Glycosaminoglycan therapy prevents TGF-beta1 overexpression and pathologic changes in renal tissue of long-term diabetic rats. *J. Am. Soc. Nephrol.*, 2000, 11, 2324-36.

[76] Reers, M; Smiley, ST; Mottola-Hartshorn, C; Chen, A; Lin, M; Chen, LB. Mitochondrial membrane potential monitored by JC-1 dye. *Methods Enzymol*, 1995, 260, 406-17.

[77] Gabryel, B; Adamek, M; Pudelko, A; Malecki, A; Trzeciak, HI. Piracetam and vinpocetine exert cytoprotective activity and prevent apoptosis of astrocytes in vitro in hypoxia and reoxygenation. *Neurotoxicology*, 2002, 23, 19-31.

[78] Justo, P; Lorz, C; Sanz, A; Egido, J; Ortiz, A. Intracellular mechanisms of cyclosporin A-induced tubular cell apoptosis. *J. Am. Soc. Nephrol.*, 2003, 14, 3072-80.

[79] Ishikawa, Y; Kitamura, M. Inhibition of glomerular cell apoptosis by heparin. *Kidney Int*, 1999, 56, 954-63.

[80] Widlak, P; Garrard, WT. The apoptotic endonuclease DFF40/CAD is inhibited by RNA, heparin and other polyanions. *Apoptosis*, 2006, 11, 1331-7.

Chapter 10

COMPARATIVE MEDICINAL PROPERTIES OF SEAWEED SULFATED POLYSACCHARIDES

Wladimir Ronald Lobo Farias[*]

Fishes Engineering Department, Federal University of Ceará, UFC,
Fortaleza, Brazil

ABSTRACT

In recent years, much attention has been focused on sulfated polysaccharides (SP) isolated from natural sources. They are found in various marine organisms and their biological activities are of great interest in the medical sciences and animal health. Anticoagulant and antithrombotic properties have been the most widely exploited SP biological's activities but these molecules also show other activities such as antitumor and antioxidant properties, inhibition of the human complement system, and immunomodulatory and antinociceptive actions. These biological activities appear to be dependent on the sulfation content and/or position of the sulfate groups. A repeating structure (4-α-D-Galp-1→3-β-D-Galp-1→) with a variable sulfation pattern was found for a sulfated galactan purified from the red algae *Botryocladia occidentalis*. The presence of two sulfate esters to a single alpha-galactose residue enhances the anticoagulant action. The antithrombotic activity of these sulfated galactans was investigated on an experimental thrombosis model and in contrast with heparin, the sulfated galactans showed a dual dose-response curve preventing thrombosis at low doses but losing the effect at higher doses. The SP isolated from the seaweed *Champia feldmannii* promotes an antitumor activity with high inhibition rates of sarcoma 180 tumor development. It was demonstrated that this polymer acts as an immunomodulatory agent, raising the production of specific antibodies. This SP was not antiinflammatory, but rather induced maximal edematogenic activity, increased vascular-permeability and stimulated neutrophil migration. The polymer was also antinociceptive and extended human plasma coagulation time by 3 times, suggesting that this molecule may be an important immunostimulant. Immunomodulatory agents such as SP from marine algae have also been widely used to minimize stress in cultivated aquatic organisms. Stress is the most powerful immunesuppressor agent on aquaculture, causing the decrease of

---

[*] E-mail: wladimir@ufc.br.

animal's natural defenses, leaving them weakened and susceptible to contaminations by pathogens. The administration of SP from the seaweeds *B. occidentalis*, *H. pseudofloresii* and *Spatoglossum schroederi* enhances *Litopenaeus vannamei* shrimp survival after stress and SP from *B. occidentalis* improved tilapia, *Oreochromis niloticus*, growth. Fish survival was also increased after the administration of SP extracted from the red marine algae *Gracilaria caudata* to *O. niloticus* post-larvae. This large range of biological activities expressed by SP from seaweed in human or animal biological systems turns these molecules important tools to promote human and animal health.

1. INTRODUCTION

Sulfated polysaccharides (SP) are anionic polymers widespread in nature and occur in a great variety of organisms mainly in macroalgae. SP extracted from green marine algae are more complex than those from red and brown algae, but they present a predominant structure and less structural variations, as compared to the high variation of sulfation pattern and the occurrence of branches and nonsulfated fucose residues in the SP of red and brown algae, respectively [1]. These SP possess a wide range of biological activities such as anticoagulant and antithrombotic activities, antiviral, antioxidant, anticancer, immunomodulatory [2-3] and also antinociceptive activities [4-6]. Anticoagulant and antithrombotic activities are among the most widely studied properties of sulfated polysaccharides and various anticoagulants have been extracted from marine macroalgae including red [7-12] brown [13-17] and in a lesser amount green seaweeds [18-20]. Marine algae SP [21-28] and alginates wich are brown algae non sulfated polysaccharides [29-33] have also been used as immunostimulants in aquaculture. Immunostimulants agents have been used as dietary supplements in cultivated aquatic organisms in order to improve innate defenses against pathogens during periods of high stress, such as grading, reproduction, transfer and handling [34-35]. The aim of this chapter is to discuss some potential medicinal uses of SP extracted from marine algae collected at Ceará State-Brazil in human and animal health.

2. EXTRACTION AND PURIFICATION OF ALGAE SP

Marine red and brown algae used in our experiments discussed in this chapter [5, 6, 10, 38, 41, 45, 46, 49, 51, 52, 53, 54] were collected at the west coast of Ceará, Brazil, separated from other species and oven dried at 60 °C. The dried tissue (2 g) was cut in small pieces, suspended in 100 mL of 0.1 M sodium acetate buffer (pH 5.0) containing 204 mg of crude papain, 5 mM of EDTA and 5 mM of cystein, and incubated at 60 °C for 24 h. The incubation mixture was then filtrated and the supernatant saved. The residue was washed with 55.2 mL of distilled water, filtered again and the two supernatants were combined. SP in solution were precipitated with 6.4 mL of 10% cetylpiridinium chloride and after standing at room temperature for 24 h, the mixture was centrifuged. The SP in the pellet were washed with 244 mL of 0.05% cetylpiridinium chloride solution, dissolved with 69 mL of a 2 M NaCl:ethanol (100:15, v/v) solution, and precipitated with 122 mL of absolute ethanol. After 24 h at 4 °C, the precipitate was collected by centrifugation, washed twice with 122 mL of 80% ethanol, and once with the same volume of absolute ethanol. The final precipitate was dried at 60 °C overnight to obtain the crude SP extract.

The crude SP extract as above (1 mg) was applied to a DEAE-cellulose column equilibrated with 0.1 M sodium acetate buffer (pH 5.0) containing 5 mM of EDTA and 5 mM of cystein. The column was developed by a step wise gradient 0.5 – 2.0 M NaCl in the equilibrium buffer. The flow rate of the column was 1.0 mL/min, and fractions of 0.5 mL were collected and assayed by metachromasia using 1,9-dimethylmethylene blue and by the phenol-H_2SO_4 reactions. Anion exchange chromatography on DEAE-cellulose separated the crude SP into fractions according to the salt concentration and fractions with the highest metachromatic activity were dialysed against distilled water and freeze-dried to be used in the biological assays. In order to evaluate SP purification, DEAE-cellulose fractions (10 µg) were applied to a 0.5% agarose gel in 50 mM 1,3-diaminopropane/acetate buffer (pH 9.0) and the electrophoresis was done at 110 V for 1 h.

3. POTENTIAL MEDICINAL PROPERTIES OF SP FROM MARINE ALGAE

3.1. Human Health

3.1.1. Anticoagulants and Antithrombotics

Plaque rupture is the main cause of human arterial thrombosis which results in the exposure of thrombogenic material, such as collagen and tissue factor, to the flowing blood. This causes the activation of platelets and the coagulation system as well as the release of vasoactive substances that induces thrombus formation and vasoconstriction, which may cause myocardial ischaemia and acute coronary syndromes [36]. Despite the large amount of new anticoagulants drugs only heparin, a glycosaminoglycan, and its low molecular weight derivatives are used in thrombosis treatment and prevention. The search for new anticoagulant and antithrombotic substances is justified by the side effects of heparin therapy like bleeding and thrombocytopenia and also the limited source of material (37-40). Marine algae have a great amount of SP and that extracted from the marine red alga *Botryocladia occidentalis* showed a high anticoagulant activity. Like heparin, they enhance thrombin and factor Xa inhibition by antithrombin and/or heparin cofactor II. These polysaccharides are sulfated galactans much more potent as anticoagulants than sulfated galactans and fucans from marine invertebrates. They occur as three fractions with increasing sulfate groups and that with higher sulfate content are more anticoagulants [10]. However, like occur with sulfated fucans and sulfated galactans from invertebrates the anticoagulant activity is not only a consequence of their charge density and the interaction of these molecules with the coagulation factors and their target proteases depends of the nature of the sugar residues and the site of sulfation and/or the position of glycosidic linkage [1].

The potent anticoagulant activity of the SP from *B. occidentalis* is related to the occurrence of 2,3-dissulfated galactose units in the following repeating structure (-4-α-D-Galp-1→3-β-D-Galp-1→). The addition of two sulfate esters to a single α–galactose residue in this SP results in an amplifying effect on its anticoagulant action [10]. A similar effect was also observed in invertebrates sulfated fucans that increases the anticoagulant activity ~38-fold from a 2-sulfated α-L-fucan to a 2,4-dissulfated α-L-fucan [1]. Another sulfated galactan with anticoagulant activity extracted from the red alga *Gelidium crinale* also presents 2,3-dissulfated galactose units like SP from *B. occidentalis*, but the different proportion and/or distribution of these units along these polysaccharides results in different anticoagulant activities which do not limit the thrombin inhibitory reaction when antithrombin is used as cofactor, but it is critical when antithrombin is replaced by heparin cofactor II [41].

The antithrombotic activity of the sulfated galactans extracted from *B. occidentalis* was investigated on an experimental thrombosis model in which thrombus formation was induced in rats by a combination of stasis and hypercoagulability. In contrast with heparin, these SP prevent thrombosis at doses up to 0.5 mg/kg body weight but at higher doses this effect is lost and also show a high anticoagulant activity and a strong induce of platelet aggregation [38]. Fucoidans extracted from the brown alga *Fucus vesiculosus* and its fractions also exhibit platelet aggregation and anticoagulant action at high doses. On the other hand, they showed a low hemorrhagic activity even at higher doses compared to low molecular weight heparin and unfractionated heparin. A similar behavior was also observed with the sulfated galactans from *B. occidentalis* which their platelet effect accounts for the lack of bleeding and compensates the drastic modification of blood coagulation [17, 38].

3.1.2. Anticancer and Immune-Modulators

Reactive oxygen species also called free radicals can directly induce the formation of cancer cells in human body since they attack membrane lipids, proteins and DNA. On the other hand, macrophages which are the immune cells of the human innate immune system are a predominant source of pro-inflammatory factors, and chronic inflammation can also be the cause of neoplastic cells formation because some classes of irritant agents cause tissue damage and inflammation, enhancing cell proliferation. Thus, natural anticancer drugs such as SP extracted from marine algae, which have antioxidant and immunomodulatory activities have gained a great importance as preventive agents in treatment of cancer [3].

The SP isolated from the marine red alga *Champia feldmannii* was effective in the inhibition of Sarcoma 180 tumor growth in mice and when the animals were treated with the SP and the chemotherapeutic agent 5-FU, at the same moment, the inhibition rate increased significantly. It was also demonstrated that the SP from *C. feldmannii* elicited the production of specific antibodies and also enhanced anti-OVA antibodies generation in the treated animals. Moreover, these SP induced a hyperplasia of lymphoid folicules and reverted the leucopenia induced by 5-FU in the treated mice, which reinforces the hypothesis that they act upon immunological stimulation [42]. These SP also presented pro-inflammatory effects in rats paw edema and peritonitis and significantly increased leukocyte count and neutrophil migration in the animal peritoneal cavity which corroborated, one more time, their immunostimulating properties. A similar pro-inflammatory effect was also observed when the SP from the red seaweed *Solieria filiformis* was used in the same paw edema model [5, 45]. In contrast, the SP isolated from the brown seaweed *Lobophora variegata* showed an anti-edematogenic effect in these acute inflammatory models which seems to occur via inhibition of nitric oxide synthase and cyclooxygenase activities [46]. The SP extracted from *C. feldmannii* showed no inhibition action on human cells proliferation *in vitro* but, in contrast, the SP isolated from the green seaweed *Monostroma nitidum* presented a high *in vitro* inhibitory activity on a human cancer cell line (AGS) with a direct effect on the cancer cells, even so they also stimulated a macrophage cell line, which suggested they also could be potent immunomodulators [42, 43]. This controversy can be due to the utilization of different methods of extraction and the use of different SP fractions in the cytotoxic assays. In fact, a SP fraction extract from the brown marine alga *Sargassum latifolium* was cytotoxic against lymphoblastic leukemia cells, while other fraction extracts, obtained by different extraction methods, had no cytotoxic effect against all the tested cancer cell lines [44]. The SP extracted from another specie of the *Sargassum* genus, *S. filipendula*, showed the highest antiproliferative activity against HeLa cells when compared with the SP obtained from others species of brown and green algae [2].

3.1.3. Analgesics

Research on analgesics compounds is a growing pharmaceutical sector of high importance and interest. The search for efficient antinociceptive drugs to relieve pain is a subject of research in the pharmaceutical industry and academic experiments since pain is a common symptom of many diseases [47]. In contrast with the well know biological activities of the SP from marine algae there is a lack of information about the antinociceptive activity of these molecules. The SP extracted from the marine red alga *C. feldmannii* showed an antinociceptive activity in the acetic acid-induced writhing test in mice. The i.v. treatment of the animals with a dose of 1mg/kg, 30 minutes before i.p. administration of 0.6% acetic acid resulted in the best antinociceptive activity. A similar effect was obtained when the same dose of the SP extracted from the red marine algae *Bryothamnium seaforthii* was used in this model of nociception, also after oral administration [5, 4]. When the SP extracted from the brown seaweed *Spatoglossum schroederi* were used in the formalin and Von Frey nociception models they showed an antinociceptive action and this was the first report of this type of biological activity from a brown alga [6]. As discussed before, the SP from *C. feldmannii* showed a pro-inflammatory action which should be a conflicting data, but morphine which is an opioid like-drug, despite of a lack of anti-inflammatory activity is a potent analgesic and this same opioid mechanism was proposed for the antinociceptive action of the SP extracted from *B. seaforthii* [5, 4]. Recently, the antinociceptive activity of various extracts from the green seaweeds *Caulerpa mexicana* and *Caulerpa sertularioides* was demonstrated using the writhing, the hot plate and the formalin-induced nociception tests, however the mechanism for their antinociceptive actions and the active principles present in the extracts of these *Caulerpa* species still need to be characterized [48].

3.2. Aquatic Animal Health

Stress is the most powerful immunesuppressor agent on aquaculture, causing the decrease of cultured animal's natural defenses, leaving them weakened and susceptible to infections by pathogens [48]. Immunomodulatory agents such as SP from marine algae have also been widely used to minimize stress in cultivated aquatic organisms. The use of these agents can improve the innate defense of aquatic animals and the resistance to pathogens during the periods of stress [35]. The intensive fish larviculture process exposes larvae to considerable stresses of chemical, physical and biological nature which could be detrimental to the culture animals [50]. Sex reversion with synthetic androgens is nowadays one of the most frequently applied techniques to produce monosex male populations of tilapia which exposes larvae to a considerable level of stress and this situation promotes a suppression of the fish's immune system [51, 50].

Immunostimulants compounds can be administered to aquatic animals by injection, in the diet or through immersion baths. However, the injection method is labor intensive and becomes impractical in small fishes [34]. The administration of the SP from *B. occidentalis* and *Gracilaria caudata* incorporated into fish ration at low doses prior the addition of the synthetic hormone resulted in an enhancement of growth and survival in tilapias, *Oreochromis niloticus*, larvae [51, 53]. Growth-promoting activity and better survival rates are considered to be indirect effects of immunostimulation and can overcome immune suppression caused by sex hormones in fish [34]. The SP extracted from the red seaweeds *B. occidentalis* and *Halymenia pseudofloresia* also improved *L. vannamei*, post-larvae and juveniles survival rates, when used in immersion baths [52, 54]. Studies reported that the better survival rates and enhanced resistance against bacteria observed when aquatic animals are injected or submitted to immersion baths with water extracts [22, 25, 26] or purified SP from seaweeds [55] are a consequence of a stimulation action of the innate immune system. In addition, purified SP from marine algae directly induced fish phagocytes activity [21, 24] and murine macrophages activity [56].

The fish immune system is very similar to that of mammals with a humoral defense and phagocytic cells like macrophages, neutrophils, and natural killer cells, as well as A and B lymphocytes [34] while shrimp immune system is different and dependent of the phagocytic activity of the hyaline haemocytes and also of the semi-granular and granular haemocytes induction [27]. Total haemocyte count, prophenoloxidase activity and respiratory burst and other immune parameters increased significantly when shrimp were immersed, injected or fed diets containing hot-water extracts from the red seaweeds *Gracilaria tenuistipitata* and *Gelidium amansii* [22, 26] and from the brown seaweeds *Sargassum fusiforme* and *Sargassum duplicatum* [23, 25]. It was demonstrated that the use of the hot-water extracts of *Gracilaria tenuistipitata* in these shrimps also showed an earlier recovery in immunity after a *Vibrio alginolyticus* injection and a protective innate immunity and up-regulation of gene expressions after low-salinity stress [27, 28]. Thus, it is of fundamental importance to the success of the aquaculture activity the development of strategies that aim to turn the aquatic animals more resistant and the use of immunomodulatory compounds such as the SP from marine algae could be an important tool [49].

CONCLUSION

In conclusion, the evident effective effects and the crescent continuous search for sulfated polysaccharides extracted from marine algae to be used in human and animal systems are resulting in a very fast discovery of a great amount of these new bioactive molecules. However, it is very important to standardize the extractions methods and accomplishment the characterization and purification of these polymers in order to better establish the relationships between their structure and various biological actions. The large range of biological activities expressed by SP from seaweed turns these molecules important tools to promote human and animal health.

ACKNOWLEDGMENTS

The author gratefully acknowledge financial support from CAPES (Brazil) and Dr Ana Maria Sampaio Assreuy and Dr Letícia Veras Costa Lotufo for all collaborations.

REFERENCES

[1] Pomin V. H., and Mourão P. A. S. (2008). Structure, biology, evolution, and medical importance of sulfated fucans and galactans. *Glycobiology*, 18, 1016-1027.

[2] Costa L. S., Fidelis G. P., Cordeiro S. L., Oliveira, R. M., Sabry, D. A., Câmara, R. B. G., Nobre L. T. D. B., Costa M. S. S. P., Almeida-Lima J., Farias, E. H. C., Leite, E. L., and Rocha H. A. O. (2010). Biological activities of sulfated polysaccharides from tropical seaweeds. *Biomedicine and Pharmacotherapy*, 64, 21-28.

[3] Wijesekara I., Pangestuti R., and Kim S. K. (2011). Biological activities and potential health benefits of sulfated polysaccharides derived from marine algae. *Carbohydrate Polymers*, 84, 14-21.

[4] Vieira L. A. P., Freitas A. L. P., Feitosa J. P. A., Silva D. C., and Viana G. S. B. (2004). The alga Bryothamnion seaforthii contains carbohydrates with antinociceptive activity. *Brazilian Journal of Medical and Biological Research*, 37, 1071-1079.

[5] Assreuy A. M. S., Gomes D. M., Silva M. S. J., Torres V. M., Siqueira R. C. L., Pires A. F., Criddle D. N., Alencar N. M. N., Cavada B. S., Sampaio A. H., and Farias W. R. L. (2008). Biological effects of a sulfated polysaccharide isolated from the marine red algae Champia feldmannii. *Biological Pharmacological Bulletin*, 31, 691-695.

[6] Farias W. R. L., Lima P. C. W. C., Rodrigues N. V. F. C., Siqueira R. C. L., Amorim R. M. F., Pereira M. G., and Assreuy A. M. S. (2011). A novel antinociceptive sulphated polysaccharide of the brown marine algae Spatoglossum schroederi. *Natural Products Communications*, 6, 863-866.

[7] Potin P., Patier P., Jean-Yves F., Jean-Claude Y., Rochas C., and Kloareg B. (1992). Chemical characterization of cell-wall polysaccharides from tank-cultivated and wil plants of Delesseria sanguinea (Hudson) Lamouroux (Ceramiales, Delesseriaceae): culture patterns and potent anticoagulant activity. *Journal of Applied Phycology*, 4, 119-128.

[8] Sen S. R., Das A. K., Banerji N., Siddhanta A. K., Mody K. H., Ramavat B. K., Chauhan V. D., Vedasiromoni J. R., and Ganguly D. K. (1994). A new sulfated polysaccharide with potent blood anti-coagulant activity from the red seaweed Grateloupia indica. *International Journal of Biological Macromolecules*. 16, 279-280.

[9] Kolender A. A., Pujol C. A., Damonte E. B., Matulewicz M. C., and Cerezo A. S. (1997). The system of sulfated $\alpha(1\rightarrow 3)$ linked d-mannans from the red seaweed Nothogenia fastigiata: Structures, antiherpetic and anticoagulant properties. *Carbohydrate Research*, 304, 53-60.

[10] Farias W. R. L., Valente A. P., Pereira M. S., and Mourão P. A. S. (2000). Structure and anticoagulant activity of sulfated galactans. Isolation of a unique sulfated galactan from the red alga Botryocladia occidentalis and comparison of its anticoagulant action

with that of sulfated galactans from invertebrates. *The Journal of Biological Chemistry.* 275, 29299-29307.

[11] Barabanova A. O., Shashkov A. S., Glazunov V. P., Isakov V. V., Nebylovshaya T. B., Helbert W., Soloveva T. F., and Yermak I. M. (2008). Structure and properties of carrageenan-like polysaccharides from the red alga Tichocarpus crinitus (Gmel.) Rupr. (Rodophyta, Trichocarpaceae). *Journal of Applied Phycology*, 20, 1013-1020.

[12] Lee S. H., Athukorala Y., Lee J. S., and Jeon Y. J. (2008). Simple separation of anticoagulant sulfated galactan from marine red algae. *Journal of Applied Phycology,* 20, 1053-1059.

[13] Grauffel V., Kloareg B., Mabeau S., Durand P., and Jozefonvicz J. (1989). New natural polysaccharides with potent antithrombic activity – Fucans from brown-algae. *Biomaterials,* 10, 363-368.

[14] Nishino T., Nagumo T., Kiyohara H., and Yamada H. (1991). Structural characterization of a new anticoagulant fucan sulfate from the brown seaweed Ecklonia kurome. *Carbohydrate Research*, 211, 77-90.

[15] Soeda S., Sakaguchi S., Shimeno H., and Nagamatsu A. (1992). Fibrinolytic and anticoagulant activities of highly sulfated fucoidan. *Biochemical Pharmacology*, 43, 1853-1858.

[16] Colliec S., Boisson-Vidal C., and Jozefonvicz J. (1994). A low weight fucoidan fraction from the brown seaweed Pelvetia canaliculata. *Phytochemistry,* 35, 697-700.

[17] Azevedo T. C. G., Bezerra M . E. B., Santos M. G. L., Souza L. A., Marques C. T., Benevides N. M. B., and Leite E. L. (2009). Heparinoides algal and their anticoagulant, hemorrhagic activities and platelet aggregation. *Biomedicine and Pharmacotherapy*, 63, 477-483.

[18] Rogers D. J., Iurd K. M., Blunden G., Paoletti S., and Zanetti F. (1990). Anticoagulant activity of a proteoglycan in extracts of Codium fragile ssp. atlanticum. *Journal of Applied Phycology,* 2, 357-361.

[19] Maeda M., Uehara T., Harada N., Sekiguchi M., and Hiraoka A. (1991). Heparinoide-active sulphated polysaccharides from Monostroma nitidum and their distribution in the chlorophyta. *Phytochemistry*, 30, 3611-3614.

[20] Matsubara K., Matsuura Y., Bacic A., Liao M., Hori K., and Miyazawa K. (2001). Anticoagulant properties of a sulfated galactan preparation from a marine green alga, Codium cylindricum. *International Journal of Biological Macromolecules*, 28, 395-399.

[21] Castro R., Zarra I., and Lamas J. (2004). Water-soluble seaweed extracts modulate the respiratory burst activity of turbot phagocytes. *Aquaculture,* 229, 67-78.

[22] Hou W. Y., and Chen J. C. (2005). The immunostimulatory effect of hot-water extract of Gracilaria tenuistipitata on the white shrimp Litopenaeus vannamei and its resistance against Vibrio alginolyticus. *Fish and Shellfish Immunology*, 19, 127-138.

[23] Huang X., Zhou H., and Zhang H. (2006). The effect of Sargassum fusiforme polysaccharides extracts on vibriosis resistance and immune activity of the shrimp, Fenneropenaeus chinensis. *Fish and Shellfish Immunology,* 20, 750-757.

[24] Castro R., Piazzon M. C., Zarra I., Leiro J., Noya M., and Lamas J. (2006). Stimulation of turbot phagocytes by Ulva rigida C. Agardh polysaccharides. *Aquaculture*, 254, 9-20.

[25] Yeh S. T., Lee C. S., and Chen J. C. (2006). Administration of hot-water extract of brown seaweed Sargassum duplicatum via immersion and injection enhances the immune resistance of white shrimp Litopenaeus vannamei. *Fish and Shellfish Immunology*, 20, 332-345.

[26] Fu Y. W.., Hou W. Y., Yeh S. T., Li C. H., and Chen J. C. (2007). The immunostimulatory effects of hot-water extract of Gelidium amansii via immersion, injection and dietary administrations on white shrimp Litopenaeus vannamei and its resistance against Vibrio alginolyticus. *Fish and Shellfish Immunology*, 22, 673-685.

[27] Yeh S. T., and Chen J. C. (2009). White shrimp Litopenaeus vannamei that received the hot-water extract of Gracilaria tenuistipitata showed earlier recovery in immunity after a Vibrio alginolyticus injection. *Fish and Shellfish Immunology*, 26, 724-730.

[28] Yeh S. T., Lin Y. C., Huang C. L., and Chen J. C. (2010). White shrimp Litopenaeus vannamei that received the hot-water extract of Gracilaria tenuistipitata showed protective innate immunity and up-regulation of gene expressions after low-salinity stress. *Fish and Shellfish Immunology*, 28, 887-894.

[29] Miles D. J. C., Polchana J., Lilley J. H., Kanchanakhan S., Thompson K. D., and Adams A. (2001). Immunostimulation of striped snakehead Channa striata against epizootic ulcerative syndrome. *Aquaculture*, 195, 1-15.

[30] Cheng W., Liu C. H., Yeh S. T., and Chen J. C. (2004). The immune stimulatory effect of sodium alginate on the white shrimp Litopenaeus vannamei and its resistance against Vibrio alginolyticus. *Fish and Shellfish Immunology*, 17, 41-51.

[31] Bagni M., Romano N., Finoia M. G., Abelli L., Scapigliati G., Tiscar P. G., Sarti M., and Marino G. (2005). Short- and long-term effects of a dietary yeast □-glucan (Macrogard) and alginic acid (Ergosan) preparation on immune response in sea bass (Dicentrarchus labrax). *Fish and Shellfish Immunology*, 18, 311-325.

[32] Cheng W., Liu C. H., Kuo C. M., and Chen J. C. (2005). Dietary administration of sodium alginate enhances the immune ability of white shrimp Litopenaeus vannamei and its resistance against Vibrio alginolyticus. *Fish and Shellfish Immunology*, 18, 1-12.

[33] Skjermo J., Storseth T. R., Hansen K., Handa A., and Oie G. (2006). Evaluation of □-(1→3, 1→6)-glucans and High-M alginate used as immunostimulatory dietary supplement during first feeding and weaning of Atlantic cod (Gadus morhua L.). *Aquaculture*, 261, 1088-1101.

[34] Sakai M. (1998). Current research status of fish immunostimulants. *Aquaculture*, 172, 63-92.

[35] Bricknell I., and Dalmo R. A. (2005). The use of immunostimulants in fish larval aquaculture. *Fish and Shellfish Immunology*, 19, 457-472.

[36] De Caterina R., Husted S., Wallentin L., Agneli G., Bachmann F., Baigent C., Jespersen J., Kristensen S. D., Montalescot G., Siegbahn A., Verheugt F. W. A., and Weitz J. (2007). Anticoagulants in heart disease: current status and perspectives. *European Heart Journal*, 28, 880-913.

[37] Arthur U. K., Isbister J. P., and Aspery, E. M. (1985). The heparin induced thrombosis-thrombocytopenia syndrome (HITTS): a review. *Pathology*, 16, 82-86.

[38] Farias W. R. L., Nazareth R. A., and Mourão P. A. S. (2001). Dual effects of sulfated d-galactans from the red algae Botryocladia occidentalis preventing thrombosis and inducing platelet aggregation. *Thrombosis and Haemostasis*, 86, 1540-1546.

[39] Pereira M. S., Melo F. R., and Mourão P. A. S. (2002). Is there a correlation between structure and anticoagulant action of sulfated galactans and sulfated fucans? *Glycobiology*, 12, 573-580.

[40] Mourão P. A. S. (2004). Use of sulfated fucans as anticoagulant and antithrombotic agents: Future perspectives. *Current Pharmaceutical Design*, 10, 967-981.

[41] Pereira M.G., Benevides N. M., Melo M. R., Valente A. P., Melo F. R., and Mourão P. A. S. (2005). Structure and anticoagulant activity of a sulfated galactan from the red alga, Gelidium crinale. Is there a specific structural requirement for the anticoagulant action? *Carbohydrate Research*, 340, 2015-2023.

[42] Lins K. O. A. L., Bezerra D. P., Alves A. P. N. N., Alencar N. M. N., Lima M. W., Torres V. M., Farias W. R. L., Pessoa C., Moraes M. O., and Costa-Lotufo L. V. (2009). Antitumor properties of a sulfated polysaccharide from the red seaweed Champia feldmanni. *Journal of Applied Toxicology*, 29, 20-26.

[43] Karnjanapratum S., and You S. (2011). Molecular characteristics of sulfated polysaccharides from Monostroma nitidum and their in vitro anticancer and immunomodulatory activities. *International Journal of Biological Macromolecules*, 48, 311-318.

[44] Gama-Eldeen A. M., Ahmed E. F., and Abo-Zeid M. A. (2009). In vitro cancer chemopreventive properties of polysaccharide extract from the brown alga, Sargassum latifolium. *Food and Chemical Toxicology*, 47, 1378-1384.

[45] Assreuy A. M. S., Pontes G. C., Rodrigues N. V. F. C., Gomes D. M., Xavier P. A., Araujo G. S., Sampaio A. H., Cavada B. S., Pereira M. G., and Farias W. R. L. (2010). Vascular effects of a sulfated polysaccharide from the red marine alga Solieria filiformis. *Natural Products Communications*, 5, 1267-1272.

[46] Siqueira R. C. L., Da Silva M. S. J., Alencar D. B., Pires A. F., Alencar, N. M. N., Pereira M. G., Cavada B. S., Sampaio A. H., Farias W. R. L., and Assreuy A. M. S. (2011). In vivo anti-inflammatory effect of a sulfated polysaccharide isolated from the brown marine alga Lobophora variegata. *Pharmaceutical Biology*, 49, 167-174.

[47] Sousa D. P., (2011). Analgesic-like activity of essential oils constituents. *Molecules*, 16, 2233-2252.

[48] Matta C. B. B., Souza E. T., Queiroz A. C., Lira D. P., Araújo M. V., Cavalcante-Silva L. H. A., Miranda G. E. C., Araújo-Júnior J. X., Barbosa-Filho J. M., Santos B. V. O., and Alexandre-Moreira M. S. (2011). Antinociceptive and anti-inflammatory activity from algae of the genus Caulerpa. *Marine Drugs*, 9, 397-318. `

[49] Lima P. C. W. C., Torres V. M., Rodrigues J. A. G., Sousa-Júnior J., and Farias W. R. L. (2009). Effect of sulfated polysaccharides from the marine brown alga Spatoglossum schroederi in Litopenaeus vannamei juveniles. *Revista Ciência Agronômica*, 40, 79-85.

[50] Valdstein O. (1997). The use of immunostimulation in marine larviculture: possibilities and challenges. *Aquaculture,* 155, 401-417.

[51] Farias W. R. L., Rebouças H. J., Torres V. M., Rodrigues J. A. G., Pontes G. C., Silva F. H. O. S., and Sampaio A. H. (2004). Enhancement of growth in tilapia larvae (Oreochromis niloticus) by sulfated polysaccharides extracted from the red marine alga Botryocladia occidentalis. *Revista Ciência Agronômica*, 35, 189-195.

[52] Barroso F. E. C., Rodrigues J. A. G., Torres V. M., Sampaio A. H., and Farias W. R. L. (2007). Effect of sulfated polysaccharides extracted from the red marine alga

Botryocladia occidentalis in shrimp Litopenaeus vannamei post-larvae. *Revista Ciência Agronômica,* 38, 58-63.

[53] Araujo G. S., Farias W. R. L., Rodrigues J. A. G., Torres V. M., and Pontes G. C. (2008). Oral administration of sulfated polysaccharides from Gracilaria caudata rhodophyta on tilapias post-larvae survival. *Revista Ciência Agronômica,* 39, 548-554.

[54] Rodrigues J. A. G., Sousa-Junior J., Lourenço J. A., Lima P. C. W. C., and Farias W. R. L. (2009). Cultivation of shrimps treated with sulfated polysaccharides of Halymenia pseudofloresia rhodhophyceae through a prophylactic strategy. *Revista Ciência Agronômica,* 40, 71-78.

[55] Yeh S. T., and Chen J. C. (2008). Immunomodulation by carrageenans in the white shrimp Litopenaeus vannamei and its resistance against Vibrio alginolitycus. *Aquaculture,* 276, 22-28.

[56] Na Y. S., Kim W. J., Kim S. M., Park J. K., Lee S. M., Kim S. O., Synytsya A., and Park Y. I. (2010). Purification, characterization and immunostimulating activity of water-soluble polysaccharide isolated from Capsosiphon fulvescens. *International immunopharmacology,* 10, 364-370.

In: Seaweed
Editor: Vitor H. Pomin

ISBN 978-1-61470-878-0
© 2012 Nova Science Publishers, Inc.

Chapter 11

STRUCTURAL AND BIOLOGICAL INSIGHTS INTO ANTITUMOR SEAWEED SULFATED POLYSACCHARIDES

Hugo Alexandre Oliveira Rocha[*], *Leandro Silva Costa and Edda Lisboa Leite*
Department of Biochemistry, University of Rio Grande do Norte – UFRN, Natal, RN, Brazil

ABSTRACT

Sulfated polysaccharides comprise a complex group of macromolecules. These anionic polymers are widespread in nature, occurring in a great variety of organisms such as mammals and invertebrates. Seaweeds are the most important source of non-animal sulfated polysaccharides. Furthermore, the structure of algal sulfated polysaccharides varies according to the species of seaweed. Thus, each new sulfated polysaccharide purified from a seaweed is a new compound with unique structures and, consequently, with potential novel biological activities. Sulfated polysaccharides are found in varying amounts in three major divisions of marine algal groups, Rhodophyta, Phaeophyta and Chlorophyta. These compounds found in Rhodophyta are manly galactans consisting entirely of galactose or modified galactose units. The general sulfated polysaccharides of Phaeophyta are called fucans, which comprise families of polydisperse molecules based on sulfated L-fucose. Heterofucans are also called fucoidans. The major polysaccharides in Chlorophyta are polydisperse heteropolysaccharides, although, homopolysaccharides also may be found. The seaweeds are an untapped source of bioactive sulfated polysaccharides and the marine pharmacology research during the last years, with researchers from several international research institutes, contributing to the preclinical pharmacology of several polysaccharides which are part of the preclinical marine pharmaceuticals. Here we have reviewed publications regarding the bioactivity of seaweed polysaccharides-rich extracts or as yet structurally uncharacterized sulfated polysaccharides that showed several promising biological activities, with emphasis on antitumor. In addition, the present paper will also review the recent progress in research

[*] E-mail: hugo@cb.ufrn.br.

on structural features and the major antitumor effect of the marine algal biomaterials in a comprehensive manner. Moreover, when possible, the relationship between structure and antitumoral action of algal sulfated polysaccharides will also be reviewed.

1. INTRODUCTION

The word cancer is of Latin origin. It means crab, probably as an analogy to its penetrating nature, which can be compared to a shellfish sinking its claws into the substrate, making it difficult to remove. It is one of society's most feared diseases, having become synonymous with death and pain. In the scientific field, the term used for cancer is malignant neoplasm. However, the reader must keep in mind that cancer is a term used to define a set of over one hundred diseases whose most common characteristics are the uncontrolled growth of abnormal cells and the ability of these cells to move from their point of origin, invading and spreading to other tissues.

Neoplasms may occur in many different tissues and tumor cells retain the characteristics of their tissues of origin. The names of resulting tumors can be traced back to these origins: sarcomas (cancers derived from mesodermic cells such as bone and muscle); carcinomas (originating from epithelial cells); adenocarcinomas (from glandular cells); lymphomas (lymphocytes); myelomas (from antibody-producing bone marrow plasma cells), etc.

Carcinogenesis (tumor formation) is a highly complex process influenced by inherited and environmental risk factors, such as diet, smoking, occupation and exposure to radiation and chemical agents. The process is composed of several steps and mechanisms, involving genotoxic events (mutation), gene expression alteration on a transcriptional, translational and post-translational (epigenetic events) level and cell survival alteration (proliferation and apoptosis) [1].

In many cases these events are cumulative as a result of not being repaired by the DNA error repair mechanism found in cells. Carcinogenesis is generally understood as consisting of three distinct sequences: initiation, promotion and progression.

Neoplasic conversion (initiation) occurs when the DNA is modified by the agent. This promotes a genetic event (mutations, chromosomal rearrangement, gene insertion or deletion and gene amplification) leading to deregulation of certain genes, in most cases in the activation of oncogenes and/or lack of expression – or product inactivation – of tumor suppressing genes. Cells are genetically altered in this phase, although it is not yet possible to clinically detect a tumor. The promotion stage involves clonal expansion of the "initiated" cells, which requires cell proliferation. The initial cell is gradually changed into a malignant cell. For this transformation to occur, prolonged continuous contact with the promoting carcinogenic agent is necessary. Suspension of contact at this stage often interrupts the process. Progression, characterized by uncontrolled multiplication, is the third and final stage. The cancer is already installed, evolving to the point where initial clinical manifestations appear.

Cells start acquiring particular characteristics during these stages. These are considered tumor cell marks (self-sufficiency in growth signals, insensitivity to growth-inhibitory (anti-growth) signals, and evasion of programmed cell death (apoptosis, limitless replication potential, sustained angiogenesis, tissue invasion and metastasis). Each of these physiologic changes acquired during tumor development represents the successful breaching of an anticancer defense mechanism hardwired into cells and tissues [2].

The most commonly applied cancer treatments are surgeries, photodynamic therapies, immune therapies and hormone therapies. However, chemotherapy treatment, applied solely or with other therapies, is still the most widely used. Due to the peculiarity of the more than one hundred cancer types and the high toxicity levels of current chemotherapy treatments for normal cells, there is an extensive search for new molecules capable of substituting those currently in use. There is also a search for molecules capable of preventing tumor formation, commonly called chemopreventive molecules. Chemoprevention is defined as the systematic use of non-cytotoxic doses of nutrients and/or pharmacological agents during the period between onset and tumor progression. This is a large time window for anti-carcinogenic agents to start acting, thus reversing or suppressing the development of pre-malignant lesions into tumors [3].

Among the natural sources studied in the search for potential antitumor molecules is seaweed. For centuries traditional Asian culture has been using seaweed extracts for this purpose. Studies show that sulfate polysaccharides are the main antitumor components found in seaweed.

Sulfate polysaccharides (SPs) from marine seaweeds are polymers with hemiester sulfate groups, covalently linked to their monosaccharide residues. The degree of substitution could vary greatly, but it always exerts a negative influence on the polysaccharides in question. This property allows SPs to link to several basic molecules, including proteins, thereby developing their activities [4].

Purification and structural characterization studies on seaweed SPs have focused primarily on their pharmacological proprieties, which results in limited structural knowledge regarding these polymers, since those that did not exhibit activities in studies are usually discarded in favor of those with pharmacological potential.

Knowledge acquired to date has led to the following conclusion: seaweeds from different taxa synthesize different polysaccharides: red seaweeds (Rhodophytas) synthesize primarily sulfated homo- and heterogalactans; brown seaweeds synthesize homo- and heteropolymers of fucose, known as sulfated fucans and fucoidans, respectively; and green seaweeds (Chlorophyta) exhibit greater SP structural heterogeneity and are rich in arabinose, glucose, xylose, galactose, mannose, rhamnose and/or glucuronic acid. Homopolysaccharides are also found in these seaweeds (Table I).

Each seaweed synthesizes its sulfated polysaccharide and displays unique structural characteristics, which are reflected in the biological, pharmacological and biotechnological proprieties of the polysaccharide (Figure 1). Furthermore, these structural characteristics can be changed by biotic and abiotic factors that the seaweed is exposed to, as well as extraction and purification methods used in an attempt to obtain sulfated polysaccharides. Recent reviews regarding seaweed SPs have been restricted to their structural characteristics and primary biological/pharmacological activities [5], [6], [7].

The main objective of this chapter is to review the most significant data related to seaweed sulfated polysaccharides with antitumor activity published in recent years.

Several studies have reported that SPs exhibit in vitro antiproliferative activities against tumor cells, as well as tumor growth inhibition in animal models [25]. Furthermore, SPs display antimetastatic activity by inhibiting tumor cell adhesion and migration to extracellular matrix molecules [26], act as metalloproteinase inhibitors [27], are antiangiogenic and/or

Table 1. Some sulfated polysaccharides from seaweeds: Major sugars and source

Source	Major sugar(s)	Reference
Phaeophyceae		
Ascophyllum nodosum	Fucose, galactose	[8]
Ecklonia kurome	Fucose	[9]
Fucus vesiculosus	Fucose	[10]
Padina gymnospora	Fucose, xylose	[11]
Sargassum stenophyllum	Fucose, mannose, galactose	[12]
Spatoglossum schröederi	Fucose, galactose, glucuronic acid	[13] [14]
Turbinaria ornata	Fucose, glucosamine	[15]
Rhodophyceae		
Gloiopeltis tenax	Galactose, glucose, mannose	[16]
Gracilaria birdiae	Galactose	[17]
Grateloupia indica	Galactose	[18]
Halymenia durvillei	Galactose	[19]
Porphyra haitanensis	Galactose	[20]
Chlorophyceae		
Monostroma nitidum	Glucose, rhamnose	[21]
Ulva rigida	Rhamnose, xylose	[22]
Codium fragile	Galactose, rhamnose	[23]
Codium isthmocladum	Galactose	[24]

immunostimulators [7]. They are also chemopreventive compounds since they exhibit antioxidant and free radical scavenging activity.

Most studies on seaweed SPs are based on the prospecting of bioactive extracts rich in sulfated polysaccharides. For example, Yamamoto et al. evaluated antitumor activity in polysaccharide-rich extracts of 20 different species of seaweed in Sarcoma 180 cells implanted subcutaneously in mice and determined that five types of seaweed showed SP with marked antitumor activity (*Laminara angusta, Laminaria japonica, Eclonia clava, Eisenia bicyclis* and *Sargassum kjellmanianum*) [25], [28]. Later, these authors purified fucans from the *S. kjellmanianum* seaweed and found that only the most sulfated were effective against L-1210 leukemic cells [29].

Many seaweed species such as the brown seaweed *Sargassum* sp, *Fucus* sp, *Aschopyllum* sp and *Laminaria* sp; the red *Porphyra* sp; and the green *Enteromorpha intestinalis* and *Ulva* sp are grown or collected in nature in large amounts for food production and extraction of products such as alginates and agar for a number of industries. Thus, thousands of tons of seaweed are processed annually and sub-products such as SP consequently do not reach their full potential. On the other hand, populations, mainly in Asian countries, use seaweed culturally in the treatment of diseases. Such factors make the SPs of these species the target of studies, given their possible activities, without undergoing initial evaluation, such as the seaweed Chondrus *ocellatus*, distributed naturally throughout the intertidal zone along the southeast China and Atlantic coasts. In ancient times, seaweeds were not only edible, but also used as medicine to cure chronic constipation and treat bone fracture. This seaweed produces large amounts of λ-carrageenan. Zhou and colleges [30] used ultrasound to obtain large populations of λ-carrageenan with different molecular weights and observed that these carrageenans showed no cytotoxic activity in sarcoma S180 and H22 heptocarcinoma cells. However, when these cells were implanted in mice to induce tumor formation, then treated with different carrageenans, it was observed that those with low molecular weight (15 and 9.3 kDa) strongly hindered tumor growth (~65%). The authors concluded that the antitumor activity of this galactan acts as immunostimulator. It has been reported that other SPs, such as the heterofucan from an edible and commercially available seaweed (*Eisenia bicyclis*) [31], a sulfated galactan from the seaweed *Champia feldmannii* [32] and heterofucans from *Sargassum thumbergii* [33] exhibit in vivo antitumor activity but no in vitro cytotoxic activity in several cell models.

Results obtained from cultivated macrophages show several indications of the immunostimulating mechanism of action of seaweed SPs. Studies with a sulfated galactan from the green seaweed *Codium fragile* demonstrate that this SP is capable of stimulating nitric oxide production and raising the kevels of proinflammatory interleukin cytokine IL-1, IL-6, IL-12 and tumor necrosis factor-α (TNF-α), consequently promoting macrophage activation [34]. The activation of cells such as macrophages, neutrophils, T cells, and NK cells is one of the main steps with which the various SPs, not only from green seaweed, but also from the red and brown varieties, perform antitumor activity. For example, when heterofucans from *Ascophyllun nodosum* and *Laminaria augustata* are in contact with RAW 264.7 macrophages, they stimulate the release of TNF-α, granulocity colony-stimulating factor (G-CSF) [35] and IL-6 as well as increasing nitric oxide (NO) levels [36]; sulfated glucan extracted from the green seaweed *Monostroma nitidum* stimulate NO and prostaglandin E production, the latter by increasing cyclooxygenase-2 activity [21].

The immunostimulating action of seaweed SPs is related to their ability to bind to cell surface receptors. Studies performed with a heterofucan from the seaweed *Cladosiphon okamuranus* show that it stimulates RAW cells to increase production of NO, IL-6 and TNF-α, since they bind to toll-like receptor 4, CD14 and macrophage scavenging receptor 1 (MSR1) and activate MAPK signaling pathways [37]. However, another fucan from the seaweed *Fucus vesiculosus* binds to the MRS1, but does not lead to an increase in NO production, occurring only when it binds to other receptors [38]. This shows that the immunostimulating mechanisms of action of each SP may occur through different pathways.

Studies using cultivated tumor cells revealed valuable information on the mechanism of the antiproliferative action. These include investigations conducted with SP-rich extract from the green seaweed *Capsosiphon fulvescens*. This extract induced apoptosis and exhibited antiproliferative activity against tumor cells by promoting caspase activation and decreasing Bcl-2 levels, as well as decreasing phosphorylation levels of the insulin-like growth factor-I receptor and the PI3K/Akt pathway [39]. In more recent studies, Costa et al. evaluated the antioxidant and antiproliferative activities of SP-rich extracts from 11 types of seaweed (1 red, 4 green and 6 brown), and observed that the *Sargassum filipendula, Dictyopetis delicatula, Caulerpa prolifera* and *Dictyota menstrualis* inhibited about 68% of HeLa cell proliferation [40].

The antiproliferative activity of SPs could be a result of their interference in the cell cycle. SP from the seaweeds *Turbinaria ornata* [15] and *Bifurcaria bifurcata* inhibit the proliferation of lung carcinoma (NSCLC-N6) by blocking its passage in the G_1 phase of the cell cycle [41]. This effect was also observed with sulfated fucans from the seaweed *A. nodosum*, which inhibit proliferation of NSCLC-N6 cell lines by blocking cell growth in the

G_1 phase [42]. When low-molecular-weight sulfated fucans were obtained from the same seaweed, they exhibited antiproliferative activity against some of the tumor lines. It was observed that these compounds inhibit colon adenocarcinoma cell (Colo320 DM) proliferation by 85% [43]. This inhibition has been shown to be reversible, depending on bovine fetal serum concentration (BFS) and was not related with fucan internalization. An alteration in cell distribution was also observed during the cell cycle. Additionally, it was demonstrated that the effect was dependent on molecular weight and sulfate clusters, as well as on the spatial orientation of these clusters [44]. This indicates that these fucans bind to cell receptors and/or growth factors to exert their effect.

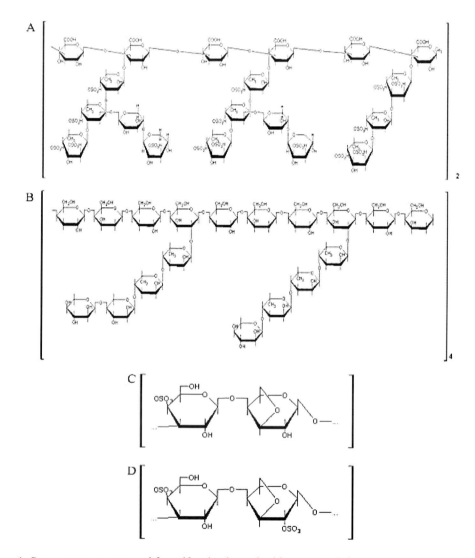

Figure 1. Some structure proposed for sulfated polysaccharides. A – Xylofucoglucuronan, a heterofucan named fucan A extracted from *Spatoglossum schröederi* [14]. B – A sulfated galactofucan named fucan B also extracted from *S. schröederi* [13]. C – Disaccharide unit from kappa carragenan. D – Disaccharide unit from iota carragenan.

The antiproliferative activity of seaweed SPs is also related to their ability to induce apoptosis in several cell lines. Studies with porphyran, an SP extracted from red seaweed of the genus Porphyra, showed that these polysaccharides reduced the phosphorylation levels of Insulin-like growth factor-I receptor (IGF-IR), which correlated with Akt activation. Thus, porphyran appears to negatively regulate IGF-IR phosphorylation by causing a decrease in the expression levels of AGS gastric cancer cells, and then inducing caspase-3 activation [45].

The most complex studies were conducted with SPs from the brown seaweed and these can give us an idea of how pro-apoptotic SPs behave. In most cases apoptosis induction is related to caspase activation. However, other proteins involved in cell survival pathways may be affected by the presence of SP in the culture medium. For example, the heterofucan from *Undaria pinnatifida* induced apoptosis in A549 human lung carcinoma cells through the down regulation of anti-apoptotic protein Bcl2 and activation of the caspase pathway. This heterofucan also down regulated p38MAPK and PI3K/Akt and activated the ERK1/2MAPK pathway [46].

There is a report that caspase activation may not be necessary for apoptosis induction by some SPs. A recently published study showed a heterofucan with pro-apoptotic activity, extracted from the seaweed *Sargassum filipendula*, denominated SF-1.5V. This SP inhibited HeLa (adenocarcinoma cervical cell) cell proliferation by inducing apoptosis induction through a caspase- independent mechanism. SF-1.5v induces apoptosis in HeLa mainly by mitochondrial release of an apoptosis- inducing factor (AIF) into cytosol. In addition, SF-1.5v decreases the expression of Bcl-2 and increases the expression of the apoptogenic protein Bax [47].

The most widely studied SP with regard to its pro-apoptotic mechanism of action is the *F. vesiculosus* fucan known as fucoidan. This SP promotes caspase-3 activation, which consequently induces apoptosis of human lymphoma HS-sultan cells. However, when these cells were concomitantly incubated with fucoidan and a caspase inhibitor, the fucoidan effect was partially offset. The authors showed that the fucoidan did not affect the p38 kinase and AKT pathway, but decreased the ERK phosphorylation and that this effect also led cells into apoptosis [48]. When this *F. vesiculosus* fucoidan was evaluated with HCT-15 human colon carcinoma cells, it was also observed that it stimulated apoptosis in HCT-15 cells by caspase activation. However, in these cells the fucoidan, in contrast to what was seen earlier, increased p38 kinase and ERK phosphorylation and blocked the PI3K/AKT pathway [49]. The data showed that this effect was responsible for the pro-apoptotic effect of the fucoidan. When the pro-apoptotic mechanism of this same fucoidan was evaluated in human promyeloid HL60 leukemic cells, it was found that, in addition to activating the caspases and increasing ERK phosphorylation levels, the pro-apoptotic effect of the fucoidan was also dependent on an increase in NO production and a decrease in cytoplasmic glutathione levels [50]. On the other hand, the fucoidan did not modify ERK phosphorylation levels in HT-29 human colon cancer cells. However, it was determined that it activated caspase, promoted the release of pro-apoptotic mitochondrial factors (cytochrome c and Smac/Diablo) to the cytoplasm and increased death receptor-5 levels. In short, the pro-apoptotic mechanism of the fucoidan in HT-29 cells is mediated through both the death receptor-mediated and mitochondria-mediated apoptotic pathways [51].

A quick analysis of the aforementioned data suggests that that the pro-apoptotic mechanism of action of the fucoidan seems to vary according to the cell line and that this might be true for other SPs. However, apoptosis is a very complex process requiring the control of several factors. To date, none of the published reports on pro-apoptotic SPs have been able to evaluate all possible factors involved in the pro-apoptotic mechanism of action of the fucoidan in a certain type of cell at the same time. When this occurs a standard is followed to confirm if the pro-apoptotic mechanisms of action of the fucoidan and other SPs are similar or not, and if they vary according to cell type [52].

The antitumor activity of SPs is also related to their ability to inhibit metastasis formation. Coombe and colleges [53] demonstrated that the *F. vesiculosus* fucoidan inhibits 13762-mammary-adenocarcinoma metastasis in rats by interfering in tumor cell passage through the vascular endothelium. This polysaccharide did not affect tumor cell adhesion to the vascular endothelium. The authors suggested that the fucoidan could be inhibiting the action of degrading enzymes involved in the invasive process. This group also demonstrated that the fucoidan was capable of inhibiting heparanase "in vitro" [54]. Heparanase, which acts on heparan sulfate proteoglycans, degrading the molecules of this glycosaminoglycan, is one of the enzymes involved in invasive and metastatic processes [55]. The heterofucan from *Cladosiphon novae-caledoniae* was capable of hindering MMP-2/9 metalloprotein activity, thus inhibiting the invasion of HT1080 human fibrosarcoma cells through Matrigel.

Compounds that inhibit cell adhesion could theoretically prevent metastasis formation. Soeda and colleges [56] reported inhibition of the binding of laminin to Lewis carcinoma cells in the presence of fucoidan from *F. vesiculosus* between 2.5 and 25 µg/mL. No effect was detected for fibronectin or type IV collagen.

Working with human adenocarcinoma cells (MCF7 and MDA-MB231), Liu and col. [57] determined that fucans from the brown seaweeds *A. nodosum* and *Laminaria brasilensis* have antiadhesive activity dependent mainly on polysaccharide structure, sulfate content and molecular weight. The authors assumed that this activity was due to the direct interaction between the fucans and the matrix molecules, thus hindering recognition by adhesion molecules, such as integrins and proteoglycans, of binding sites existing in the extracellular matrix.

Using a biotinylated fucan as probe, Rocha and colleges [26] showed that it bound specifically to fibronectin, thus hindering CHO cell adhesion to this matrix molecule. This fucan had no antiadhesive effect when laminin, collagen, and vitronectin were used as cell adhesion substrate (Figure 2). The authors also showed that this effect is not simply due to repulsion of charges, since sulfated polysaccharides from animals, such as chondroitin sulfate, were unable to inhibit cell adhesion to fibronectin. Liu and col. [58] showed that fibronectin has three SP binding sites, but sulfated polysaccharides, such as heparin, only bind to two of them, while fucans bind to three. This could be the molecular target of fucans to act as antiadhesive compounds.

Using a different approach, Cumashi and col. [59] showed that fucans from *Laminaria saccharina, L. digitata, Fucus serratus, F. distichus*, and *F. vesiculosus* strongly blocked MDAMB-231 breast carcinoma cell adhesion to platelets, an effect that might have critical implications in tumor metastasis.

Due to its antioxidant activity in several models, many SPs can be considered as chemopreventive agents, acting as antitumor agents. Unlike cytotoxic agents that damage tumor cells, antioxidants, which are generally beneficial to cells, act by preventing the onset of cancer during carcinogenesis. Oxidants, such as reactive oxygen and nitrogen species damage macromolecules, such as proteins, lipids, enzymes, and DNA.

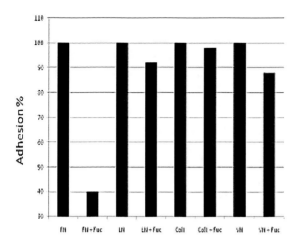

Figure 2. Fucan from *S. schröederi* inhibits CHO cell adhesion by bind the extracellular fibronectin. A – Plates were coated with protein from extracellular matrix and after CHO cells were added to plates in the presence or absence (control) of fucan (400 μg/mL). After 24 hours the percentage of cell adhesion was determinate. B – Cells were grown for three days and expose to biotinylated fucan, which was detected using streptavidin conjugated to alexa flour 594 (red). The cell nuclei were stained with DAPI. FN – fibronectin; Col I – collagen I; VN – vitronectin; Lan – laminin; Fuc – fucan. Barr = 20 μm.

It was recently demonstrated that a fucoidan from *Laminaria cichorioides* interacted with EGF to suppress EGF-induced phosphorylation of EGFR and the downstream signaling to both the ERK/p90RSK/AP-1 and JNK/c-Jun/AP-1 pathways. This is the first report on the molecular mechanism of the cancer-preventive action of fucoidan seaweeds.

Another way that sulfated polysaccharides act as antitumor agents is through their antiangiogenic activity. Sulfated polysaccharides from the seaweeds *Cladosiphon novae-caledoniae*, *Laminaria saccharina*, *Laminaria digitata*, *Fucus evanescens*, *Fucus serratus*, *Fucus distichus*, *Fucus spiralis* have shown antiangiogenic activity in several models [59].

Despite these important activities attributed to SPs as antitumor agents, little is known regarding the structural characteristics that provide them with antitumor activity. SPs are structurally very complex, since they are synthesized by organisms without any pattern strictly guiding the synthesis steps, such as occurs with proteins and nucleic acids. Some studies have reported on the importance of sulfate groups, of their amount and position, as well as the size of the SPs. In some cases, smaller molecules from the same SP were more active than their larger counterparts. Much more needs to be elucidated and the current search is to identify enzymes able to degrade seaweed SPs and produce bioactive oligosaccharides whose structures are more easily identifiable, allowing clear correlations between structure and activity to be established.

REFERENCES

[1]   J. Chen and X. Xu, *Adv. Genet.* 71, 237 (2010).
[2]   D. Hanahan and R. A. Weinber, *Cell.* 100, 57 (2000).
[3]   A. M. Bode and Z. Dong, Nat. Rev. *Cancer.* 9, 508 (2009).

[4] C. F. Becker, J. A. Guimarães, P. A. S. Mourão, H. C Verli, *J. Mol. Graph. Model.* 26, 391 (2007).
[5] G. Jiao, G. Yu, J. Zhang and H. S. Ewart, *Mar. Drugs.* 9, 196 (2011).
[6] M. Cianca, I. Quintana, A. S. Cerezo, *Curr. Med. Chem.* 17, 2503 (2010).
[7] B. Li, F. Lu, X. Wei and R. Zhao, *Molecules.* 13, 1671 (2008).
[8] E. G. Medcalf, P. Whitmen, B. Larsen, *Carbohydr. Res.*, 59, 531 (1977).
[9] T. Nshino, Y. Aizu, T. Nagumo, *Carbohydr. Res.*, 211, 77 (1991).
[10] J. Conchie, E. G. V. Percival, *J. Chem. Soc.*, 827 (1950).
[11] C. P. Dietrich, G. G. M. Farias, L. R. D. Abreu, E. L. Leite, H. B. Nader, *Plant. Sci.*, 108, 143 (1995).
[12] M. E. R. Duarte, M. A. Cardoso, M. D. Noseda, A. S. Cerezo, *Carbohydr. Res.*, 333, 281 (2001).
[13] H.A.O. Rocha, F. A. Moraes, E. S. Trindade, C. R. C. Franco, R. J. S. Torquato, S. S. Veiga, A. P. Valente, P. A. S. Mourao, E. L. leite, H. B. Nader, C. P. Dietrich, *J. Biol. Chem.*, 280, 41278 (2005)
[14] E. L. Leite, M. G. L. Medeiros, H. A. O. Rocha, G. G. M. Farias, L. F. Silva, S. F. Chavante, C. P. Dietrich, H. B. Nader, *Plant Sci.*, 132, 215 (1998).
[15] E. Deslandes, P. Pondaven, T. Auperin, C. Roussakis, J. Guezennec, V. Stiger, C. Payri, *J. Appl. Phycol.*, 12, 257 (2000).
[16] B. Lim, I. Ryu, *J. Med. Food,* 12, 442 (2009).
[17] J. S. Maciel, L. S. Chaves, B. W. S. Souza, D. I. A. Teixeira, A. L. P. Freitas, J. P. A. Feitosa, R. C. M. Paula, *Carbohydr. Pol.*, 71, 559 (2008).
[18] K. Chattopadhyay, C. G. Mateu, P. Mandal, C. A. Pujol, E. B. Damonte, B. Ray, *Phytochem.*, 68, 1428 (2007).
[19] T. A. Fenoradosoa. C. Laroche, A. Wadouachi, V. Dulong, L. Picton, P. Andriamadio, P. Michaud, *Int. J. Biol. Macromol.*, 45, 140 (2009).
[20] Z. Zhang, Q. Zhang, J. Wang, H. Zhang, X. Niu, P. Li, *Int. J. Biol. Macromol.*, 45, 22 (2009).
[21] S. Karnjanapratum, S. You, *Int. J. Biol. Macromol.*, 48, 311 (2011).
[22] M. Lahaye, B. Ray, *Carbohydr Res.,* 283, 161 (1986).
[23] J. M. Estevez, P. V. Fernandez, L. Kasulin, P. Dupree, M. Cianca, *Glycobiol.*, 19, 212 (2009).
[24] E.H. C. Farias, V. H. Pomin, A. P. Valente, H. B. Nader, H. A. O. Rocha, P. A. S. Mourao, *Glycobiol.*, 18, 250 (2008).
[25] I. Yamamoto, Y. Suzuki, I. Umenzawa, *Chemotherapy*, 28, 165 (1980).
[26] H. A. O. Rocha, C. R. C. Franco, E. S. Trindade, S. S. Veiga, E. L. Leite, H. B. Nader, C. P. Dietrich, *Planta Med.,* 71, 628 (2005).
[27] H. J. Moon, , S. R. Lee, S. N. Shim, S. H. Jeong, V. A. Stonik, V. A Rasskazov, T. Zvyagintseva, Y. H. Lee, *Biol. Pharm. Bull.*, 31, 284 (2008).
[28] I. Yamamoto, M.Takahashi, E. Tamura, H. Maruyama. *Bot. Mar.*, 25, 455 (1982).
[29] I. Yamamoto, M.Takahashi, T. Suzuki, H. Seino, H. Mori, *Japan. J. Exp. Med.*, 54, 143 (1984).
[30] G. Zhou, Y. Sunc, H. Xin, Y. Zhang, Z. Li, Z. Xu, *Pharmacol. Res.*, 50, 47 (2004).
[31] M. Takahashi, J. *Janpan. Soc. Reticuloendothel. Syst.*, 22, 269 (1983).

[32] K. O. A. L. Lins, D. P. Bezerra, A. P. N. N. Alves, N. M. N. Alencar, M. W. Lima, V. M. Torrers, W. R. L. Farias, C. Pessoa, M. O Moraes, L. V. Costa-Lutofa, *J. Appl. Toxicol.*, 29, 20 (2009).

[33] H. Itoh, H. Noda, H. Amano, H. Ito, *Anticancer Res.*, 15, 1937 (1995).

[34] J. Lee, Y. Ohta, K. Hayashi, T. Hayashi, *Carbohydr. Res.*, 345, 1452 (2010).

[35] S. Nakayasu, R. Soegima, K. Yamaguchi, T. Oda, Biosci. *Biotechnol. Biochem.* 73, 981 (2009).

[36] T. Teruya, H. Tatemot, T. Konoshi, M. Tako, *Glycocunj. J.*, 8, 1019 (2009).

[37] T. Teruya, S. Takuda, Y. Tamak, M. Tako, *Biosci. Biotechnol. Biochem.* 74, 1960 (2010).

[38] H. Do, N. Kang, S. Pyo, T. R. Billiar, E. Sohn, *J. Cell Biochem.*, 111, 1337 (2010).

[39] M. Kwon, T. Nam, *Cell Biol. Int.*, 31, 768 (2007).

[40] L. S. Costa, G. P. Fidelis, S. L. Cordeiro, R. M. Oliveira, D. A. Sabry, R. B. G. Camara, L. T. D. B. Nobre, M. S. S. P. Costa, J. Almeida-Lima, E. L. Leite, H. A. O. Rocha, *Biomed. Pharmacother.*, 64, 21 (2010).

[41] D. Moreau, H. Thomas-Guyon, C. Jacquot, M. Juge, G. Culioli, A. Ortalo-Magné, L. Piovetti, C. Roussakis. *J. Appl. Phycol.* 18, 87 (2006).

[42] D. Riou, S. Coliec-Jouault, D. P. Sel, S. Bosch, S. Siavoshian, V. Bert, C. Tomasoni, C. Sinsuin, P. Durand, C. Roussakis. *Anticancer Res.*, 16: 1213 (1996).

[43] M. Ellouali, C. Boisson-Vidal, P. Duarand, J. Jozefonviscz, *Anticancer Res.*, 13, 2011 (1993).

[44] C. Boisson-Vidal, F. Haroun, M. Ellouali, C. Blodin, A. M. Fischer, A. Agostini, J. Jozefonviscz, *Drugs Fut.*, 20, 1237 (1995).

[45] M. Kwon, T. Nam, *Life Sci.* 79, 1956 (2006).

[46] H. Boo, J. Hyun, S. Kim, J. Kang. M. Kim, S. Kim, H. Cho, E. Yoo, H. Kang. *Phytother. Res.* 25, 1082 (2011).

[47] L. S. Costa, C. B. S. Telles, R. M. Oliveira, L. T. D. B. Nobre, N. Dantas-Santos, R. B. G. Camara, M. S. S. P. Costa, J. Almeida-Lima, R. F. Melo-Silveira, I. R. L. Albuquerque, E. L. Leite, H. A. O. Rocha, *Mar. Drugs,* 9, 613 (2011).

[48] Y. Aisa, Y. Mayakawa, T. Nakazato, H. Shibata, k. Saito, Y. Ikeda, M. Kizaki, *Am. J. Hematol.*, 78, 7 (2005).

[49] J. Hyun, S. Kim, J. Kang, M. Kim, H. Boo, J. Kwon, Y. Hoh, J. Hyun, D. Park, E. Yoo, H. Kang., *Biol. Pharm. Bull.* 32, 1760 (2009).

[50] J. Jin, M. Song, Y. Kim, J. Park, J. Kwak, *Mol. Carcinog.* 49, 771 (2010).

[51] E. Kim, S. Y. Park, J. Lee, J. H. Y. Park, *BMC Gastroenterology.* 10, 96 (2010).

[52] J. Ye, Y. Li, K. Teruya, Y. Katakura, A. Ichikawara, H. Eto, M. Hosoi, S. Nishimoto, S. Shirahata, *Cytotechnology.* 47, 117 (2005).

[53] D. R. Coombe, C. R. Parish, I. A. Ramshaw, J. M. Snowden, *Int. J. Cancer,* 39, 82 (1987).

[54] C. R. Parish, D. R. Coombe, K. B. Jakobsen, F. A. Bennet, P. A. Underwood, *Int. J. cancer*, 40, 511 (1987).

[55] I. Vlodavsky, Y Friedmann, *J. Clin. Invest.*, 108, 341 (2001).

[56] S. Soeda, Y. Ohmagugi, H. Shimeno, A. Nagamatsu, *Biol. Pharmac. Bullet.*, 17, 784 (1994).

[57] J. Liu, F. Haroun-Bouhedja, C. Boisson-Vidal, *Anticancer Res.* 20, 3265 (2000).

[58] J. M. Liu, J. Bignon, F. Haroun-Bouhedja, J. Vassy, S. Fermandjian, J. Wdzieczak-Bakala, C. Boisson-Vidal, *Anticancer Res.* 25, 2129, (2005).

[59] A. Cumashi, N. A. Ushakova, M. E. Preobrazhenskaya, A. D'Incecco, A. Piccoli, L. Totani, N. Tinari, G. E. Morozevich, A. E. Berman, M. I. Bilan, A. I. Usov, N. E. Ustyuzhanina, A. A. Grachev, C. J. Sanderson, M. Kelly, G. A. Rabinovich, S. Iacobelli, N. E. Nifantiev, *Glycobiol.*, 17, 541 (2007).

INDEX

#

20th century, 36
21st century, xv, 20, 198

A

acarbose, 209
acclimatization, 58
acetaminophen, 232, 233, 248, 249
acetic acid, 59, 107, 211, 258
acetogenins, xv, 105, 197, 198, 207, 218
acetone, 70, 226
acetylcholine, 97
acetylcholinesterase, 214
acidic, 42, 84
acrylic acid, 208, 210, 211, 220
active compound, 129, 180, 206
additives, 39
adenocarcinoma, 200, 202, 209, 273, 274, 275
adenoma, 119
adhesion, 102, 111, 169, 270, 275, 276
adipose tissue, 87, 109
adiposity, 87
adjustment, 14, 156
adolescents, 31
adsorption, 105, 123
adults, 114, 158
Africa, 61
agar, x, xi, xii, 19, 21, 35, 38, 39, 40, 41, 55, 63, 72, 76, 83, 101, 119, 121, 123, 124, 128, 136, 137, 144, 146, 151, 166, 179, 193, 198, 271
age, 69, 156, 180
aggregation, 256
agility, 29
agriculture, 21
AIDS, 249

alanine, ix, xii, 2, 3, 4, 5, 7, 8, 9, 10, 11, 12, 13, 16, 82, 91
alanine aminotransferase, ix, xii, 2, 9, 16, 82, 91
alanine aminotransferase (AlaAT), ix, 2, 9, 16
aldehydes, 216
algorithm, 64, 72
ALI, 78
alimentary canal, 129
alkaloids, 41, 105, 199, 200
allergic rhinitis, 102
ALT, xii, 82, 90, 91, 93, 179
alters, 246
amines, 171
amino, ix, x, xii, 1, 2, 3, 4, 5, 6, 7, 8, 9, 10, 11, 13, 14, 15, 19, 20, 25, 33, 37, 81, 84, 106, 131, 142, 169, 174, 182, 183, 195
ammonia, 169
ammonium, 135
amylase, 87
analgesic, 185, 258
androgens, 258
anemia, 28, 34, 36, 133
angiogenesis, 104, 112, 269
angiotensin converting enzyme, 236
angiotensin II, 236, 237, 240
antibiotic, 136, 215
antibody, 103, 268
anti-cancer, xii, xiv, 81, 165
anticancer activity, 203
anticancer drug, 257
anticoagulant, xi, xiii, xvi, 50, 64, 72, 83, 101, 106, 113, 121, 125, 134, 177, 185, 246, 253, 254, 256, 261, 263, 264
anticoagulation, 208
antioxidative activity, 174, 176, 177, 187
antioxidative potential, xiv, 166, 168
antitumor, xiii, xv, xvi, xvii, 60, 61, 83, 103, 104, 116, 117, 121, 125, 136, 179, 193, 197, 201, 204,

216, 217, 253, 268, 269, 270, 271, 272, 275, 276, 277
antitumor agent, 201, 204, 276, 277
antiviral agents, 117
antiviral drugs, 61, 245
apoptosis, 103, 107, 116, 186, 204, 217, 235, 242, 244, 249, 251, 268, 269, 272, 274, 275
apoptotic mechanisms, 275
apoptotic pathways, 275
appetite, 86, 160
apples, 85
aquaculture, xvi, 21, 30, 45, 62, 132, 138, 143, 254, 258, 259, 263
aquatic habitats, 170
aqueous solutions, 41, 42, 229, 247
Arabidopsis thaliana, 10, 16
Argentina, 45, 134
arginine, 131
arsenic, 27, 159
arteries, 169
arteriosclerosis, x, xiv, 19, 165, 168
arthritis, 59, 136
ascites, 204
ascorbic acid, 27, 136, 167, 171, 174, 208
Asia, xiii, 20, 40, 59, 109, 110, 122, 126, 161, 163, 245
Asian countries, xiii, xiv, xv, 29, 40, 123, 125, 145, 165, 223, 271
Asian cuisine, xiii, 145
aspartate, xii, 82, 91
aspartic acid, 3, 16, 131
assessment, 112, 147, 154
assimilation, 142, 230
asthenia, 36
astrocytes, 111, 251
asymmetry, 15
atherogenesis, 172
atherosclerosis, 28, 93, 132, 167, 168, 169
atmosphere, 211, 213
atoms, 132, 168, 213
atopic dermatitis, 101, 112
ATP, 235, 242
ATR technique, 77
atrophy, 94
awareness, 146

B

Bacillus subtilis, 150, 200
bacteria, 10, 41, 105, 119, 123, 128, 129, 148, 149, 172, 205, 215, 259
bacterium, 193
basal forebrain, 97, 111

base, 38, 137
basement membrane, 233, 236, 239
baths, 135, 259
beef, 148
bending, 76
beneficial effect, 22, 158, 233, 237
benefits, xiii, xv, 22, 30, 52, 54, 60, 145, 146, 151, 154, 157, 160, 166, 180, 184, 223, 225, 230, 260
benign, 103
benzene, 208, 209, 210
beverages, 71, 106, 157, 188
bile, 37, 103, 104, 234
bile duct, 103, 104
bioavailability, 33, 128, 147, 152, 159, 207
biochemistry, 142
biological activities, xvi, xvii, 65, 72, 129, 166, 179, 201, 207, 208, 253, 254, 258, 260, 267
biological activity, 117, 179, 192, 199, 200, 258
biological markers, 169
biological sciences, xiii, 122, 134
biological systems, xvi, 186, 254
biologically active compounds, 22
biomarkers, xii, 81
biomass, 45, 55, 57, 247
biomaterials, xvii, 268
biomedical applications, 184
biomolecules, 85, 168
biomonitoring, 187
biopolymers, 114
bioremediation, 123, 139
biosynthesis, xv, 2, 9, 197, 198, 214, 216
biotechnological applications, 41
biotechnology, 40, 41, 188, 194
biotic, 270
Black Sea, 119
bleeding, 256
blood, 28, 38, 83, 87, 88, 90, 91, 93, 109, 113, 134, 137, 154, 155, 157, 174, 231, 236, 246, 250, 255, 256, 261
blood clot, 134
blood flow, 93, 236, 250
blood pressure, 28, 109, 231
blood stream, 91, 174
blood vessels, 174
bloodstream, 28
BMI, 155, 156, 157
body fat, 158, 159
body weight, 28, 88, 89, 93, 96, 97, 157, 256
bone, 136, 231, 268, 271
bone marrow, 268
bone resorption, 231
bradykinin, 86
brain, xii, 29, 82, 86, 96, 97, 98, 100, 111

branching, 38, 74, 76, 179
Brazil, 45, 146, 230, 253, 254, 255, 260, 267
breakdown, 16, 170
breast cancer, 31, 58, 103, 104, 115, 153, 155, 157, 159, 202, 204, 217, 224
breast carcinoma, 276
breast milk, 159
breastfeeding, 156
Britain, 135
Brittany, 29, 30, 36, 70, 143
bromination, 201
bromine, 27, 198, 199, 202, 204, 207, 214, 217
building blocks, 130
buns, 40
by-products, 60, 140, 188

C

cabbage, 37
CAD, 244, 251
cadmium, 135, 159
calcium, x, 19, 20, 24, 25, 27, 28, 29, 31, 36, 40, 42, 49, 65, 85, 102, 124, 125, 128, 151, 158, 159, 242
calorie, 40, 42, 86
Canary Islands, 185, 215
cancer, xii, xiv, 81, 103, 104, 107, 115, 116, 137, 156, 165, 168, 169, 173, 186, 201, 209, 224, 230, 245, 257, 264, 268, 269, 274, 276, 277, 279
candidates, 123, 178
capsule, 157
carbohydrate, 27, 35, 37, 49, 67, 71, 152, 179, 183, 246
carbohydrates, xiii, 2, 14, 38, 73, 121, 126, 127, 260
carbon, 107, 132, 179, 233, 246, 248
carbon tetrachloride, 107, 179, 233, 246, 248
carboxyl, 151
carboxylic acid, 73, 211
carboxylic acids, 211
carcinogen, 58, 116, 186
carcinogenesis, 58, 115, 144, 167, 170, 276
carcinoma, 103, 116, 153, 202, 204, 272, 274, 275
cardiac muscle, 93
cardiovascular disease, xii, 29, 65, 81, 152, 155, 166, 209, 231, 234, 245
carotene, 31, 59, 83, 85, 124, 132, 167, 174, 180, 181, 194
carotenoids, 31, 69, 85, 86, 132, 133, 143, 167, 170, 171, 172, 180, 181, 184, 188, 189, 193, 194, 230
carrageenan, x, xi, xii, 19, 21, 39, 43, 44, 45, 48, 60, 61, 63, 72, 76, 77, 83, 101, 104, 110, 117, 121, 123, 124, 128, 137, 143, 146, 157, 166, 179, 192, 193, 198, 261, 271
Caspase-8, 115, 116

caspases, 242, 274
catalase (CAT);, xii, 82
catalysis, 187
cation, xi, 64
cattle, 123
causal relationship, 154
cDNA, 10, 11
cell culture, 153, 186
cell cycle, 272
cell death, 69, 85, 169, 176, 269
cell line, 53, 103, 104, 107, 115, 116, 186, 202, 203, 204, 209, 215, 242, 257, 272, 274, 275
cell lines, 103, 107, 116, 186, 202, 203, 204, 215, 257, 272, 274
cell signaling, 69
cell surface, 104, 123, 272
cellulose, xi, 63, 65, 75, 129, 151, 154, 167, 224, 255
cerebrovascular disease, 53, 107
challenges, 146, 264
changing environment, 69
charge density, 256
cheese, xiii, 35, 122, 126, 148
chemical, xii, xiv, 41, 60, 76, 81, 82, 84, 85, 106, 119, 126, 128, 140, 141, 143, 160, 165, 170, 185, 187, 194, 198, 199, 210, 214, 218, 258, 268
chemical characteristics, 210
chemical properties, 60, 106, 140, 187
chemicals, 20, 75, 227, 244
chemiluminescence, 102, 191
chemopreventive agents, 190, 276
chemotherapeutic agent, 257
chemotherapy, 104, 116, 269
children, 131, 156
Chile, 21, 38, 39, 45, 46, 55, 139
China, xiv, 20, 21, 22, 27, 28, 30, 37, 41, 45, 58, 124, 128, 138, 165, 168, 193, 224, 230, 271
chitosan, 100
Chitosan, 106
chlorine, 27, 198, 199
chloroform, 211, 213
chlorophyll, xv, 137, 166, 167, 171, 224, 230
Chlorophyta, xii, xv, xvii, 21, 22, 23, 26, 61, 121, 124, 146, 147, 167, 197, 198, 201, 207, 210, 249, 267, 270
chloroplast, 224
CHO cells, 276
cholestasis, 91
cholesterol, xii, 38, 81, 83, 86, 87, 88, 91, 96, 110, 129, 137, 144, 154, 155, 158, 169, 230, 231, 234, 247
choline, 38, 137
cholinesterase, 111
chondroitin sulfate, 245

chromatography, xiii, 41, 67, 68, 122, 132, 134, 255
chromium, 230, 247
chronic diseases, 65, 69, 170, 237
chronic fatigue syndrome, 133
chronic myelogenous, 202
chronic renal failure, 96, 111, 246
circulation, 37, 152
classes, xv, 179, 197, 198, 199, 207, 210, 213, 257
cleaning, 36, 135
climate, 134
clinical application, 112
clone, 206
clusters, 273
coastal communities, 136, 138, 146
cobalt, 29
cocoa, 42, 43
colitis, 59, 107
collagen, 92, 98, 100, 239, 241, 255, 275, 276, 277
college students, 157
colleges, 271, 275, 276
Colombia, 47
colon, 29, 83, 103, 104, 116, 166, 167, 184, 186, 200, 202, 204, 273, 274
color, 23, 37, 224
colorectal cancer, 104, 116, 119
commercial, xii, 29, 30, 39, 44, 121, 126, 135, 137, 138, 205
community, 142, 151, 213
comparative analysis, 50
competition, 43, 198
complement, xvi, 141, 253
complex carbohydrates, 124
complications, 28, 32, 110, 233
composition, x, xii, 3, 4, 20, 22, 23, 28, 45, 51, 54, 55, 56, 57, 58, 73, 74, 76, 81, 106, 107, 118, 123, 126, 131, 140, 141, 142, 143, 191, 199, 210
compost, 134
composting, 134
compounds, xi, xv, xvii, 2, 27, 30, 38, 49, 52, 61, 63, 65, 69, 70, 91, 97, 102, 103, 104, 105, 123, 136, 152, 153, 166, 167, 170, 171, 172, 173, 175, 179, 182, 184, 188, 190, 191, 197, 198, 199, 200, 201, 202, 204, 205, 206, 207, 208, 210, 211, 213, 214, 215, 216, 220, 225, 242, 245, 258, 259, 267, 271, 273, 276
condensation, 244
conditioning, 125
configuration, 76, 168
congestive heart failure, 92
connective tissue, 174
consciousness, 69
conservation, 10
constipation, 29, 136, 166, 271

constituents, x, 19, 29, 51, 65, 69, 82, 110, 141, 194, 195, 230, 247, 264
Constitution, 245
construction, 172
consumers, 36, 147, 170, 224
consumption, x, xii, xiii, xiv, xv, 19, 20, 21, 28, 31, 49, 58, 64, 65, 69, 70, 71, 83, 85, 86, 98, 101, 103, 112, 122, 123, 126, 128, 134, 137, 145, 146, 147, 151, 153, 155, 156, 157, 159, 160, 165, 184, 197, 223, 224, 225
contamination, 28, 159, 170, 226, 227
control group, 154, 155, 157, 158
cooking, 25, 147
cooling, 152
copolymer, 41, 73
copper, 27, 29, 70, 71, 135, 170, 230, 247
coral reefs, 122
cornea, 192
coronary heart disease, 85, 129
correlation, ix, xi, 2, 12, 13, 31, 64, 71, 82, 99, 157, 189, 234, 263
correlations, 277
cosmetic, 138, 179, 198
cosmetics, xii, xv, 20, 64, 76, 134, 137, 166, 197, 231
cost, 41, 184
couh, 136
covering, 199
CPC, 226
creatinine, xii, 82, 93, 96
crop, 15, 49
crude oil, 138
crystals, 42
CSF, 272
cultivars, 15
cultivation, 20, 24, 37, 54, 213
culture, xiii, 3, 11, 37, 41, 48, 122, 134, 137, 144, 198, 258, 261, 269, 274
culture conditions, 11
culture media, xiii, 41, 122, 134
culture medium, 274
cures, 136, 271
Cyanophyta, 136
cycles, 10, 12, 142
cyclohexanone, 208, 209
cyclooxygenase, 103, 257, 272
cyclosporine, 245, 249, 250
cysteine, 195, 226
cytochrome, 242, 244, 275
cytomegalovirus, 104, 117
cytoplasm, 91, 122, 132, 275
cytotoxic agents, 204, 276
cytotoxicity, 110, 194, 201, 202, 203, 208, 209, 215

D

damages, 90
D-amino acids, 8, 9, 16
danger, 180
data analysis, 228
database, 10
decay, 168, 169
decomposition, 99, 170
defence, xiv, 133, 166, 168, 170, 213
deficiency, 31, 59, 101, 113, 128, 159, 180, 194
deformation, 73
degenerative conditions, 105
degradation, 60, 152, 193, 237, 238, 250
dementia, 97, 168
denaturation, 2
dengue, 104, 107, 117
Denmark, 45
deoxyribonucleic acid, 168
deposition, 94
deposits, 135, 138
depth, ix, 25
deregulation, 268
derivatives, 2, 41, 61, 169, 171, 178, 179, 184, 189, 193, 204, 208, 210, 211, 215, 219, 220, 255
desiccation, 122
destruction, 168
detoxification, 10, 12
developing countries, x, 19
diabetes, x, xii, 19, 29, 31, 65, 81, 83, 84, 85, 101, 110, 131, 155, 156, 169, 231
diabetic nephropathy, 233, 236
dialysis, 67, 144
diarrhea, 105
diastolic blood pressure, 86
diatom communities, 142
diatom frustule, 135
diatoms, 135
diet, xii, xiii, xiv, xv, 21, 31, 49, 69, 70, 71, 82, 87, 88, 89, 90, 91, 92, 93, 94, 95, 96, 97, 98, 99, 100, 121, 123, 124, 126, 127, 137, 145, 146, 147, 152, 154, 156, 157, 158, 159, 165, 223, 230, 259, 268
dietary fiber, xi, xii, 25, 33, 52, 53, 60, 63, 64, 65, 66, 81, 82, 83, 126, 128, 129, 140, 166, 184
dietary intake, 102, 147, 152, 154, 160
dietary iodine, 128, 143, 247
dietary supplementation, 235
digestibility, xi, 52, 64, 184, 230, 247
digestion, 59, 65, 86, 101, 130
digestive enzymes, 153, 179
dioxin, 91, 110
diploid, ix, 2, 3, 8, 13
direct action, 176
diseases, x, xii, xiii, xiv, 19, 31, 32, 49, 65, 69, 81, 84, 85, 132, 136, 137, 145, 153, 165, 168, 169, 180, 183, 186, 231, 235, 238, 258, 268, 271
disorder, 2, 12
dispersion, 235
distilled water, 226, 255
distribution, 14, 30, 52, 55, 122, 152, 177, 190, 220, 229, 256, 262, 273
diterpenoids, 143, 204
diversification, 45
diversity, 69, 122, 123, 199, 201, 215
DNA, 10, 12, 41, 59, 168, 169, 176, 183, 186, 191, 199, 201, 216, 232, 244, 248, 257, 268, 276
DOI, 51, 52, 54, 57, 61
donors, 189
dopaminergic, 97, 111
dosage, 157
double bonds, 132
dressings, 42
drought, 3, 8, 13, 14
drugs, 153, 199, 205, 206, 213, 242, 255, 258
dry matter, xiii, 33, 71, 122, 126, 127, 130, 131, 133
drying, 37, 39
dyes, xiii, 122, 134

E

E.coli, 105
East Asia, x, 19, 146, 153, 154
ecology, 119, 214, 218
economics, 144
eczema, 101
edema, 257
edible mushroom, 248
edible Spanish seaweeds, xi, 63, 65, 66, 67, 68, 140
editors, 49, 50, 51, 53, 54, 55, 56, 59, 60
education, 38, 156
effluents, 135, 230
egg, x, 20, 131
Egypt, 56, 121, 141
eicosapentaenoic acid, 27, 132, 167
electromagnetic, 49
electron, 168, 171, 172, 189, 242
electrons, 168
electrophoresis, xiii, 40, 41, 122, 134, 187, 244, 255
emission, 211, 242
employment, 138
emulsions, 38, 42, 152
encephalopathy, 111
encoding, ix, 2, 10
endonuclease, 244, 251
endothelial cells, 141, 169
endothelium, 275

energy, 39, 49, 57, 87, 107, 135, 155, 158, 171, 235
energy expenditure, 87
engineering, 14, 100, 112
England, 188
environment, xv, 13, 69, 122, 197, 198, 213
environmental change, 49
environmental conditions, xv, 2, 142, 166, 168
environmental factors, 45
environmental stress, 14
enzymes, x, xii, xiii, 20, 81, 86, 91, 93, 121, 126, 127, 152, 170, 175, 179, 180, 183, 184, 198, 199, 212, 214, 235, 242, 245, 275, 276, 277
EPA, 27, 84, 132, 167, 210
epithelial cells, 117, 242, 244, 268
epithelium, 94, 187
equilibrium, 255
equipment, 39
erythrocytes, 88
esophageal cancer, 202
essential fatty acids, 132, 167
EST, 10
ester, 73, 75, 76, 151, 183, 225
estrogen, 102, 103, 104, 217
ethanol, 226, 255
ethers, 207, 210
ethnic groups, 38
ethyl acetate, 174
EU, 153
eukaryotic, 2
Europe, xiv, 29, 37, 41, 52, 53, 124, 136, 140, 141, 147, 165, 195
evidence, xiii, xiv, 14, 17, 28, 65, 71, 91, 115, 145, 153, 157, 159, 160, 176
evolution, 199, 260
excretion, 91, 110, 249
exercise, 156, 157, 169
exploitation, 126, 141, 198
exposure, 36, 86, 159, 173, 183, 195, 235, 255, 268
extracellular matrix, 237, 238, 239, 270, 275, 276
extraction, 20, 30, 41, 42, 44, 45, 60, 70, 75, 107, 132, 134, 140, 257, 270, 271
extracts, ix, xiv, xvii, 27, 56, 70, 71, 82, 101, 102, 103, 105, 109, 114, 116, 118, 132, 135, 136, 137, 143, 148, 151, 166, 167, 171, 174, 175, 176, 179, 180, 182, 183, 185, 186, 187, 190, 195, 199, 207, 208, 212, 227, 230, 231, 232, 233, 234, 245, 247, 250, 257, 258, 259, 261, 262, 268, 269, 271, 272

F

families, xvii, 65, 200, 267
farmers, 45, 134
farms, 56, 139

fast food, x, 19
fasting, 155
fat, x, xii, xiii, 19, 27, 29, 49, 67, 82, 86, 91, 96, 99, 109, 121, 126, 127, 147, 152, 159
fatty acids, xv, 31, 52, 84, 87, 105, 129, 132, 143, 159, 167, 169, 197, 199, 210, 211, 219, 232
fear, 49
feed additives, 137
feelings, 158
fermentation, 135
ferric ion, 176
ferrous ion, 175
fertility, 102
fertilizers, 16, 20, 21, 125, 134, 137
fetus, 101
fiber, x, xi, 19, 20, 22, 40, 57, 63, 65, 66, 67, 82, 100, 112, 116, 129, 135, 141, 166, 186
fiber content, 129
fibers, xi, 60, 63, 83, 93, 128, 140, 236
fibrinolytic, xiii, 121, 125, 169
fibroblast growth factor, 112
fibroblasts, 104
fibrosarcoma, 275
fibrosis, 92, 98, 112, 237, 238, 239
fibrous tissue, 91
Fiji, 45
filters, 135
filtration, 42, 135, 226, 236
financial, 260
financial support, 260
fish, 1, 30, 35, 37, 40, 42, 125, 156, 157, 175, 176, 181, 186, 191, 194, 230, 258, 259, 263
fisheries, 138
flavonoids, 32, 86, 176, 177, 189
flavor, 28, 124
flavour, x, 20, 84
flexibility, 41
flocculation, 41, 152
flooding, 13
flour, 25, 42, 191, 276
fluctuations, 69, 138
fluorescence, 242, 243
foams, 43
food, x, xi, xii, xiii, xiv, xv, 19, 20, 21, 22, 25, 27, 36, 37, 38, 39, 40, 41, 43, 44, 49, 50, 51, 52, 60, 63, 65, 70, 72, 76, 82, 119, 121, 122, 123, 125, 126, 128, 129, 130, 131, 132, 137, 138, 140, 144, 145, 146, 147, 148, 151, 152, 153, 155, 157, 158, 159, 160, 166, 168, 179, 183, 184, 187, 188, 197, 198, 213, 230, 231, 245, 247, 271
food industry, x, 20, 39, 43, 44, 60, 72, 129, 130, 147, 151, 160
food intake, 158

food poisoning, 159
food production, 271
food products, xii, xiii, xiv, 40, 50, 52, 121, 122, 123, 126, 137, 140, 145, 148, 151, 152, 157, 160, 166
force, 65
formation, xiv, 12, 34, 42, 44, 85, 97, 100, 133, 153, 158, 166, 169, 170, 171, 172, 176, 194, 195, 229, 232, 235, 244, 255, 256, 257, 268, 269, 271, 275
fouling, 172, 199, 213
fragments, 12, 238, 239
France, 21, 29, 30, 36, 45, 134, 138, 139, 141, 216, 224
free radicals, xiv, 2, 34, 133, 165, 167, 168, 169, 170, 171, 172, 180, 181, 231, 232, 233, 242, 257
freezing, 2, 12, 14, 35, 39, 152, 169
freshwater, 130, 198
fruits, 65, 69, 128, 133, 151, 187
FTIR, xi, 45, 48, 53, 54, 64, 72, 73, 74, 75, 76, 77
Fucoxanthin, xv, 30, 31, 58, 59, 104, 109, 116, 166, 167, 180, 188, 193, 194
functional food, 22, 52, 57, 105, 188
fungi, 10, 41, 172, 246

glutathione, xii, 58, 81, 86, 88, 99, 170, 176, 179, 181, 182, 186, 195, 232, 274
glutathione peroxidase (GSH-Px), xii, 81, 179
glycerol, 2, 15
glycine, 10, 182, 183, 247
glycol, 39, 42
glycosaminoglycans, 225
goiter, 31, 128, 136
gout, 136
gracilis, 97, 111
grades, 40
grading, 254
graph, 238, 239, 244
grass, 23, 147
green alga, xiii, 8, 22, 25, 50, 51, 57, 122, 124, 130, 132, 137, 141, 142, 146, 159, 198, 216, 224, 258, 262
greenhouse, 212
growth, ix, xiii, xvi, 1, 2, 15, 30, 53, 54, 100, 101, 103, 104, 112, 115, 116, 118, 121, 127, 132, 135, 139, 150, 169, 183, 198, 212, 225, 230, 247, 254, 259, 264, 268, 269, 272, 274
guidelines, 155, 158

G

gametophyte, ix, 1, 3, 77
gastric ulcer, 32, 38
gastrointestinal tract, 129
gel, xiii, 35, 40, 74, 76, 122, 134, 147, 148, 187, 244, 255
gel formation, 147
gelation, 151, 158
gene amplification, 268
gene expression, ix, 2, 10, 11, 13, 17, 69, 87, 103, 116, 259, 262, 268
genes, 2, 11, 12, 14, 268
genus, x, 20, 23, 25, 27, 30, 37, 41, 58, 129, 132, 182, 186, 199, 200, 202, 205, 206, 207, 257, 264, 274
geographical origin, xiii, 122, 126
Germany, 162
germination, 230
gland, 104, 115
glial cells, 111
glioblastoma, 204
glomerulus, 94, 233
glucose, xiv, 74, 75, 87, 101, 113, 145, 155, 158, 180, 235, 270
glucosidases, 209
glutamate, 179
glutamic acid, 3, 5, 7, 8, 9, 13, 25, 28, 131

H

habitat, xvi, 122, 223
habitats, xvi, 36, 38, 198, 223, 224
hair, 100, 112
half-life, 86, 152
halogen, xv, 197, 198, 199, 201, 202, 203, 207, 210, 213
halogenation, 198, 201, 212, 214
haploid, ix, 1, 3, 13
haploid leafy gametophyte, ix, 1, 3
harbors, 138
harvesting, 21, 124, 125
Hawaii, 38, 146
hazardous waste, 139
hazards, 232
HE, 192
healing, xii, 82, 100, 117
health, x, xiii, xiv, xv, xvi, 20, 22, 28, 29, 30, 32, 38, 49, 52, 53, 56, 60, 65, 69, 83, 106, 135, 141, 145, 146, 147, 151, 153, 154, 155, 157, 160, 166, 180, 183, 193, 223, 224, 225, 230, 253, 254, 260
health condition, 225
health effects, 53
health researchers, 160
heart and lung transplant, 250
heart disease, 83, 129, 132, 167, 190, 263
heavy metals, 28, 135, 141, 159
heme, 170

hemoglobin, 174, 175
hepatic injury, 233
hepatitis, 110, 249
hepatocytes, 89, 91, 233
hepatotoxicity, 107, 233
herpes, 104, 107, 117, 136
Heterofucans, xvii, 267
heterogeneity, 92, 270
hexane, 209
HFP, 52
high density lipoprotein, 87
high fat, 87, 88
histology, 96, 97
history, 134, 138
HIV, 33, 35, 61, 104, 105, 106, 117, 118, 136, 245
homeostasis, 103, 232
Hong Kong, 118
hormone, 86, 103, 115, 157, 259, 269
host, 117, 206
hot springs, 198
human body, 152, 170, 257
human health, x, xv, 20, 22, 49, 56, 136, 197, 213, 247
human immunodeficiency virus, 104, 117, 136
human milk, 247
hyaline, 259
hybrid, 44, 77
hybridization, 48, 53
hydrocarbons, xv, 197, 199, 211
hydrocolloids, x, 19, 38, 39, 45, 54, 123, 134, 137, 143, 198
hydrogels, 100
hydrogen, 17, 71, 105, 151, 168, 171, 172, 176, 181, 182, 183, 191, 198, 199, 201, 216, 220
hydrogen abstraction, 172
hydrogen peroxide, 105, 168, 176, 183, 191, 198, 199, 201, 216, 220
hydroperoxides, 170
hydroxyl, xiv, 71, 99, 165, 167, 168, 177, 182, 183, 203, 231
hydroxyl groups, 177
hypercholesterolemia, 29, 98
hyperglycemia, 101, 113
hyperlipidemia, 234, 249
hyperoxaluria, 96, 111
hyperplasia, 91, 257
hypersensitivity, 141
hypertension, 29, 84, 85, 136, 231, 236, 250
hypertriglyceridemia, 109
hypertrophy, 93
hypothesis, 257
hypoxia, 16, 247, 251

I

Iceland, 30, 36, 124
ideal, 35, 36, 40, 43
identification, xi, 14, 53, 64, 72
illumination, 3
images, 236
immersion, 259, 262
immigrants, 230
immune defense, 34
immune reaction, 102
immune response, 104, 115, 263
immune system, 103, 257, 259
immunity, 259, 262
immunocompetent cells, 102
immunodeficiency, 35
immunomodulatory, xvi, 33, 102, 114, 130, 253, 254, 257, 259, 264
immunomodulatory agent, xvi, 253
immunostimulant, xvi, 254
immunostimulatory, 61, 262, 263
improvements, 39, 157, 158
impurities, 35
in vitro, xii, 17, 32, 59, 60, 68, 69, 70, 71, 81, 82, 97, 98, 102, 103, 105, 111, 112, 116, 117, 131, 153, 159, 184, 187, 189, 190, 193, 195, 209, 213, 214, 217, 219, 229, 231, 234, 236, 246, 251, 257, 264, 270, 271, 275
in vivo, xii, 32, 49, 81, 86, 87, 88, 97, 98, 102, 103, 108, 111, 112, 130, 131, 187, 192, 193, 201, 213, 236, 247, 271
incidence, xiv, 85, 103, 165, 224, 245
India, 39, 51, 57, 108, 143, 165, 223, 225, 226, 247
indirect effect, 259
individuals, 156, 157, 159
indoles, xv, 197, 198, 200, 215
Indonesia, 21, 38, 45, 47, 123, 174
inducer, 204
induction, 11, 104, 116, 176, 182, 259, 274
induction period, 182
industrial wastes, 135
industries, xv, 82, 166, 179, 197, 198, 245, 271
industry, xi, 20, 21, 28, 38, 39, 40, 41, 43, 49, 51, 54, 63, 76, 128, 137, 138, 140, 153, 184, 198, 213, 230, 258
infection, 56, 117, 137, 141, 169, 206
infestations, 136
inflammation, 87, 89, 91, 92, 94, 97, 102, 104, 169, 192, 233, 257
inflammatory cells, 169
inflammatory disease, 85, 187
inflammatory responses, 102
influenza, 105

Index

influenza virus, 105
ingestion, 107, 114, 133, 158, 159
ingredients, xiv, 22, 35, 43, 53, 57, 60, 125, 138, 145, 153, 160, 166, 168, 188
inhibition, xvi, 13, 87, 101, 103, 105, 112, 117, 118, 167, 174, 204, 232, 234, 236, 253, 256, 257, 270, 273, 275
inhibitor, 117, 141, 206, 219, 239, 274
initiation, 104, 167, 169, 186, 268
injuries, 94, 98
injury, 59, 85, 90, 91, 92, 94, 110, 168, 169, 179, 232, 233, 242, 245, 246, 248, 249, 250
innate immunity, 259, 262
inositol, 38
insertion, 268
insulin, 87, 101, 113, 153, 272
integrins, 275
integrity, 104, 115, 168, 233
interference, 272
interferons, 41
internalization, 273
intertidal red macroalga, ix, 1
intervention, 153, 155, 156, 157, 158, 160
intestinal tract, 187
invertebrates, xvii, 8, 170, 256, 261, 267
iodine, 24, 25, 27, 28, 36, 49, 85, 101, 103, 108, 113, 115, 127, 128, 136, 153, 159, 198, 230
ion-exchange, xiii, 122, 123, 134
ionization, 70
ions, 65, 136, 151, 230, 247
IR spectra, 55, 228, 229
IR spectroscopy, 72, 246
Ireland, 21, 30, 36, 37, 135, 147, 214, 224
iron, x, 19, 20, 25, 27, 29, 34, 36, 124, 133, 159
irradiation, 183
ischemia, 93, 97, 169, 250
Islam, 17
islands, 38
isoflavonoids, 177, 230
isolation, 61, 72, 151
isomers, 175, 219
issues, 148

J

Japan, xiii, xiv, 1, 2, 10, 12, 14, 20, 21, 22, 24, 27, 28, 30, 37, 39, 41, 45, 51, 59, 121, 123, 124, 126, 137, 138, 140, 141, 144, 147, 159, 165, 168, 184, 188, 191, 217, 245, 278
Japanese women, 156
juveniles, 259, 264

K

KBr, 73, 201, 227
Kenya, 45
Keynes, 117
kidney, xii, 16, 82, 93, 94, 95, 96, 98, 99, 103, 108, 110, 115, 233, 236, 237, 238, 239, 242, 246
kidneys, 94, 96, 98, 233, 249
kinetics, 6, 7, 11, 54, 174
Korea, xiv, 20, 21, 22, 27, 28, 37, 50, 123, 138, 147, 159, 165, 168, 230

L

lactic acid, 149
large intestine, 159
larvae, xvi, 254, 258, 259, 264, 265
Latin America, 206
LDL, 85, 87, 99, 154, 155, 169, 174, 230, 232, 234
lead, xiv, 69, 99, 129, 135, 159, 165, 168, 179, 238, 272
leaks, 91
legislation, 52
legume, 16
leishmaniasis, 206
lesions, 91, 94, 137, 233, 269
leucine, 131, 235
leukemia, 53, 103, 104, 200, 202, 257
life cycle, ix, 1, 3, 15
ligand, 116
light, 3, 54, 69, 142, 170, 182
lignin, xi, 63, 64, 66, 67
limestone, 138
linoleic acid, 174, 175, 176
lipid metabolism, 98, 180, 213
lipid oxidation, xiv, 86, 165, 168, 176, 183, 186
lipid peroxidation, 58, 70, 86, 88, 96, 98, 169, 171, 172, 176, 181, 186, 191, 231, 247, 248
lipid peroxides, 167, 232
lipids, xiv, 27, 30, 98, 142, 145, 157, 166, 168, 188, 232, 257, 276
liposomes, 181
liquid chromatography, 53
liquids, 135
Listeria monocytogenes, 150
liver, xii, 28, 82, 86, 87, 88, 89, 90, 91, 98, 99, 109, 110, 170, 180, 187, 194, 230, 233, 246, 248, 249
liver damage, 28, 89, 170
liver disease, 87, 109
liver enzymes, 230
localization, 17
longevity, xv, 123, 223
low temperatures, 2

low-density lipoprotein, 34, 71, 133, 189, 246
lung cancer, 204
lung disease, 136
lung metastases, 112
Luo, 111
lutein, 31, 132, 167
lycopene, 31
lymphocytes, 102, 176, 179, 259, 268
lymphoid, 257
lymphoma, 202, 274
lysine, 131
lysis, 141, 182

M

macroalgae, ix, xi, xii, 1, 8, 13, 16, 49, 50, 51, 56, 63, 121, 137, 142, 143, 146, 151, 166, 171, 185, 195, 208, 213, 214, 215, 217, 220, 254
macromolecules, xvii, 41, 168, 170, 179, 267, 276
macronutrients, xi, 63, 155
macrophages, 102, 115, 169, 173, 257, 259, 272
magnesium, 24, 25, 27, 28, 42, 102
majority, 198, 199
Malaysia, 38, 39, 45, 81, 123
mammal, 86
mammals, xii, xvii, 82, 87, 94, 101, 103, 259, 267
man, 68, 126, 138, 140
management, 65
manganese, 27, 29, 170
manipulation, 14, 45
mannitol, 2, 14, 15, 27
manufacturing, 39, 137, 152
marine environment, 224
marketing, 37, 124
Maryland, 130
mass, x, 19, 25, 130, 134
mast cells, 169
materials, 40
matrix, 34, 133, 148, 170, 191, 204, 237, 238, 240, 242, 250, 275, 276
matrix metalloproteinase, 237, 240
matter, iv
measurement, 174, 185, 186, 248
meat, 35, 40, 41, 54, 126, 147, 148, 149, 152
mediation, 119
medical, ix, xiii, xvi, 118, 122, 134, 136, 137, 179, 253, 260
medical science, xvi, 253
medicine, xii, 21, 110, 116, 117, 119, 121, 125, 136, 271
Mediterranean, 56, 119, 214
mellitus, 155
melting, 35

melting temperature, 35
membranes, ix, xiv, 1, 2, 131, 165, 168, 199, 210
memory, 29
mercury, 159
Metabolic, 14
metabolic pathways, xv, 69, 197, 198
metabolic syndrome, 83
metabolism, 16, 87, 91, 101, 102, 104, 109, 144, 168, 187, 189, 194, 213, 235
metabolites, xv, 49, 103, 105, 123, 139, 142, 180, 181, 182, 183, 185, 194, 197, 198, 199, 201, 203, 204, 209, 210, 211, 213, 214, 215, 217, 220, 233
metabolizing, 245
metal ion, 123, 136
metalloproteinase, 270
metals, 30, 123, 135, 159, 189
metastasis, 98, 269, 275
methanol, 70, 113, 174
methodology, 88
methylene blue, 227
Mexico, 45, 47, 48, 214
mice, 59, 107, 115, 137, 141, 185, 194, 249, 257, 258, 271
microorganism, 199
migration, xvi, 169, 254, 257, 270
milligrams, 72
Ministry of Education, 14
mitochondria, 91, 235, 242, 249, 275
mitochondrial damage, 96
mitogen, 176
MMP, 237, 238, 240, 275
MMP-2, 239, 240, 275
model system, xiv, 54, 166, 182, 186
models, 70, 153, 247, 257, 258, 270, 271, 276, 277
moderate activity, 206, 211
modernity, xiii, 145
modifications, 226
moisture, 41, 42, 152
molecular biology, 184
molecular weight, 83, 107, 174, 176, 179, 211, 212, 232, 233, 236, 249, 250, 255, 256, 271, 273, 275
molecules, xvi, xvii, 35, 49, 168, 169, 179, 181, 183, 184, 188, 212, 232, 238, 253, 256, 258, 260, 267, 269, 270, 275, 277
mollusks, 30
monosaccharide, 269
monounsaturated fatty acids, 84
Moon, 115, 247, 278
morbidity, 92
Morocco, 39, 45, 55
morphine, 258
morphology, 97
mortality, 92, 156

mRNA, 236, 238, 239, 240
mucous membrane, 36
multiples, 172
multiplication, 269
muscles, 28
mutagen, 116
mutant, 15
mutation, 268
myocardial infarction, 169
myocardium, 91, 92
myocyte, 93

N

Na^+, 180, 194
NaCl, 17, 226, 255
natural compound, 201
natural food, x, 19, 128, 151
natural habitats, 138
natural killer cell, 259
natural resources, 37, 184
NCTC, 150
necrosis, 91, 92, 93
negative effects, 170
neoplasm, 268
nephritis, 236, 250
nephropathy, 249
nervous system, 28
Netherlands, 56
neurological disease, 104, 117
neurons, 96, 97, 111
neutral, xi, 63, 66, 132, 179
neutral lipids, 132
neutrophils, 259, 272
New England, 124
New Zealand, 39
niacin, 25, 29, 30, 167
nickel, 29, 135, 230, 247
nicotine, 124
nicotinic acid, 136
nitric oxide, 17, 133, 168, 169, 232, 248, 257, 272
nitrogen, 13, 16, 135, 170, 232, 276
NK cells, 272
NMR, 53, 64, 77
North America, xiv, 37, 136, 147, 165
North Korea, 45
Norway, 21, 28, 30, 138, 196
nuclear magnetic resonance, 64
nuclei, 224, 244, 276
nutraceutical, 30
nutrient, 22, 45, 57, 59, 109, 114, 132, 142, 230, 247
nutrition, 49, 51, 53, 56, 69, 125, 128, 147, 160, 193, 195, 212

O

obesity, x, xiv, 19, 29, 30, 31, 84, 87, 88, 109, 111, 145, 165, 166
obstruction, 94
oceans, 25, 137, 211
oedema, 94
oesophageal, 187
OH, 61, 73, 168, 170, 228
oil, xi, 42, 64, 68, 124, 132, 152
oleic acid, 84
oligomers, 173
oligosaccharide, 60, 98, 112, 179, 184, 193
omega-3, 84, 87
oncogenes, 268
operations, 45
optimization, 193
organ, 91, 96, 97, 98
organic chemicals, 172
organic compounds, 122, 127, 211, 212, 220
organic solvents, 249
organism, 69, 126, 150
ornithine, 116
osteoarthritis, 31, 60, 102, 114, 131
overweight, 84, 158
ovulation, 102
oxalate, 248, 249
oxidation, xiv, 34, 70, 71, 85, 87, 88, 104, 133, 166, 169, 171, 174, 176, 180, 183, 186, 187, 189, 193, 198, 231, 232, 234, 246
oxidation products, xiv, 166
oxidative damage, 98, 168, 170, 178, 232, 235, 249
oxidative stress, xiv, 12, 32, 59, 69, 85, 86, 91, 98, 152, 165, 167, 169, 170, 174, 180, 183, 186, 187, 194, 235, 237, 248
oxygen, xiv, 13, 69, 86, 93, 165, 167, 168, 169, 170, 171, 174, 176, 180, 181, 186, 189, 190, 194, 207, 248, 250, 257
oxygen consumption, 93
oysters, 37
ozone, 212

P

Pacific, 25, 29, 50, 54, 59, 109, 110, 126, 142, 143, 161, 211
pain, 102, 258, 268
paints, xiii, 122, 134, 135
Pakistan, 56, 118, 143
palmate, xv, 143, 166, 185, 194
Panama, 48
pantothenic acid, 136
parallel, 153, 229

parasite, 36, 206
participants, 156, 158
pasta, 53, 147, 148
patents, 129
pathogenesis, 169, 186
pathogens, xvi, 198, 205, 254, 258
pathology, 169
pathways, 13, 14, 100, 272, 274, 277
PCR, 10, 11, 12, 236, 238, 239, 241
peptides, 52, 86, 130, 236
percentage of fat, 37
perinatal, 111
peritoneal cavity, 257
peritonitis, 257
permeability, xvi, 254
peroxidation, 86, 169, 175, 182, 186, 189, 248
peroxide, 99, 183
personal communication, 147, 148
Peru, 45
PGE, 64, 71, 83
pH, 65, 73, 101, 151, 226, 255
Phaeophyta, xii, xv, xvii, 14, 57, 61, 73, 106, 108, 121, 124, 139, 146, 167, 190, 197, 198, 207, 210, 213, 267
phagocytosis, 102
pharmaceutical, ix, xi, xiii, xv, 54, 63, 76, 82, 122, 134, 137, 153, 197, 198, 199, 213, 258
pharmaceuticals, xvii, 166, 231, 268
pharmacology, xii, xvii, 121, 125, 267
phenol, 171, 172, 255
phenolic compounds, xi, 64, 71, 105, 152, 153, 172, 207, 208, 209
phenoxyl radicals, 189
Philippines, 21, 24, 30, 38, 45, 46, 47, 48, 134, 138, 157
phosphate, 2, 15, 43, 179, 180
phosphatidylcholine, 249
phospholipids, 168, 171
phosphorus, 24, 29, 36, 85, 102, 135
phosphorylation, 272, 274, 277
photosynthesis, 58, 225
photosynthesize, 122
phycocyanin, 59, 107, 167
phycoerythrin, 167
phylum, xiii, 122, 126
physical properties, 198
physicochemical properties, xi, 57, 64, 68, 74, 140
Physiological, 15, 16, 17, 58, 216
physiology, 100, 112, 130, 140
phytoplankton, 142
PI3K, 272, 274
PI3K/AKT, 274
pigmentation, 167

placebo, 157
plant growth, 134
plants, x, xiii, 2, 8, 10, 13, 14, 15, 20, 25, 40, 50, 68, 83, 118, 122, 123, 124, 126, 128, 130, 133, 134, 138, 152, 170, 172, 177, 184, 224, 225, 246, 261
plasma cells, 268
plasma membrane, 248
plasminogen, 237
plastics, xv, 135, 197
platelet aggregation, 107, 256, 261, 263
platelets, 255, 276
point of origin, 268
polar, 30, 131, 171, 211, 220
polarity, 70
pollutants, 124
pollution, 134, 135
polyether, 215
polymer, xvi, 253
polymerase, 236
polymerase chain reaction, 236
polymerization, 208
polymers, xvii, 76, 172, 173, 176, 249, 254, 260, 267, 269
polyphenols, xi, xv, 27, 32, 61, 64, 65, 71, 83, 86, 101, 104, 108, 166, 171, 172, 173, 174, 181, 190, 208
polysaccharide, xii, xv, xvii, 25, 61, 64, 65, 71, 72, 73, 74, 75, 84, 97, 100, 101, 103, 105, 107, 110, 113, 114, 118, 157, 166, 167, 177, 184, 192, 213, 225, 227, 234, 236, 242, 245, 246, 248, 249, 260, 261, 264, 265, 267, 270, 271, 275
Polysaccharides, viii, 33, 55, 56, 61, 72, 73, 74, 76, 117, 128, 151, 177, 185, 225, 226, 227, 229, 231, 232, 233, 234, 235, 236, 242, 253, 267
polyunsaturated fat, 37, 51, 82, 84, 86, 98, 131, 171
polyunsaturated fatty acids, 37, 82, 84, 86, 98, 132, 171
ponds, 138
population, xv, 55, 149, 151, 154, 155, 156, 223
Porphyra yezoensis, vii, ix, 1, 2, 15, 17, 21, 27, 31, 37, 58, 102, 107, 114, 195
Portugal, 19, 28, 39, 45, 197
positive correlation, 158, 159, 174, 183
positive relationship, 174
potassium, x, 20, 27, 36, 40, 201, 227, 236, 250
potential benefits, 153
poultry, 123, 148
precipitation, 35
prematurity, 85
preparation, xiii, 25, 40, 41, 42, 73, 117, 122, 134, 184, 250, 262, 263
preservation, xi, 64
preservative, 148, 150

prevention, xii, xiv, 32, 65, 70, 71, 81, 84, 93, 108, 110, 131, 133, 157, 166, 178, 193, 231, 247, 249, 256
principles, 171, 184, 188, 191, 258
probe, 276
producers, xii, 41, 45, 64, 122, 131, 137, 200
profitability, 139
progesterone, 102
pro-inflammatory, 257, 258
project, 65
prokaryotes, 14
prolapsed, 136
proliferation, 91, 103, 104, 169, 173, 174, 183, 190, 195, 257, 268, 269, 272, 274
proline, 2, 4, 5, 8, 12, 13, 15, 17
prolyl endopeptidase, 110
promoter, 116
prophylactic, 93, 111, 172, 265
propylene, 42
prostaglandins, 34, 132, 133
prostate cancer, 30
protection, xiv, 12, 165, 168, 174, 176, 182, 249
protective role, 190, 248
protein kinase C, 250
protein kinases, 176
protein oxidation, xiv, 166, 232
protein synthesis, 232
proteinase, 209
proteins, ix, x, xiii, 1, 2, 17, 20, 27, 33, 36, 43, 51, 72, 75, 82, 84, 107, 118, 121, 126, 127, 130, 131, 140, 141, 151, 166, 168, 169, 232, 239, 257, 269, 274, 276, 277
proteinuria, 236, 240, 250
proteoglycans, 250, 275
public health, xi, 63
pulp, 43
purification, 139, 185, 255, 260, 270
pyrimidine, 168
pyrolysis, 20, 134, 135

Q

quantification, 107, 190, 214

R

radiation, 114, 169, 172, 195, 268
radical reactions, 70, 189
radicals, xiv, 166, 168, 169, 170, 171, 172, 180, 181, 182, 186, 187, 212, 231, 233
rancid, 169
random assignment, 155
rat kidneys, 247

raw materials, xiii, 30, 45, 121, 126
reactions, 168, 171, 172, 201, 255
reactive oxygen, xiv, 2, 69, 85, 97, 99, 165, 170, 175, 183, 231, 276
Reactive oxygen species (ROS), xiv, 165, 167
reactivity, 210
receptors, 104, 111, 199, 272, 273
recognition, 275
recommendations, 65, 157
recovery, 93, 232, 238, 247, 259, 262
recreational, 134
recycling, 13
Red Sea, vii, 63, 76
reference system, 50, 55
regeneration, 36
regression analysis, 71
regulations, 153
rejection, 237, 250
repair, 59, 100, 268
replication, 105, 269
reproduction, 113, 254
repulsion, 276
requirements, 28, 39, 133
researchers, xvi, xvii, 138, 154, 213, 223, 267
residues, 10, 11, 41, 73, 75, 76, 151, 169, 225, 245, 254, 256, 269
resistance, 205, 232, 236, 258, 259, 262, 263, 265
resources, xi, 50, 55, 63, 125, 144, 166, 184
respiratory syncytial virus, 117
response, ix, xvi, 2, 4, 7, 13, 16, 69, 92, 103, 158, 169, 183, 230, 253
restaurants, 29, 36
restoration, 242
retail, 138
retinol, 31, 59, 180, 194
retinopathy, 85
retrovirus, 105
reverse transcriptase, 106, 118
rheumatoid arthritis, 85
Rhodophyta, xii, xv, xvii, 15, 20, 23, 24, 26, 30, 39, 43, 50, 54, 55, 76, 108, 118, 121, 124, 136, 146, 167, 183, 190, 197, 198, 200, 207, 208, 210, 215, 267
riboflavin, 136, 167
rings, 86
risk, 22, 29, 31, 58, 65, 69, 101, 102, 103, 112, 128, 129, 132, 154, 155, 156, 167, 172, 190, 231, 234, 237, 268
RNA, 10, 12, 251
rods, xiii, 122, 134
room temperature, 151, 226, 255
Royal Society, 60, 140, 141, 160
Russia, 30, 39, 45

S

safety, 159
salinity, 3, 8, 13, 14, 15, 135, 259, 262
salmon, 115
salt concentration, 255
salt tolerance, 122
salts, 27, 40, 41, 42
saturated fat, 84, 210
saturated fatty acids, 84
scavengers, 86, 170, 178, 183, 189, 231, 233
seafood, 37
seaweed phylum, xiii, 122, 126
secrete, 169
secretion, 94, 100, 218, 237, 250
sedimentation, 41, 43
seed, 50, 59
seedlings, 16
self-sufficiency, 269
semen, 136
sensitivity, 150
serum, 52, 91, 93, 96, 102, 108, 129, 141, 154, 155, 157, 180, 194, 234, 273
sewage, 124, 135
sex, 136, 259
sex hormones, 259
shade, 226
shape, 25
shellfish, 126, 156, 268
shock, 169
shoot, 230
shores, 122
showing, 70, 73, 149, 150, 205, 206, 232
shrimp, xvi, 30, 126, 254, 259, 262, 263, 265
side chain, 207
side effects, 256
signal transduction, 13
signaling pathway, 272
signals, 269
silicon, 27
silver, 100
Singapore, 123
sinuses, 97
skeleton, 76
skin, 97, 100, 112, 136, 138, 167, 186
smoking, 268
smooth muscle, 169
society, x, 19, 231, 268
sodium, x, 20, 42, 43, 73, 75, 83, 124, 150, 158, 159, 192, 226, 233, 249, 250, 255, 263
solubility, 65, 70, 76, 114
solution, 38, 65, 73, 136, 189, 226, 255
solvents, 70, 135, 171, 192

South Africa, 144, 215, 216
South America, 45, 54
South Korea, 45, 156
Southeast Asia, 27
soybeans, 130
Spain, 28, 45, 63, 65, 70, 140, 143, 185, 215
species, xii, xiv, xvi, xvii, 2, 8, 9, 20, 21, 22, 23, 24, 25, 27, 28, 29, 30, 36, 37, 38, 39, 42, 45, 55, 56, 69, 70, 71, 72, 74, 82, 85, 86, 87, 97, 99, 105, 106, 121, 122, 123, 124, 126, 127, 129, 130, 131, 133, 134, 135, 136, 137, 143, 154, 159, 165, 167, 170, 171, 173, 176, 182, 183, 186, 198, 199, 201, 205, 207, 213, 218, 223, 224, 225, 227, 229, 230, 231, 232, 248, 255, 257, 258, 267, 271, 276
spectroscopy, xi, 53, 54, 64, 72, 78, 79
spleen, xii, 82, 97, 100, 192
sponge, 22
sporophyll, 61
sporophyte, ix, 2, 3
Spring, 46, 47, 48
stability, xiv, 166, 168, 170, 231
stabilization, 43
stabilizers, 38, 152
standard deviation, 66, 67, 68
starch, 36, 44, 66, 67, 128, 151, 152, 154, 157, 184
starch polysaccharides, 67
stasis, 256
state, 38, 125, 171
sterile, 122, 227
steroids, 41
sterols, 97, 111, 132, 143
stomach, 136, 137
stomach ulcer, 137
stomatitis, 117
storage, xi, 25, 64, 65, 128, 132, 231, 232
stress, ix, xv, xvi, 1, 2, 4, 5, 7, 8, 9, 10, 11, 12, 13, 14, 15, 16, 17, 58, 59, 69, 108, 152, 166, 168, 169, 184, 193, 195, 232, 248, 254, 258, 259, 262
stretching, 73, 75, 229
striatum, 48, 60, 61, 193
stroke, 53, 107, 169
structural characteristics, 270, 277
structural knowledge, 269
structural variation, 207, 254
structure, xvi, xvii, 41, 72, 73, 76, 108, 114, 118, 151, 171, 173, 175, 176, 177, 178, 179, 185, 192, 195, 214, 216, 246, 247, 253, 254, 256, 260, 263, 267, 273, 275, 277
substitutes, 36, 112
substitution, 209, 269
substrate, 112, 179, 268, 276
sucrose, 40

Index

sulfate, xi, xvi, 64, 65, 68, 71, 72, 73, 74, 75, 76, 77, 101, 104, 118, 250, 253, 256, 261, 269, 273, 275, 276, 277
sulfated polysaccharides (SP), xvi, 253
sulfur, 104
sulphur, 69, 199
superoxide dismutase (SOD), xii, 81, 86, 99, 176, 179
supplementation, 87, 89, 91, 93, 94, 97, 99, 155, 157, 234
suppliers, xii, 121, 124
suppression, 104, 150, 167, 175, 179, 181, 259
survival, xvi, 104, 199, 236, 254, 259, 265, 268, 274
susceptibility, 98, 149
suspensions, 38
swelling, xi, 64, 68, 235
symmetry, 73
symptoms, 56
synapse, 97
syndrome, 35, 263
synergistic effect, 49, 132, 143, 182
synthesis, ix, 1, 2, 9, 13, 14, 15, 91, 180, 238, 277
Syria, 141
systolic blood pressure, 231

T

T cell, 249, 272
Taiwan, 194
tanks, 37, 147
tannins, 172
Tanzania, 45, 47, 48
target, 13, 105, 168, 209, 236, 256, 271, 276
taxa, 270
techniques, 20, 72, 77, 258
technology, 39, 123
temperature, ix, 1, 2, 3, 4, 5, 6, 7, 8, 11, 12, 13, 139, 151
terpenes, xv, 105, 197, 198, 200, 201, 203, 205
testing, 71
textiles, 166
texture, x, 20, 23, 35, 42, 129
TGF, 98, 100, 104, 112, 237, 239, 240, 250
Thailand, 38, 174, 190
therapeutic agents, 22, 153
therapeutic use, 233
therapeutics, 199
therapy, 117, 136, 157, 230, 250, 256
thiamin, 167
thickening agents, 76, 152, 166
thrombin, 101, 219, 256
thrombocytopenia, 256, 263
thrombosis, xvi, 105, 132, 167, 172, 253, 255, 256, 263
thrombus, 255, 256
thyroid, 28, 31, 102, 103, 108, 114, 128, 157
tissue, xii, 34, 41, 74, 82, 85, 91, 98, 100, 108, 112, 133, 168, 169, 173, 179, 183, 186, 232, 233, 235, 238, 239, 241, 250, 255, 257, 269
TNF, 272
tobacco, 124
tocopherols, 132, 133, 181, 182, 231
tones, 38, 137
total cholesterol, 87, 88, 154, 156
total energy, 156
toxicity, 61, 159, 168, 170, 173, 248, 269
toxicology, 107
trace elements, xiii, 27, 29, 85, 121, 126, 127, 159
trade, 29, 45
transcription, 17
transcription factors, 17
transesterification, 132
transformation, 14, 169, 269
transforming growth factor, 104, 237
transplantation, 236
transport, 29, 242
trauma, 169
treatment, ix, 8, 11, 32, 34, 36, 60, 66, 67, 84, 109, 111, 114, 115, 133, 136, 158, 201, 209, 230, 233, 237, 242, 246, 256, 257, 258, 269, 271
trial, 111, 136, 149, 155, 157
triglycerides, 87, 91, 154, 155, 234
tryptophan, 131
TSH, 102
tuberculosis, 136
tumor, xvi, 60, 104, 115, 167, 202, 215, 225, 253, 257, 268, 269, 270, 271, 272, 273, 275, 276
tumor cells, 104, 268, 270, 272, 276
tumor development, xvi, 253, 269
tumor growth, 257, 270, 271
tumor metastasis, 276
tumor necrosis factor, 115, 272
tumor progression, 269
tumorigenesis, 186
tumors, 103, 268, 269
turnover, 237, 239, 247
type 2 diabetes, 112

U

UK, 21, 140, 141
ulcer, 144
ultrasound, 271
UN, 143, 144
under-exploited marine bio-resources, xi, 63

urea, xii, 82, 96
uric acid, xii, 82, 96
USA, 10, 12, 21, 41, 45, 53, 56, 136, 139, 147, 214
UV, 172, 173, 183, 190, 195

V

valine, 131
vanadium, 201, 212, 216
vancomycin, 205, 206
variations, xiii, 122, 126, 131, 151, 155, 184
varieties, xii, 28, 121, 123, 127, 131, 150, 154, 272
vascular diseases, 31, 131, 172
vasoconstriction, 250, 255
vegetables, x, xiii, 20, 22, 30, 37, 60, 65, 68, 69, 121, 124, 125, 127, 128, 130, 133, 140, 141, 147, 151, 156, 187
vegetation, 50
versatility, 49, 213
vertebrates, 8, 10
vibration, 73, 75, 229
vibrational FTIR-ATR spectroscopy, xi, 64
Vietnam, 38
viruses, 117
viscosity, 129, 152, 157, 158
vision, 36
vitamin A, 25, 36, 133
vitamin B1, 25, 49, 58, 133, 143
vitamin B12, 49, 58, 133, 143
vitamin B2, 25, 132
vitamin C, xii, 29, 32, 36, 58, 81, 85, 86, 133, 171
Vitamin C, 34, 83, 133
vitamin E, 25, 34, 86, 133, 143, 171, 182, 188, 189
vitamin K, 134
vitamins, x, xii, xv, 19, 20, 25, 27, 29, 30, 37, 49, 69, 121, 124, 126, 128, 132, 133, 137, 166, 167, 231
VLDL, 234
vocabulary, 36
volatile organic compounds, 220

W

Wales, 37, 124
Washington, 185
waste, 135
wastewater, 135, 247
water, xi, xiii, 2, 15, 29, 35, 38, 40, 42, 52, 64, 65, 68, 70, 71, 76, 83, 86, 100, 103, 122, 123, 124, 127, 128, 134, 147, 152, 157, 159, 170, 175, 179, 183, 198, 226, 233, 259, 262, 265
wave number, 72
weakness, 36, 128
wealth, 126, 166
weight gain, 96
weight loss, 27, 84
weight management, 109, 157, 158
welding, xiii, 122, 134
welfare, 22
well-being, 97
wellness, 225
West Indies, 38
Western countries, 22, 65, 69
WHO, 65, 84
workers, 173
worldwide, xii, 70, 121, 122, 125, 126
wound healing, 100, 112, 136, 172

Y

yeast, 149, 157, 209, 263
yield, 42, 54

Z

zinc, 29, 135, 170